European Cities & Technology

The Cities and Technology series

Pre-industrial Cities and Technology
(C. Chant and D. Goodman, eds)
The Pre-industrial Cities and Technology Reader
(C. Chant, ed.)

European Cities and Technology: industrial to post-industrial city
(D. Goodman and C. Chant, eds)
The European Cities and Technology Reader: industrial to post-industrial city
(D. Goodman, ed.)

American Cities and Technology: wilderness to wired city
(G.K. Roberts and J.P. Steadman)
The American Cities and Technology Reader: wilderness to wired city
(G.K. Roberts, ed.)

European Cities & Technology

industrial to post-industrial city

Edited by
David Goodman and Colin Chant

in association with

The Open
University

Copyright © 1999 The Open University

Published by Routledge   Written and produced by The Open University
11 New Fetter Lane   Walton Hall
London EC4P 4EE   Milton Keynes MK7 6AA

Simultaneously published in the USA and Canada by Routledge
29 West 35th Street
New York, NY 10001

Edited, designed and typeset by The Open University
Printed in The United Kingdom by Scotprint Ltd, Musselburgh, Scotland

A catalogue record for this book is available from the British Library
A catalog record for this book is available from the Library of Congress

ISBN 0 415 20079 2 hardback
ISBN 0 415 20080 6 paperback

at308bk2i1.1

This book is part of the Cities and Technology series listed on the back of the first page. The series has been
prepared for the Open University course AT308 *Cities and Technology: from Babylon to Singapore*. Details of
this and other Open University courses can be obtained from the Course Reservations and Sales Office,
PO Box 724, The Open University, Milton Keynes MK7 6ZS, United Kingdom; tel: + 44 (0)1908 653231.

Much useful course information can also be obtained from The Open University's website: http://www.open.ac.uk

Contents

Introduction

This textbook is part of a series about the technological dimension of one of the most fundamental changes in the history of human society – the transition from rural to urban ways of living. The series[1] is intended first and foremost as a contribution to the social history of technology; the urban setting serves above all as a repository of historical evidence with which to interpret the historical relations of technology and society. The main focus, though not an exclusive one, is on the social relations of technology as exhibited in the physical form and fabric of towns and cities.

The main aims of the series are twofold. The first is to investigate the extent to which major changes in the physical form and fabric of towns and cities have been stimulated by technological developments (and conversely how far urban development has been constrained by the existing state of technology). The second aim is to explore within the urban setting the social origins and contexts of technology. To this end the series draws upon a number of disciplines involved in urban historical studies – urban archaeology, urban history, urban historical geography and architectural history. In so doing, it seeks to correct an illusion created by some past historical writing – the illusion that all major changes in urban form and fabric might be sufficiently explained by technological innovations. In brief, the series shows not only how towns and cities have been shaped by applications of technology, but also how such applications have been influenced by, for example, politics, economics, culture and the natural environment.

The wide chronological and geographical compass of the series serves to bring out the general features of urban form which differentiate particular civilizations and economic orders. Attention to these differences shows how civilizations and societies are characterized both by their use of certain complexes of technologies, and also by the peculiar political, social and economic pathways through which the potentials of these technologies are channelled and shaped. Despite its wide sweep, the series does not sacrifice depth for breadth: case-studies of technologies in particular urban settings form the bulk of the material in the Readers associated with each textbook.

Definitions

The series calls upon a diverse set of interpretations, models and approaches from the social history of technology, from urban historical and geographical studies, and from archaeology and architectural history. In a wide-ranging series such as this, it seems appropriate to introduce theoretical issues when called for by a particular topic. There is, however, a place in this introduction for discussion of the series' two main variables.

In this series, 'technology' is interpreted broadly to cover all methods and means devised by humans in pursuit of their practical ends, thereby including relevant developments in science, mathematics, public health and medicine. But however broad this interpretation, there are still distinctions to be drawn between, say, technology (the means of building) and the built environment (the product of building technologies); or between technology (the means of

---

1   The three textbooks and Readers in this Cities and Technology series – all published by Routledge, in association with The Open University – are listed on the back of the first page of this volume.

achieving human ends) and society (perhaps the most ambiguous term of all, but in one sense the summation of human ends in the form of a set of religious, moral and political values). It may turn out to be difficult to draw a sharp line between these concepts; but the same can be said of night and day, or indeed, the modern city and the countryside, and we cannot do without such concepts in our dealings with the world. Similarly, without the distinctions proposed between technology, society and the built environment, we cannot hope to think clearly about the questions this series raises. In practice, the series focuses on the implications for urban form and fabric of a well-defined range of innovations, above all those in: agriculture and food-processing; military technology; energy; materials; transport; communications; water supply, sanitation and other developments in public health and social medicine; production processes; and building construction, including representational and measurement techniques and engineering science.

The contribution of these technologies is analysed at differing angles of relevance to the history of urban form and fabric. Innovations in agricultural, military, industrial and transport technology are linked with broad developments in the history of urbanization, including the origins of urban settlements, the changing relationship between town and country, and the increasingly specialized nature of cities within systems of cities. At the core of the series are the technologies most intimately involved in the processes of city-building – construction techniques, intra-urban transport, energy systems, water supply and sanitation, and communications networks. The series also attends to developments in science, technology and public health stimulated by a variety of urban crises.

If 'technology' is a slippery term, so too is 'city', the other great historical variable of this series. It should be emphasized at the outset that no particular store is set by the distinction between cities and towns, a distinction often made in a culturally specific way – for example, the peculiarly British criterion that cities have cathedrals. In this series, cities and towns are seen as part of an urban continuum, and references to cities should usually be read as covering all settlements of an urban degree of complexity and specialization. This needs to be stressed, because in a series as wide-ranging as this, the largest cities of a given period and region tend to claim the most detailed attention. But this is already to presuppose a distinction between cities and villages, the type of permanent human settlement which developed alongside agriculture, beginning some 11,000 years ago. What differentiates towns or cities from villages, apart from the brute facts that the former are physically bigger and contain more people?

The very fact that 'cities' have been transformed throughout history is what makes them compelling objects of historical investigation; but it also makes any abstract definition elusive. In many earlier periods and places, the city, densely populated and built up within its defining defensive walls, was sharply demarcated from the surrounding countryside and its small agricultural settlements. In the modern era, the old fortification rings which defined the city, especially in Europe, became redundant, partly through developments in artillery. They sometimes metamorphosed directly into circular roads, becoming part of the transport system which facilitated the outward diffusion of the city into the countryside, to the point where some recent commentators have begun to question whether the term 'city' is becoming outmoded. Cities, clearly, are dynamic entities, subject to great changes, not only expanding but declining and contracting, and even at times being reduced to rubble or ashes. For all we know, they may be a phenomenon of certain successive configurations of human society only, and be destined to disappear, just as there were none in human history, it seems, until little more than 5,000 years ago. There is no neat definition of a city, as the following quotation indicates:

> the city may be defined initially as a community whose members live in close proximity under a single government and in a unified complex of buildings, often surrounded by a wall. Since, however, this definition would also cover many villages, military camps, religious communities and the like, the city may further be described as a community in which a considerable number of the population pursue their main activities within the city, in non-rural occupations. But other communities, such as a monastery or small factory surrounded by the dwellings of its workmen, might be similarly characterized. A third characterization may therefore be that the city is a community which extends at least its influence and preferably its control over an area wider than that simply necessary to maintain its self-sufficiency.
>
> (Hammond, 1972, p.8)

The problem with this process of definition and redefinition is that in order to cover marginal exceptions, the definition gets loaded with various historical attributes of certain cities that are not part of the *meaning* of words such as 'city' or 'urban'. The search for a perfect definition, intended not only to cover all types of city in history but also to expose those settlements with bogus urban credentials, is surely a futile one, and this series eschews it. The main point that is emphasized is that a significant or dominant proportion of the population is engaged in activities other than agriculture: government, religion, administration, law, education, finance, manufacture, commerce, entertainment and so on. As it turned out, those involved in the non-agricultural occupations made possible by improved agricultural productivity generally clustered in relatively densely populated and built-up settlements; the reasons for this are considered at appropriate points in the series.

But a criterion based on economic specialization seems too disembodied, too dissociated from the physical urban reality, to capture the full meaning of 'city'. It is important therefore to add that the activities which distinguish towns and cities from villages become associated with particular spaces and structures: forums, squares and parks, streets, bridges and railways, markets, factories and shopping malls, temples, town halls and theatres; and that these spaces and structures often become emblematic of a given town or city. Always remembering that the urban built environment is designed by human beings for human purposes, we suggest that it is, apart from geographical location itself, the most definitive aspect of a city's identity, however dramatically it may be transformed by war or disaster. Emphasis on the changing spatial form and physical fabric of cities, and the ways in which they emerge from human activities, is therefore no arbitrary selection from the multiple historical phenomena of urbanization.

Conventions and acknowledgements

Short numbered passages – readily identifiable by the different typeface used and the grey rectangle beside the page number – have been printed at the end of Chapters 2 and 4 of this book. These are extracts from published works, and in some cases bibliographical references have been edited out of them without ellipses in the text; ellipses do appear where other portions of text have been omitted. Where references have been retained in such extracts, the full source is included in the bibliographic list at the end of the chapter.

Chapter authors have sought to avoid what are now widely accepted as sexist expressions in their own writing, but these remain without editorial comment where they occur in quotations from existing sources. As a general rule, measures have been given in metric rather than imperial units; where non-metric measures appear in existing sources, metric equivalents have generally been added. It may be that many readers have until now negotiated life's vagaries without having at the forefront of their minds the difference between the metric 'hectare' and the imperial 'acre'. It might therefore be helpful to point out that a hectare is equivalent to a square with sides of 100 metres. A hectare is nearly two and a half times an acre, which is equivalent to a rectangle with sides of the old furlong (220 yards – about 200 metres) and the old chain (22 yards – about 20 metres).

The editors and authors of a book such as this, which provides part of the print backbone of a mixed-media Open University course, are indebted to an unquantifiable degree to innumerable colleagues in all areas of the institution. Special thanks are due to Denise Hall, the course manager, without whose unsurpassed ability in her role, and total commitment to the course, there would be no series of books; to Linda Camborne-Paynter and her colleagues in Course Management, without whose expertise in the conversion of authors' drafts to first-rate electronic text, the textbook deadlines could never have been met; to Jonathan Davies and Sarah Hofton of The Open University's Design Studio for their skill and imagination in the exterior and interior graphic design, and to their colleagues Ray Munns and Michael Andrew Whitehead for their painstaking and creative work on numerous maps and illustrations; to picture researchers Celia Hart, Anne Howard and Paul Smith, for their unstinting dealings with the appropriately variegated cultures and technologies of illustration archives; and last but by no means least, to the dedicated team of Open University editors, Hazel Coleman, John Pettit and Jane Wood, whose close and rigorous reading of the textbook chapters has saved readers from numerous opacities, solecisms and awkward expressions. The rigorous and constructive comments of colleagues on a variety of university and Arts Faculty committees have informed the thinking of the Course Team during the course's gestation. Thanks are due in particular to Tim Benton, Tony Coulson, Colin Cunningham, Janet Huskinson, Anne Laurence and Bernard Waites for comments and advice at this stage. Help given on individual chapters has been acknowledged *in situ*. The Course Team has been fortunate indeed to have the benefit of the appropriately broad knowledge and experience of its external assessor, Anthony Sutcliffe, who has been a most constructive critic, always tolerant of academically respectable interpretations that diverged from his own.

Colin Chant
The Open University

Reference

HAMMOND, M. (1972) *The City in the Ancient World*, Cambridge, Mass., Harvard University Press.

Part 1
CITIES OF THE INDUSTRIAL REVOLUTION (TO 1870)

Chapter 1: THE ONSET OF INDUSTRIALIZATION

by David Goodman

1.1 Industrialization and urban growth

Population surge

Beneath the complex social, political, economic and technological changes that so profoundly affected Europe in the eighteenth and nineteenth centuries, there lay a dramatic population growth, a human increase of such magnitude that it was unprecedented in the history of the world. The figures – still imprecise estimates for the eighteenth century, but thereafter increasingly exact records of the first censuses and registrations of births and deaths – tell their story (Anderson, 1987, p.73; Armengaud, 1973; Mitchell, 1981, pp.34, 86–8). The population of France soared from some 18 million in 1715 to 26 million in 1789. Italy's 11 million in 1700 rose sharply to 16 million by 1770. And in Spain also there was a marked population growth from some 5 million at the beginning of the eighteenth century to around 11 million at its end. The population of Europe as a whole is estimated to have risen from about 120 million in 1700 to 190 million by 1800, 266 million in 1850, and 400 million by 1900. With the population explosion came a pronounced urbanization, much more evident in western than in eastern Europe. In 1800 there were twenty-three cities of over 100,000 inhabitants in Europe; by 1900 there were 135 such cities.

Nowhere were these accelerating demographic trends more emphatic than in Britain. Here, the population of England and Wales rose from some five million in 1700 to nine million by 1800. The censuses of the nineteenth century show that Britain's population doubled and almost doubled again (see Table 1.1). And it was in Britain of the mid-nineteenth century that the transition first occurred from a majority of rural to a majority of urban dwellers (a conclusion

Table 1.1 Population of nineteenth-century Britain

Date of census	Population
1801	10,501,000
1851	20,817,000
1901	36,000,000

that followed from the definition of a town in the census of 1851 as a place
with 2,000 or more inhabitants). Britain's urbanization was on a scale
unmatched anywhere else in Europe: eight cities of over 100,000 in 1851, and
twenty-nine in 1901. And London, the 'unique city', with 2.7 million inhabitants
in 1851, was the largest city the world had ever seen. London had always been
Britain's largest city, but the list of largest provincial cities showed a revealing
change. Cities that in the eighteenth century had been high in the order of
urban population, such as York (an ecclesiastical and administrative centre)
and Norwich (a county and market town, and a long-established
manufacturing centre of worsted cloth), were replaced in the hierarchy by
rapidly rising centres of new industry and ports. Now London was followed by
Liverpool, Glasgow, Manchester and Birmingham, all showing spectacular
population growth (see Table 1.2).

Table 1.2 Population growth in four British cities

	1775	1801	1851
Liverpool	–	80,000	376,000
Glasgow	–	77,000	329,000
Manchester	c.30,000	70,400	303,000
Birmingham	–	74,000	233,000

The cause of Britain's remarkable population surge has for decades been a
matter of lively controversy for historians. It used to be commonly assumed
that the principal cause must have been a sharp fall in the death-rate, resulting
from the conquest of disease through advances in medicine – especially the
protection from the scourge of smallpox by the eighteenth-century
introduction of inoculation, and later by vaccination. But now the widely
accepted conclusion of persuasive research is that a rising birth-rate is the
explanation. According to this argument – and there is a connection with cities
and technology – women in the mid-eighteenth century married earlier and
had more babies, partly because of the greater prospects for employment for
them, their husbands and their children in the expanding urban manufacturing
centres, and partly because of falling food prices. The booming birth-rate in
the countryside provided cheap immigrant labour for the industrializing towns.
And it provided the towns with more than that. For, without this injection of
population from the countryside, Britain's urban growth could not have been
maintained. The reason was the higher death-rate in the deteriorating
environment in cities of the late eighteenth and nineteenth centuries –
overcrowding and insanitary living conditions. This was the urban penalty that
offset the advantages of employment. In 1801, babies born in a British city of
population above 10,000 had a life expectancy of thirty-one years compared
with forty-one in the countryside.

Industrial 'revolution'?

It is worth reflecting on whether Britain's population boost stimulated urban
technology and industrialization. That is dismissed by Hudson (1992, pp.159–61),
who even suggests that the effect may have been negative: rapid population
growth can result in declining per capita income, and the presence of a ready
supply of cheap labour may retard the introduction of labour-saving
technological innovations. This is just one of many unresolved questions
surrounding the Industrial Revolution. There is disagreement over when the
Industrial Revolution began: a conventional starting-point has been around
1760 or 1770, but some economic historians refuse to see an Industrial
Revolution before 1820 because, they insist, it was only then that Britain's
industrial economy began to take off. The weakness of this purely economic

criterion is that it ignores the earlier spate of key technological innovations, which occurred in the late eighteenth century and whose implementation sparked off the great industrial changes – the various inventions of textile machinery, new iron metallurgy and steam-engines. More radical revisionism comes from historians who deny the validity of the term 'Industrial Revolution', for what sense is there in calling this industrial change 'revolutionary' when it took 150 years to complete, since it is commonly supposed to have continued from its late eighteenth-century beginnings throughout the nineteenth century? But these expectations of some overnight revolution in industry, like some political *coup*, are inappropriate in the realm of technology, where change has long periods of gestation and implementation, eventually producing dramatic results. So, despite the carping criticism, the use of 'Industrial Revolution' is currently stronger than ever and it is retained in this chapter.

Amidst all the controversy, one fundamental fact is generally acknowledged: industrialization, that outstanding characteristic of the modern world, began first of all in Britain, a country richly endowed with coal and iron ore, and with available capital for investment. Then, with the assistance of British expertise, it spread to Belgium, France, the German states and eventually to other parts of Europe and beyond, transforming the face of cities everywhere. But why did industry, with its new technologies, now come – more than ever before – into the cities? After all, until the great changes of the Industrial Revolution, the most typical location of industry in Britain and elsewhere had been the countryside. In England, the principal source of wealth – from the late Middle Ages to the mid-eighteenth century – had been the predominantly rural, woollen textile industry. It was characterized by the domestic or putting-out system: an urban entrepreneur purchased raw wool and supplied it to scattered rural peasants, to men, women and children who, working in their homes, used simple skills to perform the preliminary stages of textile-manufacture, often a part-time supplement to their agricultural labour. East Anglia, the Weald of Kent and the West Country had been among the most important areas of production. Now all of these declined in industrial importance, eclipsed by the industrial towns of the Midlands and the North. An urban site for industry had the disadvantage of higher land prices. So what made it worthwhile for the new captains of industry to operate in towns? One explanation is that an urban industrial plant concentrated the work-force and so yielded the manufacturer the precious prize of constant supervision and quality control. That had always eluded entrepreneurs in the countryside, a serious weakness of rural production with its dispersed labour-force, though this did not apply to those early, water-powered textile mills established in rural areas. But the greatest advantage, which made it worth paying the price for an urban site, was proximity to the markets.

The urban centres of the Industrial Revolution were generally not new towns. Birmingham, Manchester and Glasgow were all towns of small to moderate size with centuries of manufacturing tradition behind them before they became the prime centres of the Industrial Revolution and experienced an associated population explosion. But a few towns (around twenty) were creations of the Industrial Revolution: new technology brought them into being from nothing, or almost nothing. Though anything but typical, these new towns are worth studying for the stark depiction of the forces of technology at work in moulding urban form.

1.2   *From rural hamlet to urban industrial complex*

Merthyr Tydfil is an outstanding example of a town whose creation was linked to the innovative technology of the Industrial Revolution. Merthyr's rise was intimately associated with the introduction of mineral fuel – coke – in the iron industry. That industry had for centuries relied on a vegetable fuel, charcoal, produced by controlled combustion of wood. Since charcoal is a fragile material, readily broken into useless dust, it could not be transported far. Consequently, operations to smelt iron ore were conducted close to the fuel source, in wooded regions. The traditional centres of British (and European) iron-production were therefore in the countryside – especially in the Forest of Dean and the Weald of Sussex, areas possessing an abundance of timber, iron ore and water power for the operation of the smelting-furnaces. (The necessary blast of air was provided by the opening and closing of bellows, mechanically performed by harnessing them to a rotating water-wheel.)

Historians used to believe that this iron industry was a voracious consumer of timber, causing the depletion of England's forests and, already by the seventeenth century, creating serious fuel shortages that forced the migration of ironworks to untapped woodlands elsewhere. It was argued that the ironmasters, those entrepreneurs who combined technical knowledge with keen awareness of the response of prices to supply and demand, transferred their plant to remote rural areas, particularly avoiding towns (above all, London) because of their competitive demand for domestic fuel. None of this is now certain: research over the last few decades has undermined the previous orthodoxy. (The debate is conveniently summarized in Harris, 1988, pp.19–29.) London and many other English towns were not in fact competing with the ironmasters for charcoal-fuel: the home fires of citizens had long been fed by coal from the mines of Northumberland and Durham, shipped from Newcastle. Nor is it correct to see the ironmasters as devouring large quantities of mature, structural timber of the kind used for building houses and ships. Instead, they relied on coppices – thickets of young trees, planted over a cycle of around twenty years. These provided timber in convenient sizes for charcoal-manufacture, and at an economic price. Why, then, did the ironmasters have to migrate at all? The main reason seems to have been the shortage of land for planting coppices: landowners might receive income from hiring out their estates for growing coppices, but the potential for much greater profit lay in using land for pasture or growing crops. This meant that only agriculturally unproductive terrain was sold for coppicing. And as the iron industry grew, the pressure on this marginal land became so intense that it is thought 'inconceivable' that the remarkable late eighteenth-century expansion of the British iron industry, one of the most dramatic developments of the early phases of the Industrial Revolution, could have occurred on the basis of charcoal-fuel (Harris, 1988, p.26). A large-scale iron industry was made possible only by the substitution of mineral fuel (coke, derived from coal) for charcoal. Coal was cheaper and already the established fuel in the glass, brewing and lead industries. But coal contains impurities, notably sulphur, which remained with the product, spoiling its quality. Brewers were dismayed by the effect of sulphur on the flavour of beer, and ironmasters recognized that the same impurity brought imperfection to their pig-iron. It was then discovered, first in the brewing industry, that a preliminary roasting of the coal to produce coke brought marked improvements. (Roasting removed the sulphur impurity, converting it to sulphur dioxide, a gas that escaped into the atmosphere.) And in 1709, in Shropshire, Abraham Darby had performed successful trials on the smelting of iron ore with coke, producing cast-iron pots of superior quality. Slowly the new method was adopted, coming to Merthyr Tydfil in 1760. Once the switch to coke occurred, the nature of the iron industry was completely altered. Now the fuel was immediately available (no time-lag in waiting for coppices to grow) and practically inexhaustible. Now production on a large scale could be concentrated on a site such as Merthyr, endowed with both iron ore and the rich coal-seams of the great South Wales coalfield.

How this generated a new town in Wales is powerfully portrayed in a fascinating essay by Evans (1994).[1] Although his discussion contains traces of outmoded assumptions (extensive deforestation by ironmasters in eighteenth-century England, heavy demand for charcoal-fuel for the domestic fires of London's multitudes), these can be discounted: they have no effect on the standing of his argument about the way in which technology shaped Merthyr. Instead of a dispersed and mobile work-force, conditions at Merthyr favoured the settlement of large numbers of coal-miners and ironworkers. So, within two generations, the site experienced a quite dramatic population increase from well under a thousand in 1750 to over 11,000 by 1811. And it was all due directly to technological change: a coke-smelting iron town had sprung up, almost from nothing.

Merthyr, deep into the nineteenth century, was still a town almost devoid of amenities, devoid even of public buildings. It was rudimentary also in the complete absence of control in the building of the rising town. Here we come to the real substance of the argument. First, there is the general point that Merthyr's development into a fully functioning town was hindered by the absence of an urban tradition in Wales: there were no cities here. (Merthyr in 1801 was the largest town in Wales.) Second, and of particular interest to us, Evans relates Merthyr's uni-dimensional urban development to the technological tradition. The industrialists who first installed blast-furnaces at Merthyr had an outlook coloured by their previous experience of the charcoal-fuelled industry. They had been accustomed to providing shelter on rural sites for a small labour-force – of the order of a dozen households – and would not contemplate involvement in housing hundreds of workers and their families. So instead, the new scale of housing demand was met by petty speculators who produced a formless urban mess. The lack of urban amenities is attributed to the structure of the new iron industry – a few wealthy capitalists and a large

Figure 1.1 Pen-y-Darren ironworks, Merthyr Tydfil, *c.*1850, by Samuel Homfray (courtesy of National Museums and Galleries of Wales)

---

[1] Extracts from Evans are reproduced in Goodman (1999), the Reader associated with this volume.

proletariat, a social structure that restricted the demand for diversity of urban functions. Evans does not argue that the iron industry was everywhere necessarily associated with such a sharply divided society, just that Merthyr was peculiar in having very few middle managers. And for him, the failure to achieve a rounded urban development is attributable to 'the very nature of the town's iron industry, wedded to capital goods production and incapable of stimulating ancillary urban functions' (1994, p.14).

Merthyr Tydfil 'took off' after the decision to introduce on its site the new coke-smelting technique for the production of iron. Here too, at the Cyfarthfa works, the first extensive implementation occurred of Henry Cort's puddling process for the manufacture of wrought iron. Cort himself had been brought there in 1787 to teach the workers the new process. Now coal was used instead of charcoal to refine smelted iron into wrought iron. With abundant coal and iron ore, Merthyr by 1850 had become the world's biggest producer of iron, with an annual output of over 200,000 tonnes. From a mere hamlet in 1750, it had by 1851 grown into a town with a population of 46,378, crowded into a wholly unplanned area around the blast-furnaces and mines (see Figure 1.1). Evidence to a parliamentary committee on education in 1839 included a vivid description of the face of Merthyr:

> The surface of the soil is frequently blackened with coal, or covered with high mounds of refuse from the mines and the furnaces … Volumes of smoke … stunted and blackened trees.
>
> (quoted in Young and Handcock, 1956, pp.964–5)

Middlesbrough

Another industrial giant was created even more rapidly than Merthyr. Nowhere in the industrializing world was there anything to compare with the birth and accelerating population growth of Middlesbrough. It occurred in a later phase of the Industrial Revolution, seventy years after the genesis of Merthyr, in the era of the first railways. Middlesbrough was created by the changing policy of railway managers. The world's first freight railway had been designed to carry coal from Darlington (the point of concentration of produce from the surrounding pits of County Durham) to Stockton, a port on the River Tees, for shipment to London. But within a few months of its opening in 1825, the Quaker managers of the Stockton and Darlington Railway Company discovered that the Tees at Stockton was too shallow for heavily laden colliers. So they planned an extension of the railway line to a point a few kilometres downstream where the water was deeper. This bleak and marshy terrain was to be the site of an industrial city of wholly unforeseen size and importance. In 1829, when 200 hectares of this land were bought by the five Quakers who called themselves the 'Middlesbrough Owners', the total population – a few farmhouses – was a mere forty. This was to be a planned town with a symmetrical grid layout (see Figure 1.2) and houses displaying 'uniformity and respectability' (Briggs, 1968, p.245), another world compared with Merthyr. From the mid-points of the four sides of a small central square (the market), sprang the rectilinear main streets (North Street, West Street and so on); the rest of the grid was formed by the division of the land into rectangular housing lots that were put up for general sale. A few hundred metres from the centre was the river and terminal of the railway company. At the end of the line, the company had by 1831 erected 400 metres of staithes, a wharf system for loading coal on to berthed vessels (also visible in Figure 1.2). Here there was capacity for six ships. When the coal train from Darlington arrived, wagons of coal were placed on a cradle and hoisted up by steam power on to the platforms of the staithes. From this height the coal slid down chutes into the holds of the waiting ships.

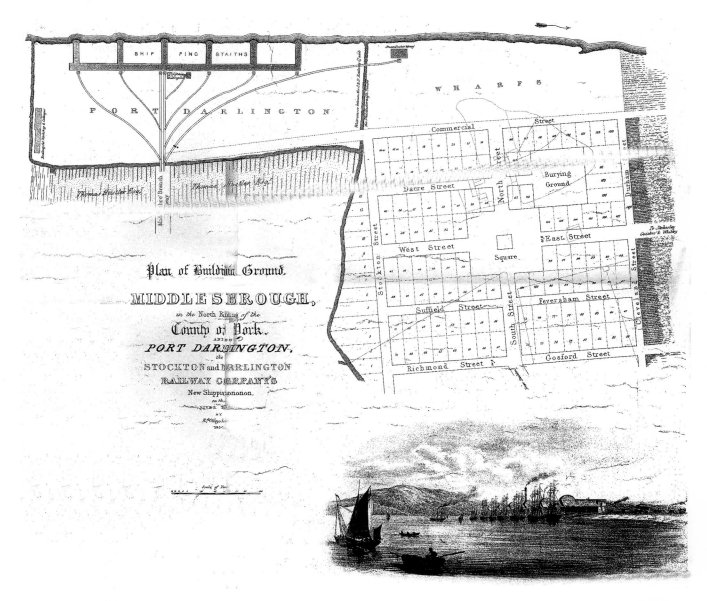

By 1840 Middlesbrough's population had leapt to over 5,000, and alleys began to spoil the purity of the original grid plan. But the greatest change was yet to come. In 1840 an iron-foundry had already been established by two immigrants, attracted by Middlesbrough's coal resources; they were John Vaughan, formerly employed in the Dowlais ironworks near Merthyr, and Henry Bolckow, a German accountant and entrepreneur. At first the local iron ore proved disappointingly sparse, but then, in 1850, tests revealed the richness of ore in the surrounding Cleveland Hills at Eston. This was a godsend for Middlesbrough: its coal-shipping trade had slumped now that coal was being transported on the new national rail network. Middlesbrough became a boom iron town with a surrounding landscape of tall blast-furnaces. By 1871 it was producing over two million tonnes of pig-iron and its population had soared to almost 40,000; by 1891 the population was 75,532.

Asa Briggs has portrayed Middlesbrough as a frontier town with the atmosphere of a US or Australian gold-rush, a rough, male-dominated place with large numbers of immigrant workers eager for a share of the prosperity: they included Irish labourers and even some from the USA and the East Indies (Briggs, 1968, pp.247–50). Yet there was also philanthropy here, some of it

Figure 1.2 *Top* 1830 design for Middlesbrough, by Richard Otley, showing the planned grid layout of wharfs. *Bottom* engraving of the first ships to arrive (Moorsom, 1976, p.10; from the archive of the owners of the Middlesbrough Estate, courtesy of Cleveland Archives: U/OME/8/4)

Quaker benevolence. In 1841, government of the town had been transferred by Act of Parliament from the Middlesbrough Owners to a group of improvement commissioners, given powers to manage street paving, drainage and lighting. They included Henry Bolckow, who rose to become Middlesbrough's first mayor and MP. From the profits of his ironworks, he gave the city its first public park and a new school (both in 1868). John Vaughan, also an ironmaster, succeeded him as mayor and then as MP. From the 1860s the citizens could benefit from a choral society, a society for the cultivation of arts and science, and a public library.

The population explosion of industrializing Middlesbrough could not be fitted into an extension of the original grid; that was prevented by the property boundaries of the outlying agricultural land, and by the railway line. Instead, working-class housing spread along the railway line and in the vicinity of the multiplying ironworks, which were situated on the periphery on cheap land and close to the railway. The better-off lived in villas on the other side of the railway line. And here, too, new municipal buildings arose; the original town centre – between river and railway line – had been abandoned (see Figure 1.3).

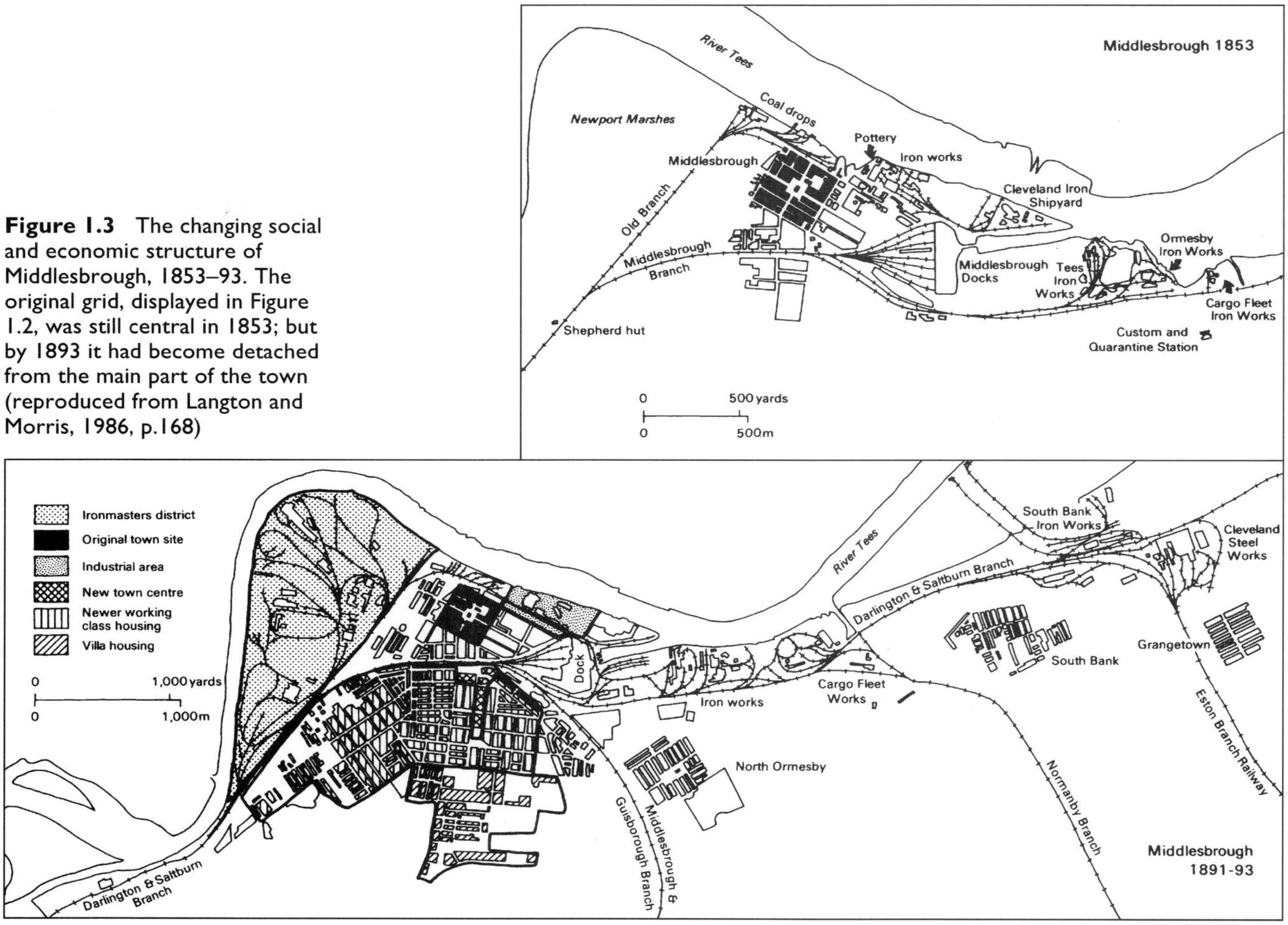

Figure 1.3 The changing social and economic structure of Middlesbrough, 1853–93. The original grid, displayed in Figure 1.2, was still central in 1853; but by 1893 it had become detached from the main part of the town (reproduced from Langton and Morris, 1986, p.168)

1.3 Technology and the urban death-trap

Medieval German citizens declared that they breathed the air of urban freedom, Victorian town-dwellers that they were inhaling the fumes of prosperity and full employment. 'We are proud of our smoke', the mayor of Middlesbrough proclaimed, in a speech to commemorate the opening in 1887 of the new town hall by the Prince and Princess of Wales – proud because it showed that 'even those in the humblest station are in a position free from want'. The words brought cheers of approval from the assembled audience (Briggs, 1968, p.263). And elsewhere in industrial Britain, complaints about air pollution were frequently overriden by fears of lost prosperity. In the 1830s the corporation of Liverpool did force the removal of a flourishing alkali-works because the chimneys were filling the city with foul-smelling gases and spreading acidic fumes for miles around, damaging crops in the surrounding countryside. But the owners of the works resumed operations a few kilometres away in the town of St Helens; and when, after years of expanding production, an aristocratic landowner complained of pollution, the local newspaper rushed to the defence of urban employment:

> Noxious as are the vapours, St Helens cannot be said to be unhealthy. The large amount of high-priced labour which these works provide would cause the inhabitants to rise as one man to resist by every legitimate means any attempt on the part of the legislature to pass any bill which would have the effect of crippling so important a branch of the trade of this district.

(quoted in Wohl, 1983, pp.216–17)

Exactly the same point was made that year (1862) by the Parliamentary Select Committee on Noxious Vapours. But in the end, apparently swayed more by the chemical destruction of aristocratic estates than the effects on public health, the committee called for the appointment of official inspectors to monitor the alkali industry and impose fines on works that produced acidic fumes beyond a set limit. It soon encouraged the implementation of new chemical technology, involving the efficient elimination of noxious by-products (Wohl, 1983, pp.227ff.). But the Alkali Acts of the 1860s and 1870s were exceptional in their effectiveness. All other industrial air pollution was left largely unchecked. A succession of by-laws to control urban smoke had been in operation from the 1840s. But the wording contained loopholes: the owners of factories were ordered to install improved furnaces which 'as far as practicable' eliminated smoke. Consequently the law was difficult to implement and prosecutions were rare (Wohl, 1983, pp.220–25).

In the first nation to experience the Industrial Revolution and its effects, city smoke aroused much less concern than the sharp deterioration in urban sanitary conditions. It was here that technology would make one of its greatest contributions to the modern city; so much so that the historian of Victorian cities was led to reflect that

> perhaps their outstanding feature was hidden from public view – their hidden network of pipes and drains and sewers, one of the biggest technical and social achievements of the age, a sanitary 'system' more comprehensive than the transport system.

(Briggs, 1968, pp.16–17)

Merthyr – with hardly any sanitary facilities – again allows us to see the issues in the clearest light. It lacked sewers and had only a few primitive latrines. It did not have a continuous supply of uncontaminated drinking-water: streams and the main river were the sources of water for washing, cooking and drinking, and these were polluted by industrial waste and sewage. Faced with a shortage of clean water, acute competition arose between industrial and domestic users. Local opposition to improvement was motivated by fear of increases in the rates. But the intervention of central government, in the form of the General Board of Health (established 1848), began to be felt. A delicate balance was, however, maintained between the extent of that intervention and the autonomy of the local board of health, whose members included men with vested interests in local industry. In the end, an elaborate sanitary system was planned and executed by expert engineers. Decisions were made on the choice of materials for the many kilometres of pipes and sewers. Both topography and geology had to be considered. The reservoir had to be built on solid terrain, not on geological faults. Gravity flow of water from an upper reservoir was the simplest method but, for some elevated districts, steam-engines were needed to pump up the water supply. The removal of sewage raised the difficulties of finding a suitable site for the outfall and then accomplishing the final stage of treatment – again a technological problem for expert consultants. Funding for all this came from loans, repayable on the rates (Grant, 1988/9).[2] None of this was peculiar to Merthyr. Therefore it is helpful to bear it all in mind when considering other cities of the Industrial Revolution.

A powerful impetus for improving the health of cities through the technology of public works came with the publication, in 1842, of Edwin Chadwick's *Report on the Sanitary Condition of the Labouring Population of Great Britain*. The campaigning author was a principal official of the Poor Law Commission, an institution which, since Elizabethan times, had administered the relief of poverty through funds raised by taxes on parishes. Now Chadwick presented a mass of evidence, collected by an army of local investigators, including doctors, which more powerfully than anything before demonstrated the shocking sanitary conditions that were part of everyday life in rural districts and, worst of all, in the overcrowded towns of industrial Britain. An abundance of filth was filling Britain's air with poisonous vapours – miasmas, the source of the terrible epidemic and other diseases currently afflicting the population, or so Chadwick and a large sector of medical opinion believed. Chadwick called for urgent action by government which was failing in its responsibility to protect the health of the population. The remedy seemed clear enough to him: the filth must be removed from dwellings and streets by an efficient water supply, flushing the refuse through a multitude of drains and sewers.

Anticipating protests that such a scheme would be enormously expensive, he assured a contrary result: profits would come from diminishing mortality and the consequent lowering of the burden on the parish to support destitute widows; he estimated that his policy would increase the life-expectancy of the working classes by at least thirteen years. And further economic gain, offsetting the initial heavy cost of laying sewers and connecting pipes to houses, was promised in his idea of using the flowing sewage as agricultural fertilizer. He insisted that the success of his plans, as well as the winning of the initial confidence to invest in them, depended on the employment of experts who possessed 'the science and skill of civil engineers'.

But the government took no immediate action, beyond ordering another investigation under the Royal Commission on the Health of Towns, which confirmed much of Chadwick's *Report*. It was only the terrifying approach of Asiatic cholera in 1847 that goaded the government into passing the Public Health Act 1848. The Act's concentration on the technologies of water supply and sewer-construction reflected Chadwick's thinking. But there the resemblance ended. Although a General Board of Health was created, there was to be no national system of hydraulic engineering organized by central government, and little in the way of compulsory improvements. Instead, the improvement of water supply and sewers was left to the new local boards of health, institutions that were not made mandatory until one-tenth of the

---

[2]   Extracts from Grant's paper are reproduced in Goodman (1999), the Reader associated with this volume.

ratepayers petitioned for one. Only where the death-rate rose above twenty-three per 1,000 did central government force a town or rural district to establish a local board. And even then there was no compulsion to introduce sanitary engineering; it was left to local authorities to decide whether or not to use the powers they had been given. But the strong hand of central government was felt in one way: although insanitary, dilapidated dwellings remained, no new houses could be built within the jurisdictional area of a local board unless they were provided with efficient drains.

Chadwick's dream – that technology for public health would be implemented as a nation-wide plan enforced by central government – was too drastic for the British political tradition. While it is a simplification to portray that tradition as *laissez-faire*, consistently shunning government intervention and jealous of local independence, those sentiments were real enough. When, in 1853, the House of Commons debated a bill to make vaccination against smallpox compulsory, Sir George Strickland protested that this smacked of continental compulsion and abandoned England's way of voluntary action, whereby the people were left to 'their own good sense' (Burn, 1964, p.157). Nevertheless he was ignored, and the Vaccination Act passed. And in the other crucial area of sanitary engineering, the power of central government gradually became more intrusive. In 1858 stronger central direction resulted when the Privy Council became responsible for public health, with John Simon as its energetic medical officer. Although he disliked Chadwick's interventionist plans, which seemed to him like 'papal forms of civil government', he was determined to introduce 'the rudiments of sanitary civilization' into Britain's 'savage life' (quoted in Wohl, 1983, pp.95, 149). His department made progress towards that goal by administering loans to local authorities for sanitary engineering works recommended by its inspectors. By 1871, millions of pounds of such loans were being authorized by a new body, the Local Government Board, headed by a minister responsible to Parliament. This central government agency offered expert advice on the technicalities of hydraulic engineering to some 1,500 local health authorities.

From the mid-1860s there was a notable trend towards the municipal take-over of private water companies. This development was due to growing opinion that the rapidly increasing demands for water could only be satisfied by a change of management: for efficient supply, public control had to replace the existing, unregulated, private companies. Municipalization was accompanied by another notable innovation – the provision of a continuous supply from a purer source, sometimes many kilometres away in remote countryside. That stimulated dam technology and a heavy demand for cast-iron pipes for urban distribution. It has been argued that the self-interest of influential businessmen also drove water supply towards municipalization. They wanted pure drinking-water to sustain a healthier work-force and pure, soft water for industrial use in their textile factories. Eventually the business and propertied classes would also benefit from the new water supplies at high pressure that protected their property by providing a more effective fire-fighting jet (Hassan, 1985).

The removal of human excrement, produced in unprecedented quantities by the swelling urban populations, was the most urgent problem of all. Still in 1900, the cesspool or a system of public collection of pails predominated over water-closets. But real progress was gradually achieved. From the middle of the nineteenth century, experts in hydraulic engineering influenced the adoption of house drains made of stone pipes, glazed to render them impervious. Experts made a further contribution in showing that brick sewers of oval section were preferable for drainage because the flow was faster than in circular sewers. But house drains and sewers only transferred the disposal problem to the local river, which became polluted with sewage, foul-smelling and deadly to anyone who drank from it. Engineers and chemists proposed various solutions. In one method, chemicals were added to precipitate the faecal matter held in suspension or solution. The resulting sludge was collected and spread over large areas of rural land on the outskirts of cities to form a sewage farm. In Nottingham the sewage farm grew grass to feed grazing cattle and pigs, in effect a recycling process that put the treated sewage back into the food chain. Another method, still in use, filtered the sewage through a succession of layers of different materials – first sand, then gravel, the purer water finally trickling through and collecting on the impervious lowest layer of tiles. The slime collecting on the sand was removed. Filtration was a slow process, and was speeded up in the 1880s by the application of compressed air. By this time Britain's largest cities were endowed with a copious and continuous water supply, pure enough to drink, and flowing fast enough to flush the new sewers.

Two matters of interpretation remain. Why were the urban authorities slow to adopt the new technology of sanitary engineering which offered so much? A common view of historians has been to blame the reluctance to pay, vested interests, and a determination to protect local government from the unwelcome intrusion of Parliament at Westminster. This picture is beginning to change as a result of other interpretations, much more sympathetic to these Victorian local authorities. On this view the technology of hydraulic engineering was difficult to understand in its detail, especially given the need to assess numerous competing projects put forward by engineers and scientists. These technicalities, along with uncertainty on the effectiveness of expensive schemes, caused understandable delays; the marvel is that so much was achieved (Hamlin, 1988).

Did these hectic years of sanitary engineering produce the desired result? No one denies that the urban death-rate fell sharply in the second half of the nineteenth century. But is this to be attributed to purer water and the new efficient sewers? The connection used to be taken for granted, part of the triumphant march of progress of modern technology. But this, too, is not as simple as was once supposed. The several improvements were not introduced simultaneously, the final step of sewage treatment coming decades after the provision of house drains; so the full benefits would not have been felt for a long time, or so it is argued. And it is alleged that by 1850, when the improved water supplies began to function, cholera and typhoid, the main water-borne diseases, had already receded (Hassan, 1985, pp.543–4). Despite these reservations, most historians continue to assume that sanitary engineering contributed to the perceived reduction in urban mortality. They can be no more precise than that, because of an insuperable difficulty: in these decades of falling death-rate in the cities, there was not only the spate of sanitary engineering projects, but also a marked improvement in nutrition. It is impossible to isolate the effects of these two types of progress. All one can do is acknowledge that populations who were better fed had a greater resistance to disease, and that this, too, has to be part of the explanation of the falling death-rate in mid-Victorian cities.

1.4   Mechanization and new building types

The Industrial Revolution left a permanent imprint on the complexion of cities. Still today the scars and the embellishments are clearly visible. These physical features include new building types, manifestations of hectic bursts of new technology, erected at the centre or periphery of a city.

Factories and warehouses

Urban factories of the new textile industry first appeared in Derbyshire (Cromford, Belper and Derby – the earliest: 1721), Nottingham and Preston. Their imposing forms were dictated by the mechanization of processes that until then had been dominated by manual operations. Now that was all changed by a spate of eighteenth-century inventions, gadgets and machines, which speeded up the preliminary carding (the combing and disentangling) of fibres, and the subsequent stretching, twisting and spinning into yarn. Arkwright's water frame for spinning (1769), his carding engines and roving machines (for drawing out and twisting the fibres), and Crompton's mule for spinning (1779), constituted a new technology that transformed Britain's cotton industry, one of the principal motors of the Industrial Revolution. This new machinery was bulky, heavy and produced strong vibrations in its continuous, violent operations. The building in which it was to be accommodated would require a strong structure and a large floor area. And for textile-production in all of its phases to be concentrated on this site, the building must be designed to facilitate the whole sequence of operations, from the initial work on the raw material to the subsequent processing stages that ended with the spinning of yarn. A single-storey structure occupying an extensive area would not have been practicable, whether erected in town or country, because of the inefficient distribution of power in such a factory. The power of a water-wheel or steam-engine would then have to drive the textile machinery by long, connecting belts. But the further the machine from the central power source, the greater the power losses from friction. And quite apart from this inefficiency in power supply, a sprawling single-storey structure in towns would have been prohibitively expensive because of the high cost of land. The answer was a multi-storey factory with specialized machinery suitably distributed on separate floors, and all closely connected to an installed power source, a water-wheel or, later, one of Watt's steam-engines. Another conspicuous feature of the textile factory, the large area of windows, met the essential requirement of maximum daylight for the work-force (see Figure 1.4).

The typical eighteenth-century textile mill was five or six storeys high, with strong brick walls, and wooden floors laid on wooden joists supported by timber beams. The traditional use of timber, the saturation of the floors with oil dripping off the machinery, the presence of quantities of cotton or other textile material, and the installation of hundreds of candles or oil lamps to provide lighting for nightwork, together produced ideal conditions for a conflagration. Several of the early mills were soon to be totally destroyed by fire. One response was to introduce fire-extinguishing equipment. At his cotton mill in Stockport in 1819, Thomas Orrell installed a water-cistern in the roof from which pipes were brought to each floor; the water was pumped by the factory's steam-engine (Tann, 1970, p.139). A more effective solution (less common because of the expense) was a revolutionary change in building materials – the substitution of iron for timber to provide a 'fireproof' factory. The transition first occurred in Derby where, in 1792–3, William Strutt, an engineer, built his cotton mill to a design that has been seen as 'one of the most important technical innovations in building since medieval times' and 'the starting-point of a structural revolution which leads to the skyscrapers of

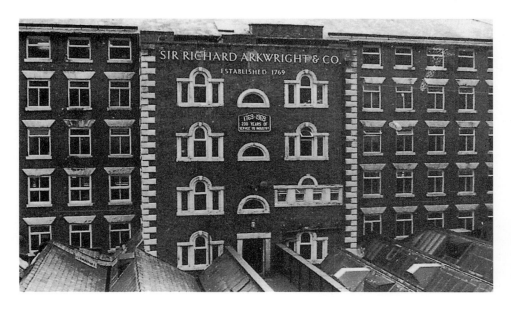

Figure 1.4 Cotton-spinning mill, Cromford, Derbyshire, 1783: six storeys, brick and scores of windows (reproduced from Pevsner, 1979, p.276; photograph: Dr Jennifer Tan)

Figure 1.5 Six-storey cotton mill, designed by William Strutt and erected in Derby, 1792–3; the first multi-storey 'fireproof' building. The floors consisted of brick arches, springing from heavy beams supported by two rows of solid cast-iron pillars (adapted from Skempton and Johnson, 1962, p.176)

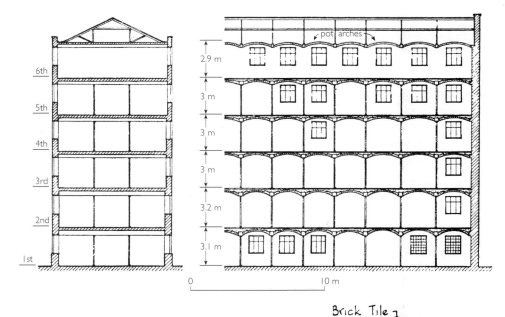

Figure 1.6 Strutt's fireproof construction as used in mills at Derby, Milford and Belper. It had a number of interesting features: solid, star-section, cast-iron pillars; timber beams protected by plaster; brick-arch floors; and wrought-iron tie-rods (reproduced from Skempton and Johnson, 1962, p.177)

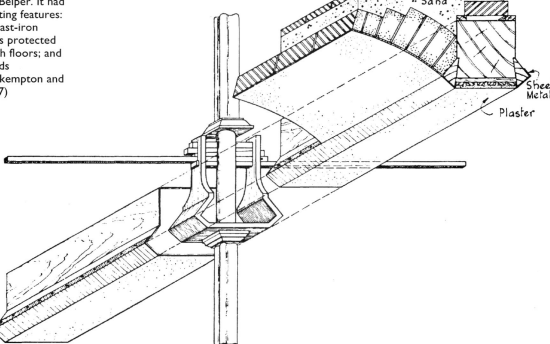

Chicago a century later' (Skempton and Johnson, 1962, pp.176, 178); but this view of the history of building progress is a simplification amounting to technological determinism. Here, in Derby, little more than a hundred metres from the spacious old market-place, arose a six-storey cotton mill with floors supported by rows of cast-iron pillars. The floors, in place of inflammable wood, consisted of shallow, brick arches, and the ceiling of the top storey was made of hollow, fireproof earthenware pots. Some timber remained in the structure – large timber beams coated with plaster for protection from fire, and exposed timber roof-trusses, a vulnerable part of the design (see Figures 1.5 and 1.6). The next step, the elimination of all timber, was taken in the construction of a flax-spinning mill in Shrewsbury in 1796–7. Cast-iron beams were the new feature here and, together with the cast-iron pillars, constituted a five-storey metallic factory: the brick arches of the floors were the only other material (see Figure 1.7). The design was by

Figure 1.7 Marshall, Benyon and Bage's flax-spinning mill, Shrewsbury, 1796–7 (reproduced from Pevsner, 1979, p.277; photograph: West Midlands Photographic Services, Shrewsbury)

Charles Bage, an engineer who had just worked out the rudiments of what may have been the earliest theory of the strength of cast-iron columns. And soon after this, he performed pioneering experiments on the torsional strength (the degree of resistance to the strains generated by twisting) of iron (Skempton and Johnson, 1962, p.179).

By the 1820s, English factories, built on these plans, were rising as high as nine storeys, bringing amazement to visiting Prussian officials on a government mission to observe at first hand the world's most advanced industrial nation. One of them, Karl Friedrich Schinkel, an architect, excitedly recorded in his notebook that 'the miracles of the new age are the machine and the buildings for it, called "factories"' (quoted in Pevsner, 1979, p.277). And in Manchester in 1826, he sketched an urban landscape of towering factories (see Figure 1.8). But these were not yet true skyscrapers: they had an iron frame, but the outer brick walls were still load-bearing. Only when the iron beams carried the walls do we have the metal skeleton that characterizes the US skyscrapers of the 1880s.

Figure 1.8 Mills in Manchester, 1826; sketch by K.F. Schinkel (reproduced from Pevsner, 1979, p.277)

Figure 1.9 Warehouse, 46–54 Mosley Street, Manchester; built in 1845 to Edward Walters' design (reproduced from Hitchcock, 1954, plate xii.14; photograph: J.R. Johnson)

As the cotton mills and other factories expanded output, so the need grew for storage facilities, capacious and fireproof. Warehouses, incorporating the cast-iron frames developed in mills, became an increasingly familiar part of the urban scene (see Figure 1.9). And from around 1850, cast-iron beams in iron-framed buildings began to be replaced by wrought iron because this material was much stronger in tension.

How effective was the fireproofing of the new structural iron? According to one expert at the time, C.F. Young[3] (a mid-Victorian civil engineer), iron pillars and beams, far from reducing urban conflagrations, had resulted in worse fire damage. That, he argued, was because of the entirely misplaced confidence in the resistance of iron to fire. As evidence, he could refer to several recent disasters. His discussion included passing remarks on the improved fire-protection that concrete might provide, and even hints of some return to timber (large wooden beams can smoulder slowly in a fire). This text is also interesting for the evidence it supplies of the increasing risks of city fires from the growing accumulation and storage of inflammable materials, the consequence of an expanding chemical industry (Young, 1866, pp.57–63).

And indeed, contrary to what may at first be supposed – it was widely supposed by Victorians – iron is not fireproof. When subjected to strong heat, iron loses its strength because of uneven expansion. Also the use of cold water in fire-fighting caused the hot iron to crack because of uneven cooling. Victorian buildings with exposed cast-iron or wrought-iron pillars and beams collapsed when blazes spread to them. This happened at Gateshead in 1854 when iron-framed warehouses were totally destroyed in a fire that reached Newcastle, causing many casualties and an estimated loss of property of over one million pounds. It happened again in 1861 at London's Tooley Street wharfs and warehouses, where a conflagration, regarded as the worst since the Great Fire of 1666, burned for fifteen days with damage to the tune of one and a half million pounds. And three years later, another 'fireproof' building, a warehouse for carpets in Gresham Street in the heart of the City of London, was turned into a red-hot, crumbling shell with total destruction of precious textiles.

A growing sense of insecurity in the cities and of dissatisfaction with the new generation of fireproofed building types was unmistakable in the evidence given to the House of Commons Select Committee on Fire Protection (1867) by builders, surveyors to insurance offices and merchants. One witness, C. Fowler (an architect and district surveyor), looked back for a solution: traditional oak beams would withstand a fire better than iron girders. But others looked forward to new materials, to the use of concrete, widely adopted for floors of buildings in Paris and already imitated in some London houses on the Duke of Bedford's estate. It was clear that the exposed iron frames of warehouses and factories required some covering material to protect them from fire. Testing of a variety of materials became intensive. Bricks and masonry were tried; but in the 1880s and 1890s it was found that the most effective fireproofing could be achieved by casting concrete around steel.

Another development within factories was to have lasting consequences for the urban environment. In the 1790s William Murdock, a technician employed at the Boulton and Watt steam-engine manufactory at Soho, Birmingham, achieved the first large-scale production of coal gas and used it to illuminate the works. Here was an illuminant that gave a better light than candles or oil and, once installed, was cheaper; it was a very attractive proposition for factory owners. Between 1805 and 1811, hundreds of gaslights were installed in a large cotton mill in Salford, in another near Halifax, and in two flax mills in Shrewsbury. After the factories, the new coal-gas technology lit up city streets and public buildings. By the 1820s, lampposts were an established feature in most of Britain's large

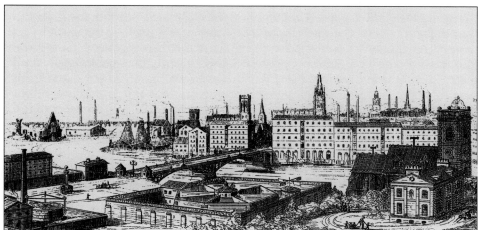

Figure 1.10 Urban decline over 400 years, as perceived by Pugin. *Top* Catholic town in 1440. *Bottom* the same town in 1840 (Pugin, 1841, p.104, facing)

towns (and spreading to many towns of Europe and the USA). Gas lighting created a new urban night-life as well as a means for securing greater law and order. By mid-century almost any British town with a population exceeding 2,500 had its gas company; gas-works and gas-holders became a common part of the urban fabric.

The changing face of the Victorian city was not always welcomed as a sign of progress and a source of municipal pride. The new building types brought utter despondency to Augustus Pugin (aesthete and architect, who gave Gothic detail to the new Houses of Parliament, rebuilt after their destruction by fire in 1834). A Catholic, he condemned the urban architecture associated with advancing technology and science as a manifestation of deepening moral degeneracy, a further victory for that same 'destructive or Protestant principle' that had wrought devastation over the centuries. And to drive the message home, he introduced a plate (see Figure 1.10) comparing two models of cities, in the second edition of his highly influential book, entitled *Contrasts: or, A Parallel between the Noble Edifices of the Middle Ages and corresponding buildings of the Present Day; showing the present decay of taste* (1841). One half of the plate, an engraving with the caption 'Catholic town in 1440', depicts a medieval, walled city dominated by the soaring spires of several churches. In the other half of the plate, 'The same town in 1840' is barely recognizable as such. Ironworks, factory chimneys, warehouses, a gas-works, gas-lamps on posts along the now iron bridge fill the unbeautiful scene, which is completed by a lunatic asylum and a 'Hall of Science', here presented under the epithet 'socialist' (a term then imbued with utter contempt).

Railway stations

Pugin's caricature of the industrial city curiously lacks a building type that had already become one of the most conspicuous products of the mechanizing age, even its most potent symbol – the railway station. By 1875, the journal *Building News* could speak of stations as 'the cathedrals of the nineteenth century'. But the earliest stations were unremarkable and would not yet have warranted that effusive description. The world's first passenger railway connected industrial Manchester to the port of Liverpool in 1830. Liverpool's first station had a single platform for arrival and departure, the locomotives brought into position by turntables. An ordinary timber roof covered the rails; it was supported by a porch and a wall (see Figure 1.11), and had apertures to let in light.

What turned stations into the architectural and engineering marvels of the age was the new industrial technology. The achievement was in the roof, a breathtaking composite structure of iron and glass. Here the strength of iron eventually permitted spans unequalled in the history of architecture, and the arched iron functioned as ribs to support glass, a transparent roof admitting light while providing weather protection. In a first stage, small panes of glass were used. The real change came in 1847, when the glass-manufacturing firm of Hartley (in Sunderland) invented a new rolling process, producing large, strong sheets of plate glass. The full effect was visible at New Street Station, Birmingham (1854). The contemporary excitement can still be felt in the newspaper engraving published in the week of the opening (see Figure 1.12). The semicircular roof – almost 340 metres long, twenty-five metres high, and with a span of sixty-five metres – filled the station with light, and beautiful flowing lines. This daring span eliminated intermediate supporting pillars: in a recent accident, at Southwark Station, a train had collided with a roof support, bringing down much of the roof. Now, at New Street, the riveted-iron span rested on pillars at its extremities, well away from the trains. At St Pancras (1868) the peak was reached with an arched roof-span of seventy-three metres, erected with the help of a vast timber scaffolding (see Figure 1.13, p.20). And next to the station, the huge hotel erected in neo-Gothic style still stands as a London landmark. It had hydraulically operated lifts for passengers and their luggage. Hotels had, since the 1840s, been built by railway companies for the convenience of their clients. Along with the stations, they formed conspicuously novel features of cities, functioning as overnight lodges, restaurants and meeting places which enriched urban social life (see Figure 1.14).

Figure 1.11 Crown Street Station, Liverpool, 1830 (with permission of the National Railway Museum/Science and Society Picture Library)

Figure 1.12 Birmingham's new central railway station, at New Street, 1854 (*Illustrated London News*, 3 June 1854, vol.24, p.505)

Figure 1.13 St Pancras Station, London: the roof is being erected with the aid of movable timber scaffolding; 1868 (reproduced from Simmons, 1968, plate 5; photograph: Butterley Company and Geoffrey Platts)

Figure 1.14 Charing Cross Hotel (1864) in London still forms the façade of the railway station (photograph: by permission of the George Washington Wilson Collection, Aberdeen University Library)

The railway termini established new gateways to cities. The monumental entrance to Euston Station (1835–9) has been interpreted as a symbolic statement of conquest by the railway-builders, conquest of long distances as well as an invasion of the greatest city in the world. Arriving and departing travellers were confronted by a colossal triumphal arch (see Figure 1.15).

1.5 Cities and the transport revolution

Figure 1.15 Euston Station, London: the monumental entrance (1835–9), since demolished (reproduced from Pevsner, 1979, p.225)

The first phase of Britain's transport revolution occurred on water. The hectic years of canal-building began around 1760, and by 1830 over 6,000 kilometres of new inland waterways had been cut. Few canals were built after that, because the railways then began their take-over of transport with a far more revolutionary technology. The canals had played an important role in the beginnings of the Industrial Revolution, above all in the bulk carriage of coal and iron for the rising industrial centres, at rates far below the cost of road transport. This was where canals came into their own. And through this bulk carriage of low-value merchandise, canals assisted urbanization, by supplying the basic necessities of life. The growing demands of rapidly expanding urban populations for coal-fuel and building materials (sand, gravel, lime, slates, timber and bricks) were satisfied by barge traffic on the canals. A few new canal towns also arose, including Goole (Yorkshire), Grangemouth (central Scotland) and Runcorn (Cheshire), none of them large.

Once the railways came into operation, canals could not compete with the speedy haulage of the steam locomotive, and activity on them dwindled – a slump also attributable to the aggressive tactics of railway companies intent on damaging canal business. Now the great railway age began. It would have profound effects on the development of Britain and subsequently many other nations. Above all, the railways would leave a lasting mark on cities.

John Kellett's standard work, *The Impact of Railways on Victorian Cities* (1969), continues to have much of value.[4] First, there is his comparative approach, a powerful tool in history for separating changes that are common from changes that are peculiar to individual cases. That is all potentially valuable in making progress towards explanations linking cause and effect. Here the aim is to use comparisons to discern how the arrival of railways tended to affect cities in general, and how particular circumstances prevailing in individual cities introduced an additional influence on the course of events. Second, his discussion makes explicit recognition of the difficulty of isolating the railway from all other developments that might have affected the growth of cities at the time. Third, his

[4] Extracts are reproduced in Goodman (1999), the Reader associated with this volume.

analysis is notable for its split of the period into different decades. That is a warning against unbridled generalizing: what you say about the railways may apply in the 1830s, but no longer in the 1850s. On the other hand, if carefully phrased, generalizations about the railways will hold for the entire Victorian period – for example, the constant importance of economic motives.

This brings us to the due attention that Kellett gives to economic matters. Economic conditions often have to be considered in the history of technology, and especially in making sense of the railway age. For in Victorian Britain – unlike continental Europe, with its general pattern of state ownership – railway-building was undertaken by private companies seeking profits for themselves and their shareholders; it was a business enterprise involving huge capital investment for locomotives, rolling stock and lines. And most expensive of all, land had to be purchased for building the lines, stations and works. The cost of that land soared, the nearer the site to a city centre. Kellett finds that the search for cheap land – and the minimizing of obstacles to land-purchase by negotiating with a single, large landowner – more than anything else dictated the route railway lines took into the cities.

Railways and urban form

New towns were created by the railways. This was the result of the change of policy whereby railway companies increasingly turned away from purchasing locomotives and rolling stock and began to manufacture these themselves; it was the only way to ensure punctual delivery of equipment, until then the object of intense competition among rival companies. Some of the new railway company workshops were located in existing towns with a long-established manufacturing tradition (Derby, Wolverhampton, Manchester, Glasgow), adding to their prosperity. But the site of other workshops was determined by strategic locations, such as junctions on the rail network. These nerve centres of railway operation were remote and sometimes unpopulated. Yet this was where the works and large numbers of workers were needed. To attract skilled artisans, the railway companies built houses for them, which – together with the extensive construction or repair works – constituted the nucleus of a new town, a sudden creation. Crewe 'came into being on a cold, wet day, on 10 March 1843' (Drummond, 1995, p.1), when the Grand Junction Railway Company settled 221 of its employees and their families on an isolated site – there were just a few scattered farmhouses in the area – where trunk railway lines from Birmingham, Manchester, Liverpool and Chester happened to converge. Crewe would become the world's most advanced railway-works, specializing in locomotive-manufacture, as well as one of the world's busiest stations. From a rural population of a few hundred in 1840, the expanding new railway town reached almost 20,000 by 1872.

According to the eminent railway historian Jack Simmons, railway towns were unplanned, for two reasons. First, the very nature of railway work, 'largely in the open [and] obtrusive with coal smoke and shunting noise', discouraged the idea of a 'neat, well-ordered town'. Second, railway companies were accountable to their shareholders, and so were not free 'to buy land and put industry in a planned new town' (Simmons, 1986, p.193). This argument will not stand up to the evidence. At Crewe the governing railway company imposed a grid plan around the central works, building and allocating houses for a clear hierarchy of personnel. Closest to the works, and constantly shaken by the drop of the massive steam-hammer of the forge, were the neat, red-brick cottages of the artisans. Further out lay the detached mansions of the engineers. Further out still, were the palatial villas of the company's officers: at the ends of their gardens were specially built platforms giving access to the railway line. The company provided urban amenities – gas and water supply, a public bath, a railway hospital, a church with a tin notice-board in the company colours, three schools, and a mechanics institute where apprentices could attend evening classes to learn how to use tools and the hearth. The rhythm of life in the town was dictated by the factory steam-gong, sounding the start and finish of the day (Drummond, 1995, p.13).

The same grid plan was given, and in the same year as Crewe, to another new railway town, just a mile away from the market town of Swindon. Again the spot chosen for a railway-works was the point of convergence of lines, in this case lines from Bristol, Gloucester and London. In 1843 the Great Western Railway Company housed 300 workers in eight parallel streets; soon after, the paternalistic company provided a church, schools, a park, gas and water supplies and a mechanics institute. At first a maintenance works, New Swindon soon expanded into a vast complex where locomotives, wagons, coaches and rails were manufactured. Many of the work-force came from the industrial north. Traditional old Swindon kept its distance, but functioned as a retail service centre for railway workers, as well as housing some of them. The two towns were eventually united in 1897 (Turton, 1969; see Figure 1.16).

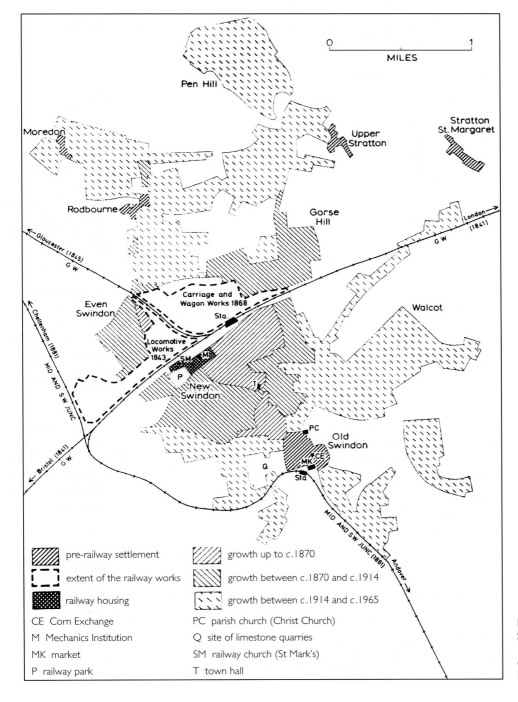

pre-railway settlement

extent of the railway works

railway housing

growth up to c.1870

growth between c.1870 and c.1914

growth between c.1914 and c.1965

CE Corn Exchange

M Mechanics Institution

MK market

P railway park

PC parish church (Christ Church)

Q site of limestone quarries

SM railway church (St Mark's)

T town hall

Figure 1.16 The growth of Swindon (adapted from Turton, 1969, p.130; with the permission of the publishers, Graphmitre Limited, Tavistock, Devon)

Cities old and new were profoundly affected by this revolution in transport. The railway has been seen as 'perhaps the most important single influence on the spatial arrangement of the Victorian city' (Morris and Rodger, 1993, p.22). What exactly does that mean? In general terms, it means that British cities by the 1850s were

> uniformly super-inscribed by the gigantic geometrical brush-strokes of the engineers' curving approach lines and cut-offs, and franked with the same bulky and intrusive termini, sidings and marshalling yards.
>
> (Kellett, 1969, p.2)

Barriers to urban expansion were introduced in the form of railway viaducts. The effect was particularly severe at Wolverhampton, where long viaducts, a railway-works and a tangle of line junctions on the town's northern outskirts, and more lines to the east, all prevented house-building in these districts well into the twentieth century. In Glasgow and Manchester, hundreds of hectares of city land were bought up by the railway companies, so constraining the direction of subsequent urban expansion. Elsewhere, conspicuous urban improvements followed the arrival of a railway. This happened in Huddersfield. The whole town, a thriving centre of the West Riding's woollen industry, was owned by an aristocratic family, the Ramsdens. They foresaw prosperity for their estate and the inhabitants if the town was connected by rail to London and other markets for its manufactures. A deal was struck with a railway company whereby, in exchange for operating a line in Ramsden territory, the company agreed to pay considerable sums towards urban development. By 1854, streets in the vicinity of the elegant, new station had been created, widened, paved and provided with sewers and drains (see Figure 1.17). A carefully planned square was laid out in front of the station with a fine neo-classical hotel (Whomsley, 1974). Little demolition resulted from these plans, but in other cities it was considerable: municipal authorities sometimes encouraged demolition, because the railway companies could execute the task for them of clearing slums. But this sanitary improvement had a negative side: the displaced poor, uncompensated, were forced to resettle in dilapidated areas of new squalor.

Railway stations acted like magnets, attracting retail shops, places of refreshment and the urban traffic of horse-omnibus routes. But residential housing tended to be repelled.

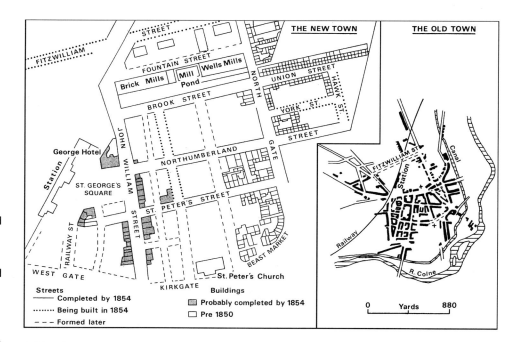

Figure 1.17 Huddersfield and the coming of the railway. By 1850 the town was a junction served by three railway lines; from 1854 its station stimulated extensive urban development, apparent from a comparison of these two maps (reproduced from Whomsley, 1974, p.204)

Freight traffic, a principal focus of attention for the profit-hungry railway companies, quickened the pulse of urban industrial activity. The railways brought to cities coal-fuel and raw materials for manufacture, and took out their industrial products for distribution to markets. A new building type, huge railway goods stations and warehouses, became conspicuous on the urban scene. In Liverpool, a warehouse stored raw materials arriving for the Lancashire textile industry and received finished products for shipment to the world (see Figure 1.18). And beneath Manchester's London Road passenger station lay a goods station, which in mid-century was handling some 2,000 tonnes of merchandise a day; filled wagons were raised by a steam-operated crane to the level of the railway (Simmons, 1995, p.44).

The transformation of urban economic life by the coming of the railways is particularly evident at Cardiff, which by the 1870s had grown into a great coal-shipping port. In the eighteenth century, coal from the Rhondda Valley had been brought to Cardiff on the backs of mules, and later by barge on the Glamorganshire Canal (1798), in all some 100,000 tonnes of coal a year. In 1801 the population of Cardiff was no more than 1,870. It all changed dramatically in 1841 with the opening of the Taff Vale Railway, connecting Cardiff to the Rhondda coal-mines. That coal was found to be of excellent quality for fuelling the world's steamships. Hectares of extensive new docks greatly expanded Cardiff's port facilities; by 1874 it was exporting over 34 million tonnes of coal a year (Simmons, 1978, pp.294–5). Its population had soared to 33,000 by 1861, and to 164,000 by 1901. This was now a prosperous city with spacious thoroughfares.

Brighton is another example of dramatic transformation. Before the railway came, it was an established health resort patronized by royalty and aristocrats in the winter season. From London it was a five-hour journey by coach, a vehicle that could take no more than fourteen passengers. The London to Brighton railway, opened in 1841, inaugurated fast, mass travel and converted Brighton into a summer seaside resort. By mid-century, excursion trains were bringing day-trippers in their tens of thousands. A new street was cut to facilitate access to the station. Other new streets with villas were the residences of newly settled businessmen, who commuted daily to London, returning to the seaside for dinner (Simmons, 1986, p.236).

Towns left unconnected to the rail network sometimes suffered. Ilchester, a Somerset coaching town, lost a previously flourishing commerce as its coaching-inns closed down, and traffic moved elsewhere to the London–Exeter rail route (Simmons, 1986, p.283). But caution is needed here. Simmons gives much food for thought in his reflections that railways did not always generate trade and population growth; after Bath and Cambridge were connected to main lines in the 1840s, their populations fell: here the railways facilitated migration of people and businesses to London. And railways could slow down the Victorian movement of population from the countryside to urban industrial centres, by stimulating agricultural industry through rail links to rural areas, as in Wiltshire (Simmons, 1986, pp.16, 19).

Figure 1.18 Goods station and warehouse, Crosshall Street, Liverpool (photograph: by permission of the National Railway Museum/Science and Society Picture Library)

Railways and city food supply

The bonds tying town to country were strengthened and extended by the growth of Britain's railway network. So were established – in the 1840s, 1850s and 1860s – new lines of communication for food supply, reaching further into once-remote farming regions, enlarging the area of supply to sustain the expanding population of the largest cities. Now agricultural produce could be carried much faster and cheaper than ever before: it was as cheap to carry goods for hundreds of kilometres by rail as to cart them thirty kilometres by road. Lincolnshire potatoes arrived in bulk at the King's Cross terminus and, to store them, the Great Northern Railway Company erected dozens of covered warehouses. And at Manchester's Oldham Road station, the Lancashire and Yorkshire Railway Company opened a potato and carrot market to sell the food it had brought in. It was part of a trend towards the creation or relocation of large city-markets in the vicinity of railway stations.

The revolutionary new means of accumulating urban food-stocks permitted the growth of Victorian cities. It was not a matter of relieving famine. In western Europe, famine had disappeared long before the coming of the railways (Ireland's potato famine of 1845–6 was quite exceptional). But in British India and the Far East, famine was still a regular occurrence. And as the British took their railway technology to the other side of the world, so they brought with them an awareness of the power of the new transport system for rapid distribution of food. In 1869 Tokyo was in the grip of famine and, to prevent its recurrence, the British ambassador, Sir Harry Parkes, called for the building of a short railway (some thirty kilometres long) to connect the city to the port of Yokohama, from where bulk food imports might be rapidly carried by trains to reach a starving population. Such a strategy came easily to Parkes, whose origins were in one of the most industrialized of Britain's regions and a railway nerve centre, the Black Country; his father had been an ironmaster in Walsall. So was built (in 1872) the first of Japan's railways, an investment against famine.

In Britain the importance of the railways for urban food supply was not for the conquest of famine but the distribution, on an unprecedented scale, of highly perishable foodstuffs, which improved the diet of Victorian citizens. Urban meat supplies increased in two distinct ways. First, the traditional droving trade went into sharp decline. Previously, cattle from the great rearing regions of Scotland had been moved on the hoof all the way to Norfolk pastures for fattening, and then sent in droves to London for sale at Smithfield market. Now cattle, sent from Scotland in railway wagons, arrived with less loss of weight and in better condition. New cattle markets, close to the railway stations where the animals arrived, opened in Liverpool, Wolverhampton, Bristol, Shrewsbury and Islington (London). Second, from the 1850s there was an increasing trend to dispatch to the cities dead meat of animals slaughtered in the countryside. The slower steamships and roads might continue to deliver live cattle to towns; but only the speed of the railway, unaffected by most weather conditions, could maintain the freshness of dead meat brought in from a distance.

The same applies to fish, an increasingly important part of the Victorian city-diet supplied by the new transport technology. Before the railways, fish was carried short distances by road, but the cost of horse-drawn transport made it a luxury, and the speed of carriage was insufficient to guarantee freshness on arrival. The exception was heavily cured herring, which was conveyed slowly inland and in bulk to be affordable for the urban poor. From mid-century an important trend was initiated by the railways, culminating in the large-scale distribution of fresh cod, sole, haddock, plaice and turbot to cities far from the coast, and by the 1890s the establishment of an urban working-class institution – the fish and chip shop. The change did not happen overnight; there was nothing automatic or inevitable about this use of the railways to bring about a supply of fresh fish for the nation's cities. Initially the railway companies were not interested in developing a volume traffic in low-price fish; instead, their policy of amassing profits from the carriage of high-value merchandise was at first continued, and the companies did little to offer attractive rates and conditions of carriage. An important first step was taken by the Manchester and Leeds Railway in 1841, when negotiations with east-coast fishermen in Flamborough and Hull concluded with a reduced freight cost. A shop, opened in Manchester, was soon selling fresh fish daily at a price ($1\frac{3}{4}$ pence per pound) low enough to bring crowds of poorer customers. By 1845 this railway company was bringing to Manchester around eighty tonnes of fish a week. But elsewhere, fish supply by rail was hindered by the companies' refusal to bear the risk of carrying perishable fish, which they eventually conceded in the 1850s (Robinson, 1986). This spirit of railway enterprise and increasing co-operation with the producers led to the creation of the port of Grimsby: in the 1850s, the Manchester, Sheffield and Lincolnshire Railway Company built houses to rent out to fishermen, attracting them from Hull. And offers of free rail travel were extended to Grimsby's fish merchants to encourage them to seek new markets inland. By the 1870s, Grimsby had a thriving fleet of over 300 fishing vessels.

Milk was another perishable that eventually became a major item of Victorian rail supply, but again the innovation was not immediate. Before the development of the railways, milk from country farms could not be carried more than about fifteen kilometres by horse-drawn vehicles without becoming spoilt by souring. Consequently several of the largest cities had their own cows, grazing in fields and housed at night in cowsheds. But these buildings were dirty and lacking in ventilation; so there were frequent outbreaks of contagious diseases (foot-and-mouth, pleuro-pneumonia and rinderpest) among town cows. An urban cattle plague of 1866 closed down most of the town dairies, and some historians have seen this as the turning-point that established large-scale traffic of railway-borne milk from the countryside to the towns. Rural Ayrshire provisioned Glasgow; Leicestershire supplied Newcastle; Cheshire farms supplied Manchester and Liverpool; Aylesbury and, later, Wiltshire became sources of Londoners' milk. But the milk did not always arrive in good condition. Milk containers on trains were not at first well designed: the lack of a tight stopper caused the milk to churn into butter. And only when refrigeration techniques were introduced in the 1880s could railway milk remain fresh in hot weather (Fussell, 1966, pp.300ff.). The quantities of liquid milk available for urban supply depended also on the willingness of farmers to produce: that depended on whether or not it was more profitable to convert the milk into butter and cheese. At Wensleydale a railway, opened in 1878, was rarely used to transport whole milk until a decade later when the price of locally sold butter and cheese had fallen (Hallas, 1991). Another complication, suggesting a negative aspect to demand, comes from an official survey of the 1860s into national diet. Here a doctor gave evidence that many mothers were ignorant of the nutritive value of milk for their children after weaning (Burnett, 1966, p.143). Nevertheless, by the 1880s railway milk had become a feature of the economy of almost every large Victorian town, though at two pence per pint it was still not yet within the reach of all the inhabitants.

Other consequences of railways

The distribution of foodstuffs on Victorian railways represented a certain standardization of diet. When, in 1841, Huntley and Palmer's factory opened in Reading, attracted there soon after the town had been placed on a main railway line, its biscuits became familiar to households up and down the country. The same applied to the beef extract, Bovril. Standardization was brought by the railways to urban life in other ways as well. Their bulk transport of building materials began to introduce a certain sameness to city buildings in many parts of Britain. By the 1860s local quarries fell into disuse as developers drew on new centres of mass brick-production, situated beside railway lines in the vicinity of Peterborough, Bletchley and Bedford. And once the railways reached west Wales, a new national source of cheap slates for roofing became available (Simmons, 1995, p.348).

Standard British time was achieved only with the railways. For the railway system, it was essential for punctuality and safety of operation. In 1847 Manchester's municipal council decided to adjust its clocks to London time. But it wasn't until 1852 that Exeter agreed to do the same; the dean at last put the cathedral clock forward by fourteen minutes, to observe the time signals arriving from Greenwich by the new telegraph system that ran beside the railway line. Clocks with standard time became prominent at railway stations. The companies supplied their servants with watches and signal boxes with clocks, so stimulating the manufacture of chronometers. The language of the railway timetable, 'the six-thirty-five', was adopted by city businessmen; the railway had become the regulator of urban time (Simmons, 1995, pp.346–7).

The invasive railway brought damage to the urban environment, increasing smoke pollution and introducing a shrill whistling that aroused a protest movement among residents of Primrose Hill (in north-west London) in 1860. More serious, the historic fabric of cities was threatened as the railways penetrated closer to their centres. Parliament and municipal corporations had the power to stop the path of a railway, but much was allowed to be destroyed. Medieval buildings were demolished for the railways entering Bradford-on-Avon and Flint, and York's medieval walls were breached before controls were imposed. A successful conservation campaign by the residents of Perth preserved the city's beauty spot, a view of the River Tay, from the proposed building of a railway station on the site (Simmons, 1995, pp.159ff.). But in London a railway bridge across Ludgate Hill so ruined the view of St Paul's (see Figure 1.19) that one mid-Victorian could record:

> That viaduct has utterly spoiled one of the finest street views in the metropolis; and is one of the most unsightly objects ever constructed, in any such situation, anywhere in the world.
>
> (quoted in Simmons, 1978, p.303)

But the idea that the advance of the railway through Victorian cities signified the irresistible march of an impersonal force of technology is unacceptable. Kellett's study concludes that we must 'turn away from technology' if we are to understand the effects of railways on the Victorian urban fabric. Instead, he finds that the most important influence on the way individual cities reacted to the revolutionary transport was 'the framework of land titles'. The local urban landowners were, on this view, more important than the railway managers as agents of change (Kellett, 1969, p.421).

Let us now consider one of the most difficult questions in the history of the railways – their connection with the appearance of those characteristic offshoots of modern cities, the suburbs. The identification of a genuine difficulty in history, such as this, has considerable importance – first because it takes us to the frontiers of knowledge and sets a problem for research, and second because it confronts us with the true complexity of the past. The essential thing is not to ignore the difficulty: if we do, an unacceptably simplified history will result. Instead, the difficulty needs to be recognized and confronted, and its existence made explicit in argument. The difficulties that prevent a simple explanation of Victorian suburban growth as a mere consequence of the railways (or any other transport) are vividly brought out in the fine introduction by F.M.L. Thompson to his edited collection of essays, *The Rise of Suburbia* (1982).[5]

---

[5]   Extracts from Thompson's introduction are reproduced in Goodman (1999), the Reader associated with this volume.

Figure 1.19 The bridge over Ludgate Hill, London (photograph: by permission of the George Washington Wilson Collection, Aberdeen University Library)

The chronology of change is crucial to the argument: suburbs came into existence before the establishment of a railway service, before even the earlier horse-omnibus service. That means that the railways were not necessary for the creation of the inner suburb (up to six kilometres from a city centre). Nor were they sufficient to generate an outer suburb (over eight kilometres from the centre): there could be a long time-lag between the opening of a railway station, in places such as Bexley, and the formation of outer suburban dormitories. So again, there is no technological determinism here. Instead, in this argument, human choice dictates change – the desire to imitate the rural lifestyle of the upper classes or the preference of rural migrants to settle in suburbs rather than in the centres of cities. At most, railways and other forms of transport facilitated suburban living; they did not create it (Thompson, 1982). Such is the prevailing current of historical interpretation.

In the next two chapters the spotlight will fall on some of the largest cities of the nineteenth-century industrial world. We will meet again some of the general features of change discussed in this chapter. But this time we are going to investigate how the patterns of change were influenced by conditions peculiar to individual cities.

References

ANDERSON, M.S. (1987, 3rd edn) *Europe in the Eighteenth Century, 1713–1783*, London and New York, Longman.

ARMENGAUD, A. (1973) 'Population in Europe, 1700–1914' in C. Cipolla (ed.) *Fontana Economic History of Europe*, Glasgow, Collins, vol.3, pp.1–76.

BRIGGS, A. (1968) *Victorian Cities*, Harmondsworth, Penguin Books.

BURN, W.L. (1964) *The Age of Equipoise: a study of the mid-Victorian generation*, London, Allen and Unwin.

BURNETT, J. (1966) *Plenty and Want: a social history of diet in England from 1815 to the present day*, London, Nelson.

DRUMMOND, D.K. (1995) *Crewe: railway town, company and people, 1840–1914*, Aldershot, Scolar Press.

EVANS, C. (1994) 'Merthyr Tydfil in the eighteenth century: urban by default?' in P. Clark and P. Corfield (eds) *Industry and Urbanization in Eighteenth-century England*, Leicester, Leicester University Press, pp.11–19.

FUSSELL, G.E. (1966) *The English Dairy Farmer, 1500–1900*, London, Cass.

GOODMAN, D. (ed.) (1999) *The European Cities and Technology Reader: industrial to post-industrial city*, London, Routledge, in association with The Open University.

GRANT, R.K. (1988/9) 'Merthyr Tydfil in the mid-nineteenth century: the struggle for public health', *Welsh History Review*, vol.14, pp.574–94.

HALLAS, C. (1991) 'Supply responsiveness in dairy farming: some regional considerations', *Agricultural History Review*, vol.39, pp.1–16.

HAMLIN, C. (1988) 'Muddling in Bumbledom: on the enormity of large sanitary improvements in four British Towns, 1855–1885', *Victorian Studies*, vol.32, pp.55–83.

HARRIS, J.R. (1988) *The British Iron Industry, 1700–1850*, Basingstoke and London, Macmillan.

HASSAN, J.A. (1985) 'The growth and impact of the British water industry in the nineteenth century', *Economic History Review*, vol.38, pp.531–47.

HITCHCOCK, H.R. (1954) *Early Victorian Architecture in Britain*, New Haven and London, Yale University Press, vol.2.

HUDSON, P. (1992) *The Industrial Revolution*, London, Edward Arnold.

KELLETT, J.R. (1969) *The Impact of Railways on Victorian Cities*, London, Routledge and Kegan Paul.

LANGTON, J. and MORRIS, R.J. (1986) *Atlas of Industrializing Britain, 1780–1914*, London and New York, Methuen.

MITCHELL, B.R. (1981, 2nd edn) *European Historical Statistics, 1750–1970*, London, Macmillan.

MOORSOM, N. (1976) *Middlesbrough As It Was*, Nelson, Hendon Publishing.

MORRIS, R.J. and RODGER, R. (eds) (1993) *The Victorian City: a Reader in British urban history, 1820–1914*, London and New York, Longman.

PEVSNER, N. (1979) *A History of Building Types*, London, Thames and Hudson.

PUGIN, A.W.N. (1841, 2nd edn) *Contrasts: or, A Parallel between the Noble Edifices of the Middle Ages and corresponding buildings of the Present Day; showing the present decay of taste*, London, Charles Dolman (first published 1836).

ROBINSON, R. (1986) 'The evolution of railway fish-traffic policies, 1840–66', *Journal of Transport History*, vol.7, pp.32–44.

SIMMONS, J. (1968) *St Pancras Station*, London, Allen and Unwin.

SIMMONS, J. (1978) 'The power of the railway' in H. Dyos and M. Wolff (eds) *The Victorian City: images and realities*, London, Routledge and Kegan Paul, vol.2., pp.277–310.

SIMMONS, J. (1986) *The Railway in Town and Country, 1830–1914*, Newton Abbot, David and Charles.

SIMMONS, J. (1995, rev. edn) *The Victorian Railway*, London, Thames and Hudson.

SKEMPTON, A.W. and JOHNSON, H.R. (1962) 'The first iron frames', *Architectural Review*, vol.131, March, pp.175–86.

TANN, J. (1970) *The Development of the Factory*, London, Cornmarket Press.

THOMPSON, F.M.L. (ed.) (1982) *The Rise of Suburbia*, Leicester, Leicester University Press.

TURTON, B. (1969) 'The railway towns of southern England', *Transport History*, vol.2, pp.105–35.

WHOMSLEY, D. (1974) 'A landed estate and the railway: Huddersfield, 1844–54', *Journal of Transport History*, vol.2 (new series), pp.189–213.

WOHL, A.S. (1983) *Endangered Lives: public health in Victorian Britain*, London, Dent.

YOUNG, C.F. (1866) *Fires, Fire Engines and Fire Brigades*, London.

YOUNG, G.M. and HANDCOCK, W.D. (eds) (1956) *English Historical Documents, 1833–1874*, London, Eyre and Spottiswoode.

Chapter 2: BRITAIN'S INDUSTRIAL NORTH – TWO URBAN CASE-STUDIES: MANCHESTER AND GLASGOW

by David Goodman

2.1 Manchester

Modern Manchester is a pure product of the Industrial Revolution. It is one of the best examples – certainly the world's earliest – of a city largely developed by the forces of technological change and industrialization. Here there was nothing of that unidimensional growth exhibited in Merthyr Tydfil, which failed to acquire the characteristics of a fully fledged city. The reason for the difference is that, unlike Merthyr (or Middlesbrough), Manchester did not arise suddenly from nothing. It has recently been revealed by archaeologists – they had taken the opportunity of working on the central city-site devastated by the IRA bomb in 1996 – that Manchester already existed in the Roman period as a fortified hill-town occupying the surprisingly large area of sixteen hectares. Medieval Manchester was an unimportant ecclesiastical centre and a small town with a weekly market. But the city acquired a new importance in the mid-sixteenth century, when the manufacture of cotton cloth and other textiles was first introduced. By the beginning of the seventeenth century, when its population is thought to have been around 2,000, Manchester had a thriving textile industry, notable for the production of fustian, a hard-wearing cloth woven from linen warp and cotton weft (the raw cotton came from the Levant, via London). By 1620 fustians were not only being sold in Manchester's new cloth-market, but exported to the Continent as well. And the new prosperity soon brought permanent benefits to the city in the form of charitable bequests from successful merchants: Humphrey Chetham's will of 1653 established Manchester's Chetham's Hospital (today a school) and Chetham's Library, one of Britain's oldest public libraries. The activities of another of these seventeenth-century textile merchants already reveals the embryonic beginnings of what was to become the most distinctive feature of Manchester's textile industry. Joshua Browne (1651–94), today remembered in the name of a street in central Manchester, owned a dye-works and a textile warehouse in the city. He was in the business of manufacturing and selling fustians. His raw material, Levant cotton wool purchased from London merchants, was distributed to weavers in Manchester; but most went to his agents in Bolton and Blackburn, to be woven by workers there or merely sold to weavers in Oldham (Carter, 1962, pp.124–5). Here were the first threads of a manufacturing connection between Manchester and a few surrounding Lancashire towns situated within a radius of some forty kilometres. This is what Defoe meant when, in his visit to the region in the 1720s, he reported that cotton-manufacturing had spread from Manchester, its centre, to Bolton. It was the consequence of the spurt in textile-manufacturing over recent decades, 'the Manchester trade we all know' (Defoe, 1962, pp.261–6).

A city at the centre of a web

So far there had been no new technology in Manchester's textile industry, merely continuation of the traditional techniques of hand-spinning and weaving. Nor had the face of the city yet changed much; an illustration of Manchester and Salford around 1760 still shows some surviving medieval character: the parish church dominates, and there are open spaces with trees close to the centre (see Figure 2.1). There is no hint yet of the great changes which would soon transform this serene image beyond recognition. From around the 1770s, the adoption of new technology brought the beginnings of the Industrial Revolution to south-east Lancashire. Manchester would never be the same again, nor would the surrounding region, which it came to dominate. Manchester's predominance was not merely another instance of a city's control over its rural hinterland, a phenomenon frequently exhibited in the past, all over Europe. Something else occurred on a scale that was unprecedented: Manchester, by the end of the eighteenth century, had forged numerous surrounding towns and villages into an intimately interlocking, industrial system under the central control of the city's entrepreneurs. This development so impressed Léon Faucher, a French visitor to Lancashire in the 1840s, that he was led to evoke an image of Manchester as 'a diligent spider, placed in the centre of the web' (Faucher, 1844, p.15).

It was technological change, from the 1770s, that had stimulated the generation of an industrial region centred on Manchester. A spate of inventions mechanized the processes of spinning and weaving. Arkwright's water-frame spinning-machines now offered the prospect of unprecedented volumes of textile-production, but the water power needed to drive them was not to be found in Manchester. The factories had instead to be installed beside fast-flowing streams in the countryside. One manufacturer found what he needed

Figure 2.1 Manchester and Salford, *c.*1760; Salford appears in the foreground (courtesy of Local Studies Unit, Central Library, Manchester)

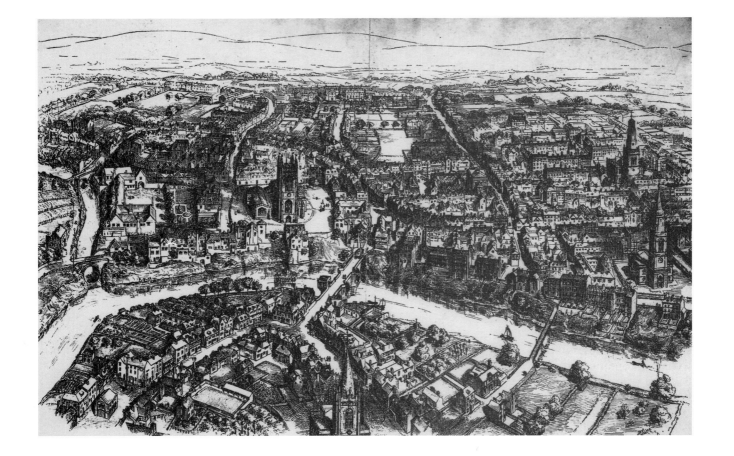

in Tyldesley, a rural hamlet sixteen kilometres to the west of Manchester. He built a six-storey spinning mill, its machinery powered by the local stream. A second mill followed and there was an influx of workers who settled there. By the early nineteenth century Tyldesley had lost its former slow tempo of cheese-production at its two farmhouses and half a dozen cottages, and was converted into an industrial dependency of Manchester with a population of some 2,000. Exactly the same urbanization happened, for the same reasons, at Altrincham, in rural Cheshire, eight miles south-west of Manchester. But far more important than these were the textile towns of Bolton, Bury, Rochdale, Oldham, Blackburn and Preston. All of these had a long tradition of textile-manufacture – Bolton's went right back to the Middle Ages. Now they were all tightly woven into Manchester's industrial web, a process facilitated by the building of a series of canals that connected most of the above-named towns to Manchester. By the beginning of the nineteenth century they had become satellites of Manchester, receiving raw material and delivering cloth to the central depositories of Manchester's merchants. Another canal, an extension of the Bridgewater Canal, which connected Manchester to Runcorn on the Mersey, played a vital role: imports of American raw cotton arriving in Liverpool could now be readily shipped to Manchester. These canals strengthened Manchester's control over distribution of raw material and marketing of products. Even more so from 1830, when some of the world's first railway lines were built in this region.

James Watt's new steam-engines allowed Manchester to overcome its industrial disadvantage of lack of water power. Watt had made dramatic improvements to the existing steam-engine: his introduction in 1769 of a separate cylinder to condense the steam made the working cycle of engines more efficient; and special gearing and other new features patented in 1782–4 made possible for the first time continuous, rotative power (see Figure 2.2 overleaf), replacing the intermittent power supplied by the reciprocating (up and down) action of previous steam-engines. Once installed in Manchester, Watt's engines had far-reaching consequences. As John Kennedy, a partner with James McConnel in the operation of Manchester's largest cotton mill – by 1833 it employed a work-force of 1,551 spinners – put it:

> About … 1790 Mr Watt's steam-engine began to be … introduced into this part of the kingdom, and it was applied to the turning of … various machines. In consequence of this, water-falls became of less value; and instead of carrying the people to the power, it was found preferable to place the power amongst the people.

(quoted in Vigier, 1970, p.92)

But the rise of textile-manufacturing in the city's new factories did not mean disintegration of Manchester's industrial web. The lower land-prices and lower wages of other Lancashire towns account for their continuing attractiveness to Manchester's entrepreneurs as sites for industrial operations. Expansion occurred throughout the region. Rural villages became towns, existing towns became larger. The result was urbanization on an unprecedented scale; by 1850 two-thirds of the population of Lancashire and Cheshire lived in thirty towns. And Manchester had itself become an industrial city, with numerous cotton-spinning mills, and a centre for dyeing and other textile-finishing processes (see Figure 2.3, p.35). In 1821 there were as many as sixty-six textile mills in the city and a quarter of the nation's spindles operating in cotton-spinning. And the power-loom had made Manchester an important centre of weaving: together with nearby Stockport, Manchester, by the 1820s, had half of Britain's power-looms. The weaving was performed on the same sites as spinning, representing a new integration of manufacturing processes by the expanding cotton firms.

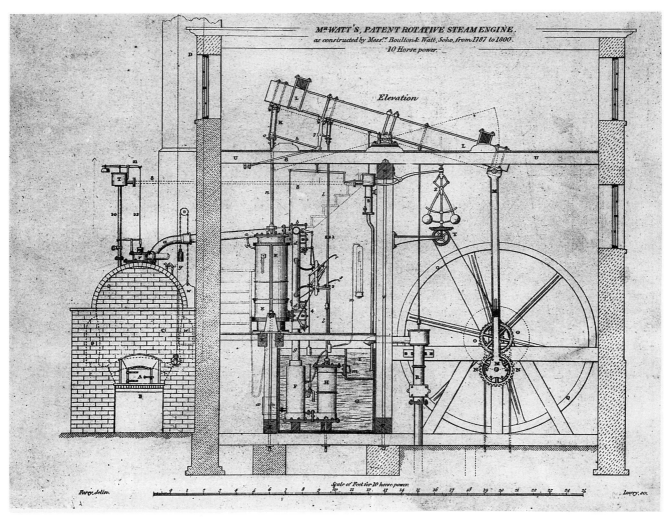

Figure 2.2 Watt's fully developed steam-engine (1784), showing the separate condenser (F), 'sun-and-planet' gearing (N, M, O) and a governor (Z) (courtesy of Science and Society Picture Library/National Museum of Science and Industry)

Immigrant workers from rural Lancashire, Wales and Ireland, attracted by the new opportunities for employment, swelled the city's population, already on the increase from a rising birth-rate. From 70,000 inhabitants in 1801, Manchester's population soared to 303,000 in 1851; by 1891 it was 505,000. But it is important not to exaggerate employment in the city's textile industry. The statistics show a surprisingly low proportion of adult males employed in this way: fewer than 25 per cent in 1841 worked in textiles, and of these only some 12 per cent in factories; the rest were engaged on the finishing processes. But the figures are higher for females: in the same year, more than 40 per cent of women over the age of twenty were at work with textiles, two-thirds of them in the city's factories. The number of children working in these factories is less well established (Walton, 1990, pp.362–3). Bolton and Oldham fit the image of an urban, textile labour-force much better than Manchester.

'Cottonopolis,' Cotton City, is the name historians used to give to Manchester – catchy but misleading. It is usual now to contrast the monolithic form of industry in satellite towns such as Oldham (though even there, it was not all cotton-spinning) with the diversification of Manchester. Apart from cotton mills, Manchester, from the end of the eighteenth century, also had a rising engineering industry. True, some of this can still be seen as an adjunct of the textile industry: the works of Sharp, Roberts and Company, established in 1821 in Faulkner Street, Manchester, were largely concerned with the manufacture of machinery for spinning and weaving. But the engineering activity was much wider than this. Sharp and Roberts had other factories (built in 1839) in Oxford Street and Great Bridgewater Street where machine-tools and locomotive

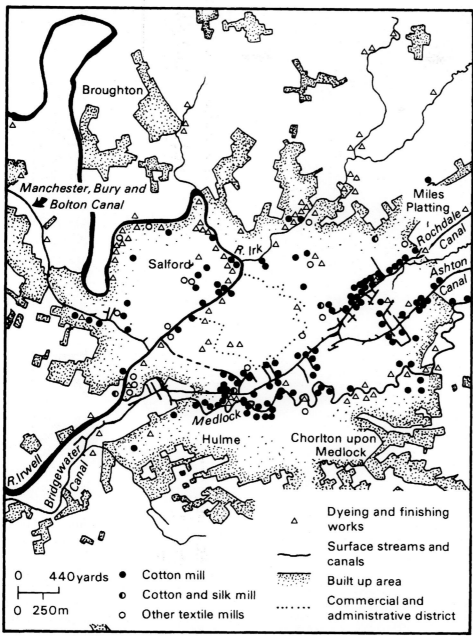

Figure 2.3 Sites of textile-production in Manchester, 1844–9 (reproduced from Langton and Morris, 1986, p.109; with the permission of the publisher)

Broughton

Manchester, Bury and Bolton Canal

Miles Platting

Rochdale Canal

Ashton Canal

R. Irk

Salford

Medlock

Hulme

Chorlton upon Medlock

R. Irwell

Bridgewater Canal

0   440 yards
0   250m

△ Dyeing and finishing works

● Cotton mill
◉ Cotton and silk mill
○ Other textile mills

Surface streams and canals

Built up area

Commercial and administrative district

engines were produced (Love, 1842, p.63). And from the 1830s Joseph Whitworth's Chorlton Street metallurgical-works became famous for its output of armaments and precision machine-tools, as well as for a road-sweeping machine, alleged to have made some of the city streets among the cleanest in the country (see Figure 2.4 overleaf). A particularly large engineering-works inspired the author of a guidebook to Manchester in the 1840s to include a detailed description of its operations and rational layout of workshops, which on the ground floor alone extended to some 150 metres. This was the plant of Nasmyths, Gaskell and Company, situated on both the Bridgewater Canal and the Liverpool–Manchester railway line at Patricroft, six kilometres west of the city centre, and soon to be absorbed in the outskirts of Manchester's expanding built-up area (see Figure 2.5 overleaf). Here 500 men were employed in the whole gamut of engineering activity: designers in the drawing office, pattern-makers preparing wooden models of metal products, foundry workers, fitters, filers and machine-erectors (Love, 1842, pp.59–62). The products were locomotive engines, a great variety of machine-tools and Nasmyth's steam-hammer.

Figure 2.4 Cleaning-up Manchester with Whitworth's recently invented road-sweeping machine of 1842. As the sketch below right shows, the contraption employed a series of brushes on an endless belt, operated by gearing (Whitworth, 1847, p.437)

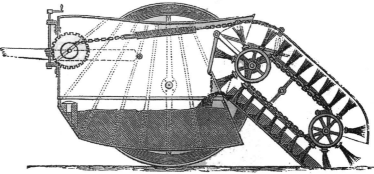

Figure 2.5 The extensive engineering-works at Patricroft (courtesy of Local Studies Unit, Central Library, Manchester)

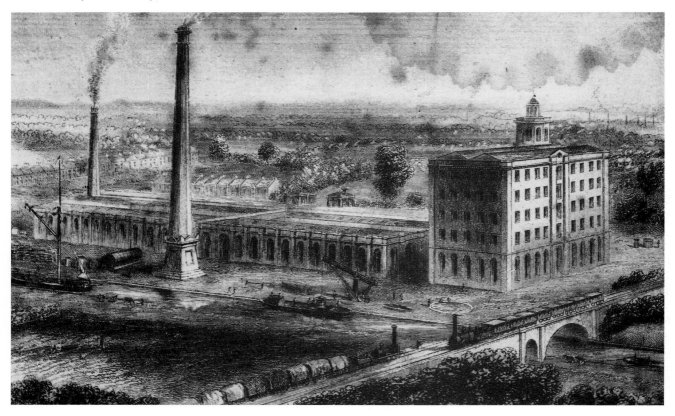

The changing face of the city

Directly or indirectly, the appearance of Manchester was profoundly altered by the technological change of the late eighteenth and nineteenth centuries. Manchester's River Irwell, a tributary of the Mersey and once a river for catching salmon, was by the 1790s declared destitute of fish, the result of pollution from the numerous dye-works (Challoner, 1959–60, p.44). In 1808 a visitor from Rotherham described the river as 'black as ink' (Briggs, 1968, p.89). And blackened also, by 1800, were the city's buildings, the effect of the smoking factory chimneys that would become an increasingly conspicuous feature of the city profile viewed from a distance (see Figure 2.6). A photograph of the mid-nineteenth century shows smoke-blackened buildings around the polluted Irwell and, above all, the obliteration of Manchester's medieval centre: the Perpendicular cathedral (the church had become a cathedral only in 1847) is now surrounded by cotton mills and dwarfed by one of the chimneys (see Figure 2.7).

Figure 2.6 Manchester from Kersal Moor, c.1840 (courtesy of Local Studies Unit, Central Library, Manchester)

Figure 2.7 River Irwell, 1859 (courtesy of Local Studies Unit, Central Library, Manchester: photograph: George Grundy)

Figure 2.8 Murray's mill, Ancoats, 1798–1801 (photograph: © Crown copyright.RCHME)

Manchester's spinning mills dominated the areas of urban space where they were erected. They were from four to eight storeys high, with each floor devoted to a particular part of the manufacturing process. The building type is well illustrated by Murray's mill in the city's central industrial district of Ancoats (see Figure 2.8). Built in 1798–1801, it is Manchester's oldest surviving mill – one of two spinning mills constructed around a canal basin into which supplies could be brought by boat.

The adoption of the power-loom led to the appearance of another building type, which became increasingly familiar in Manchester from the 1820s. Unlike the traditional hand-loom, whose comparatively gentle operation presented no threat to the structure of an ordinary house, the power-loom (see Figure 2.9) was more violent in its alternate rapid accelerations and haltings of the shuttle. It generated powerful forces and vibrations. And, of course, it required a power source. The solution commonly adopted by Manchester's cotton firms was to attach a series of long, ground-floor sheds to their multi-storey spinning mills. The low buildings were less likely to suffer damage from the strong vibrations and could harness the power of the adjacent mill, but they took up more land than the only alternative: a multi-storey structure. The roof, designed to admit the maximum amount of daylight, formed a characteristic saw-tooth pattern (see Figure 2.10). Some idea of the size individual sheds could assume comes from the records of Manchester's Chorlton New Mill, where in 1829 one shed housed 600 power-looms (Giles, 1993; Sissons, 1994). But to perceive today the full visual effect such extensive sheds once produced in Manchester, one must go fifty kilometres north-east to Saltaire, on the outskirts of Bradford. Here, in 1853–71, the philanthropic entrepreneur Titus Salt built an idealized industrial township which displays the best surviving example of a multi-storey spinning mill flanked by dozens of long, low weaving-sheds, specially built to house the power-loom (see Figure 2.11).

Figure 2.9 Power-loom, 1835 (courtesy of Science and Society Picture Library/National Museum of Science and Industry)

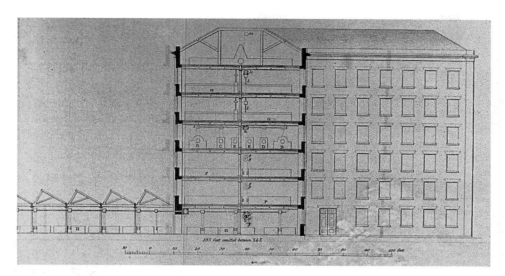

Figure 2.10 Orrell's cotton mill, Stockport, as illustrated in Ure (1836), showing shed-extensions for power-looms; on the upper floors of the mill, carding and spinning were performed (courtesy of Local Studies Unit, Central Library, Manchester; photograph: Mike Levers)

Figure 2.11 Industrial building forms influenced by function: Saltaire mills (near Bradford), showing the multi-storey spinning mill and low weaving-sheds housing power-looms (photograph: © Crown copyright.RCHME)

Figure 2.12 Middle Warehouse, Manchester, 1810 (courtesy of Neil Richardson; copyright S. Wilkinson)

Still more imposing than the mills in Manchester's urban space were those other building types generated by textile technology, the warehouses. Their function was not as simple as one might suppose; that has become increasingly clear from research over several decades (Lloyd-Jones and Lewis, 1988, pp.32–5, 44–50; Smith, 1953–4, pp.62–5). As textile output and trade increased, Manchester's manufacturers and merchants invested in warehouses, large and small. They began to appear in the late eighteenth century; by 1830 there were almost 1,000 in the city. Some, like those erected around the city's central canal basins, were unadorned industrial structures. Of this type, the Middle Warehouse, built in 1810, still stands; conspicuous on its waterfront side are two huge apertures for the passage of materials, leading to an internal hoist (see Figure 2.12). But from 1840, warehouses looking like palaces of Renaissance Italy began to dominate the architecture of Manchester's principal streets (see Figures 2.13 and 2.14); they declared the success and wealth of an urban manufacturing and merchant elite who sold to the world.

The inflammable stock in some of these warehouses was protected from fire by a system of internal water supply, pipes and hoses, an idea promoted by the engineer William Fairbairn, whose *Application of Cast and Wrought Iron to Building Purposes* (1854) devoted a whole chapter to the fireproofing of warehouses. But Manchester's warehouses were not just for the storage of raw cotton, spun yarn and finished cloth. They provided

> convenient premises in which cloth from country manufacturers, Manchester manufacturers, or even from handloom weavers could be accepted, weighed, sealed and placed ready, either for despatch to the finishers or for sale, after finishing, to the retailers in the United Kingdom.
> (Smith, 1953–4, p.63)

They served also as showrooms to display samples of finished cloth for retailers to inspect and order. And they were the site of public auctions of cottons and textile machinery. More surprising, warehouses 'played a pivotal role in the system of manufacturing' (Lloyd-Jones and Lewis, 1988, p.47). First, because they functioned as the point of quality control: expert artisans within the warehouse scrutinized the degree of perfection of products. Second, warehouses in the period 1750–1815 were centres of a putting-out system: yarn went from warehouse to hand-loom weavers in and around Manchester to return as cloth. The firm of S. and S. Stocks, manufacturers and dealers in cotton, even employed twenty hand-loom weavers in their own warehouse (Lloyd-Jones and Lewis, 1988, pp.48–9). This continuing employment of hand-loom weavers is partly explained by the initial inability of the power-loom to produce anything beyond coarse cloth; until its design was improved, finer fabrics could be woven only by hand. But even after that was solved, hand-loom production remained attractive as a cheap alternative.

Manchester's textile-manufacturers could not do without the multipurpose warehouse in some form: large, small, or even shared. The marked effects of their proliferation in the city were already noticeable in the late eighteenth century, when, one after another, the houses in Cannon Street, High Street, Market Street and other principal streets were converted into warehouses.

Figure 2.13 Britannia Buildings, a Portland Street warehouse built in 1851 (courtesy of Local Studies Unit, Central Library, Manchester)

Figure 2.14 James Brown, Son and Company warehouse, Portland Street, Manchester (*Illustrated London News*, 14 May 1853)

Already the central residential area was being transformed into the commercial zone of a manufacturing base. It was a trend that would intensify during the nineteenth century, causing George Saintsbury (literary critic and historian) to declare in the 1880s that 'hardly in London itself is there a more utter absence of residential life than on Sunday exists for a mile or so around the Exchange at Manchester' (Kellett, 1969, p.296).

This 'dead heart' of the city was just one feature of Manchester's fabric created by the industrial entrepreneurs. To escape the smoke of their mill chimneys the manufacturing elite moved to residences to the south, away from the prevailing easterlies and westerlies, to Withington, and eventually still further, to Knutsford in rural Cheshire. Their less fortunate employees, working in factories 'twelve hours in the day in an enervating, heated atmosphere, frequently loaded with dust or filaments of cotton', returned to hovels in Manchester's eastern districts, notoriously insanitary, 'pestilential streets, in an atmosphere loaded with the smoke and exhalations of a large manufacturing city' (Kay, 1833, pp.24, 27). Technology and trade had produced a city of extremes of poverty and wealth. The merchants and manufacturers dominated the life of the city. In 1821 five manufacturers founded the *Manchester Guardian*. In 1830 the same elite promoted the Liverpool and Manchester railway. And within the Exchange, the city's finest building, adorned with Ionic columns and seen at the time as 'the parliament of the lords of cotton' (Briggs, 1968, p.107), they negotiated the deals on which their economic dominance was won.

Sharp spatial variations in Manchester's mortality-rate reflected the very different living conditions of its inhabitants. In 1840 the outlying middle-class suburb of Broughton had an estimated death-rate of 15.8 per 1,000; at Ardwick, closer to the centre, it was 28.6; and closer still, in congested districts such as Ancoats, it was 35.2, the consequence of a built environment of maximum housing density, ideal conditions for the spread of tuberculosis and other infectious diseases (Flinn, 1965, p.13; Pooley and Pooley, 1984). But in the very centre, where housing density had been greatly reduced by warehouses, mortality was low.

Industrialization had induced considerable expansion from the city centre. In the 1770s Manchester was compact, its built-up area limited to a circle of radius 370 metres. By 1820 the city had spread almost to fill a circle of one and a half kilometres radius, with much greater areas devoted to industry and warehouses. Expansion would gather momentum throughout the nineteenth century. By 1850 a continuous eleven-kilometre stretch of buildings joined Manchester to Oldham, and a regular horse-omnibus service operated between the two towns. At the same time, houses and factories, spreading nearly ten kilometres to the east, filled up the spaces between Manchester, Ashton-under-Lyne, Dukinfield and Stalybridge. And in all directions the extension of Manchester's tentacles was invigorated by a multiplicity of railway lines.

For a feel of the pulse and appearance of Manchester in the 1840s, along with general comment on the inappropriateness of cities as sites for manufacturing industry, we have the vivid writing of Léon Faucher (1803–54), a French economist, journalist and Liberal politician who visited England in 1843 to study its society. His account of Manchester remains an invaluable primary source (see Extract 2.1 at the end of this chapter).

The imprint of the railway

In 1825 Ann Atherton, the heir to a family estate laid out as a sedate residential area half a kilometre to the west of the city centre, feared the consequences of revolutionary transport technology. She and her father had protected the estate with covenants prohibiting manufacturing in order to preserve an enclave of tranquillity in a zone of expanding industrialization. And now everything was threatened by a projected railway line with a terminus on her estate. In her petition she protested that 'the whole character of the neighbourhood' would be destroyed. 'The great and offensive nuisance' of passing locomotives and their concentration at the depot would be incompatible with the 'comforts' of the residents, forcing them to move out, and eventually 'inferior houses' would be built, reducing the value of existing property. The dispute ended eleven years later with a victory for the railway company. The terminus, Liverpool Road, was built and soon surrounded with poor housing. It would act as a barrier to the westward expansion of Manchester's business district (Kellett, 1969, pp.154–5).

Less difficulty was encountered with other railway approaches to the city. It is sometimes said that Manchester was unusual (compared, say, with Liverpool or Cardiff) in having no dominant landowners. Yet considerable acreage in the city was owned by the Earl of Derby and the Earl of Dulcie, and their speedy agreements with railway companies expedited Manchester's conversion to a railway communications centre. But there were nine rival railway companies, which produced a proliferation of stations in the city and competition reaching violent levels – fighting for the possession of train-sheds and even the ejection of ticket clerks through the windows of rival booking-offices (Kellett, 1969, pp.160–62).

Manchester's termini were on the edge of the central district and an inconvenient distance apart. How could they be joined? The expense and engineering difficulties of a tunnel were rejected, and in 1845 Parliament authorized the construction of a junction line, the Manchester South Junction Railway. This trans-urban railway had profound effects on the city's appearance, introducing a viaduct that crossed thirty streets and reached a height of thirty-six metres at Stockport. Dilapidation resulted. Some of the area below the viaduct was let as stables, and the unlit arches attracted undesirables (Kellett, 1969, p.16). Dilapidation also occurred at Ancoats, a district where demolition of property to build a railway line and sidings generated a slum on the east side of the city. Here there was not a single straight street that led to the city centre; instead a maze of zig-zagging, narrow streets was the product of the railway line's dominance (Kellett, 1969, p.343). And Gorton, also to the east, was covered with over sixteen hectares of sidings and locomotive-works. But railway engineering boosted Manchester's economy as well as damaging its appearance.

Did the railway influence the siting of Manchester's industry? Historians' views differ. Simmons sees the coming of the railway as a new stimulus to the city's industrial expansion; Kellett only as a reinforcement of locations already settled in the canal era (Kellett, 1969, p.350; Simmons, 1986, p.127). But Kellett argues that Manchester's textile industry was profoundly altered by the railways, even seeing manufacturers' expectations of a consequent surge in profits as the main motive for the local interest in the new transport technology (see Extract 2.2 at the end of this chapter).

According to Kellett, the railways boosted Manchester's textile industry in several powerful ways. The faster transport quickened the movements of purchasers of yarn ('the small masters') and travelling salesmen. The improved communications (this includes the telegraph, which was closely associated with the railways) clarified demand from distant markets, so that large volumes of products could be placed on railway wagons, thereby reducing the pressure on urban space already saturated with warehousing. The same enhanced communications reduced the risks of unpredictable consumer demand for a product subject to changes in taste. And the railways created an integrated system of textile-production in a group of Lancashire cities connected to and dominated by Manchester.

It is this last assertion that seems to me questionable. It is overstated, because of the word 'created'. After all (as indicated at the beginning of this chapter), this industrial web already existed decades before the railways arrived. Kellett would have done better to say that the railways intensified the functioning of an already operating industrial network of cities.

Commuting to work by train (or by horse omnibus) was the pattern of life for the very few, even at the end of the nineteenth century. Although some workers are said to have used the suburban railway in the 1840s to take them to cotton warehouses at Stretford and Chorlton (Simmons, 1986, p.113), fares were prohibitively expensive for the working class. There were continuing hopes that a reduction in fares would allow workers to live in healthier surroundings, solving Manchester's terrible housing conditions, but all efforts to persuade the railway companies to do this proved fruitless (Kellett, 1969, pp.358–9).

But if the railways failed to make much impression on the mobility of the great majority of Manchester's inhabitants, was their food supply improved? The tendency of research into this aspect of Manchester's railway era is to caution against technological determinism. Change came, but slowly, and it depended on conditions other than the availability of a new form of rapid transport. The Liverpool and Manchester Railway Company was quick to invest in cattle wagons (see Figure 2.15) but was disappointed by its share of livestock traffic: still in 1850, three-quarters of incoming cattle reached Manchester from Liverpool by road. The reason was that cattle lost little weight in the traditional droving. Pigs, however, were slower walkers and lost much of their weight in transit. There was therefore real interest in speedier pig deliveries, and by the 1840s some 80,000 pigs a year were being brought by rail from Liverpool to Manchester. This number increased considerably after the introduction of the steamship on the Irish Sea in 1838 greatly reduced the time of the sea-crossing from Dublin to Liverpool; from then on, pigs fattened in Ireland became a common source of protein in Manchester. But the traffic in dead meat was negligible until the development, around 1880, of refrigeration techniques (Scola, 1992, pp.46, 51f.). Similarly, the lack of ice-manufacturing techniques was one of several obstacles to the immediate general establishment of bulk transport of fish by rail. But gradually, fish enriched Manchester's diet. When the railway connected Manchester to Morecambe, Britain's richest shellfish grounds became accessible to the the city's teeming population, which could now feed on expensive lobsters or the cheaper shrimps and oysters (Scola, 1992, pp.126f.).

Again, with milk supply, there was no sudden resort to rail transport and, as late as the 1890s, some 30 per cent of Manchester's milk was still arriving by cart. Unlike Liverpool, with its surrounding countryside overwhelmingly devoted to arable farming, Manchester had numerous milk-producing farms within a radius of a few kilometres. That explains the unimportance of town dairies in Manchester compared with Liverpool. Before the railways, much milk arrived in Manchester by the new canals – those Cheshire farmers living close to the Bridgewater Canal switched from the prevailing cheese-production to liquid milk for the expanding urban market now within reach. Then, from the 1840s, the railways took an increasing share of this freight traffic. Until the refrigeration techniques were introduced, rail milk was just as susceptible to souring as that delivered by slower forms of transport. But in Manchester the difficulties were minimized by the short distances of the supply line: usually no more than thirty kilometres to the Cheshire farms. By 1869 railway milk was arriving in sufficient quantities to justify the building of a special milk-depot (Scola, 1992, pp.71–9).

Figure 2.15 Cattle being transported on the Liverpool and Manchester Railway, 1831 (courtesy of Local Studies Unit, Central Library, Manchester)

Water from the hills

There were no food shortages in industrializing Manchester; only poverty
restricted tens of thousands to a basic diet. The real urgency arose in water
supply. The multiplication of the city's population, the increasing requirements
of bleaching-works and dye-works for pure water, and the need to extinguish a
growing number of blazes in factories and warehouses imposed demands that
far exceeded the existing supply. That supply had, since 1809, been in the hands
of an inefficient joint-stock enterprise, the Manchester and Salford Waterworks
Company. Although the company built new reservoirs at neighbouring Gorton
in the 1820s, their 7,000 cubic metres of daily supply were never enough. By the
1840s, less than a quarter of Manchester's inhabitants were supplied and for only
a few hours in the day; the majority had to make do with polluted rivers and
wells. Many industrial plants were left unsupplied and the premises of others
increasingly at risk, with only an intermittent, low-pressure mains delivery to
fight fires (Hassan, 1984).

In 1844 the company approached John Bateman, an engineer who had improved
Bolton's waterworks, for advice on what should be done in Manchester. His
report would establish the future course of hydraulic engineering in the city.
Dissuading the company from its plans to bore deeply into Gorton's sandstone
to reach water – that would entail the expense of a constantly operating steam-
engine to pump up the water – he turned instead to gravity flow from the distant
Pennines, thirty kilometres east of Manchester. Those abruptly rising hills were
'the first to interrupt the clouds progress, carried by the westerly winds' and, if
the rain was collected in reservoirs, Manchester might have water 'nearly as pure
as it comes from the heavens', for there was neither coal-mining nor agricultural
manure in the virgin hills to pollute it. Bateman had been encouraged by his
measurements of rainfall in the area – part of that move to systematic recording
with rain-gauges which began in the 1840s. He estimated that the city required
13,500 cubic metres of water a day for domestic use, 11,500 for steam-engines in
factories, and another 7,000 for dye- and bleach-works. Revenue from these
industrial consumers would help to pay for the proposed project. They would
benefit from an abundance of pure water and reduced fire-insurance rates,
because 'a warehouse or mill can instantly be deluged by a pipe controlled by a
valve'. And reliable supplies of pure water would attract still more manufacturers
to the city – a prospect of sharply rising prosperity (Bateman, 1884, pp.39–49).

The company reacted favourably, prepared plans, and a bill was brought to
Parliament to authorize the necessary construction. It was strongly opposed by
mill owners who relied on Pennine streams now to be tapped for reservoirs. The
company backed away and resumed its attempts to sink a well at Gorton. But
Manchester Corporation was impelled to become involved. The alarming fire-
damage in the city, growing concern over public health (statistics showed that
Manchester had the second-highest death-rate in the kingdom), the economic
advantages of soft Pennine water (tests indicated an annual saving in soap to the
tune of £28,000) and a growing civic sense of the importance of public control
over a fundamental resource all motivated the corporation to buy out the
waterworks company and execute the project in the Pennine valley of
Longdendale. In 1845–51 all the necessary parliamentary approval had been
secured. Agreement had been reached to compensate mill owners and the Duke
of Norfolk, whose land would be taken for reservoirs. Bateman was appointed
corporation engineer in charge of the project, the scale of which was indicated
by tenders for the supply of 7,720 tonnes of cast-iron pipes. Work began on
three large dams, one thirty metres high, with discharge pipes connecting to an
inclined tunnel, 2,700 metres long, two metres in diameter, and lined with a type
of hydraulic cement, dropping one and a half metres every half a kilometre for
the gravity flow. The water fell, via more reservoirs and pipes, down to the
outskirts of Manchester.

This was technology at a distance, in the service of the city. And much of it was new. The firm of William Armstrong in Newcastle, specialists in innovative hydraulic engineering, invented a self-acting valve to safeguard against bursts in pipes subjected to the very high pressure in Manchester's waterworks. When the velocity of flow exceeded a certain value, a valve automatically closed and an alarm bell rang in a supervisor's house. The discharge pipes, made in both Manchester and Renfrew, were cast in a new way (vertically, with their sockets downwards) which was found to increase their strength. And this strength was tested at high pressures imposed by a press. After Bateman invited his 'chemical friends' to find an effective anti-corrosion treatment for the pipes, Angus Smith devised a process that would become standard for the prevention of rusting in water-pipes: coating them with hot tar. The pipes were first treated with linseed oil to form a retentive medium for the tar. They were then heated, to open 'the pores of the iron', and slowly lowered into the tar by a travelling crane on a gantry. After immersion, there formed a surface coating so smooth that it maximized water flow as well as providing decades of protection from rust (Bateman, 1884, pp.139–44).

Pennine water began to enter Manchester in 1851, but persistent engineering difficulties delayed completion until 1884. By then a wholly new supply organization was in place, reflecting the completeness of corporation control. The city was divided into water-districts. Warehouses and other valuable blocks of buildings were surrounded with fire-hydrants – at least one every two metres. City plumbers were appointed by the corporation and issued with official instructions. All internal fittings had to be tested and stamped. Every tap was weighed to guarantee the requisite quantity of metal:

> Now nothing can be used which does not bear the mark of having been tested and approved at the Manchester Water Office. Previously there was much inferior work. The inhabitants therefore get good articles and workmanship.
>
> (Bateman, 1884, pp.150–51, 180)

Bateman later brushed aside the suspicion (today supported by much evidence) of a correlation between soft drinking-water and higher urban mortality: 'It is economical, so I recommend it' (Bateman, 1884, p.192).

The project had cost £2.6 million, and supplied a maximum of 5,300 cubic metres of water a day. There were real gains. It stimulated the expansion of urban bleach-works and dye-works. Fire-damage was reduced as the continuous high-pressure supply enabled water from fire-hoses to reach blazing upper storeys. And by 1880 many of Manchester's homes were supplied. But how far this improved the health of the populace is less clear. It is even argued that there was deterioration, because the new supply encouraged the provision of water-closets at a time when sewers and sewage disposal were yet to be perfected. And still in 1900, Manchester's poorest had to go to public baths and wash-houses for their water (Hassan, 1984, pp.35–6, 39–41).[1]

What is clear is that there was a continuing struggle to keep pace with Manchester's exploding population and water demands. Longdendale proved to be only a temporary solution. Already in 1875, Bateman was warning of 'catastrophe' if immediate action was not taken. It was resolved in 1895 when the yet bigger Thirlmere project brought the extra water Manchester needed all the way from the Lake District. Engineering difficulties were not the only problems encountered here. The plans had generated the clamour of a vigorous environmental campaign, with protests that an area of outstanding beauty was being sacrificed to satisfy a monster city 120 kilometres away. In the end Parliament sanctioned the project, but only on condition that the aqueduct was built underground and restorative landscaping performed.

---

[1]   See the extract from J.A. Hassan and E.R. Wilson in the Reader associated with this volume (Goodman, 1999).

2.2   *Glasgow*

A distinctive Scottish urbanization?

When, in 1603, Elizabeth I died without progeny, the Tudor line came to an end, and the succession to the throne of England passed to the Scottish royal house of Stuart. Elizabeth's aunt, Margaret, had married James IV of Scotland, and it was their great-grandson, already king of Scotland as James VI, who now also became James I of England. This first union of England and Scotland was therefore a mere personal, dynastic union of crowns, not the formation of a new unified state. The Scots were the losers. They were deprived of the presence and political patronage of their king who, apart from one fleeting visit, never saw Scotland again. But a second, stronger, fusion of the two kingdoms occurred with the union of parliaments in 1707 and the creation of a new state, Great Britain. Now the Scots were represented at Westminster, they paid the same taxes as the English, and they competed for the same offices of state. Above all, there was now a common market, with Scots entitled to profit from the lucrative Atlantic trade with the American colonies – hitherto a jealously preserved English monopoly, which the Scots breached only by smuggling. New Scottish wealth from the trade in American tobacco would be the initial spark that ignited the rise of Glasgow. The union of 1707 would also stimulate Scottish urbanization by the transfer of English technology: from the late eighteenth century, close links were forged between lowland Scotland and industrial Lancashire.

The union was not complete – Scotland retained its distinctive kirk, legal and educational systems. But what of its growing cities (see Figure 2.16 overleaf)? Were they in any way distinctive? The case for a peculiarity in the character of nineteenth-century Scottish cities can be gleaned in the following passages from a discussion by R.J. Morris:

> The distinctive forms of Scottish urban government were part of the forces which shaped the development and experiences of these urban places. The Royal Burgh had an ancient origin but still formed the basis of Scottish urban authority in the nineteenth century. The seventeenth-century Dean of Guilds Court was crucial for regulating the built environment ... a powerful agency enforcing stringent building regulations in the industrial cities.
>
> The supply of housing to all but the highest income groups in Scottish cities was dominated by the tenement. This was a high-rise building of some three to four storeys. Each storey would contain several flats or 'houses' which were reached by a common stair, sometimes with galleries. Scottish law provided for the supply of building land through feuing. This was a once-and-for-all perpetual rent charge agreed on the initial transfer of occupancy and building rights. Thus the initial holder of ground rights had a motive for withholding land to gain the maximum return from the feu. This and the high legal cost of property transfer pushed up the initial cost of land and increased the motive for the intensive use of that space – hence the density of building ...
>
> When the full impact of urban industrial growth hit Scotland between 1820 and 1850, Scotland already had a structure of locally specific power structures – the royal burghs ... No major city grew within a web of parish, manorial and ad hoc institutions like Manchester or Birmingham. As urban problems ... the problems of fire, disease and traffic congestion, became clear, solutions based upon a structure of urban authority became dominant ...
>
> For a large part of Scotland's urban population, urban 'improvement' was directed by a series of local acts obtained by the major cities ...
>
> Emerging from all this is the sense of a Scottish perception of urbanisation which increasingly identified the urban place with higher levels of authority and order than was expected by the English. The Scots accepted and expected strong, positive forms of locally specific authority in their towns. The town was a place of order.

(Morris, 1990, pp.83–91)

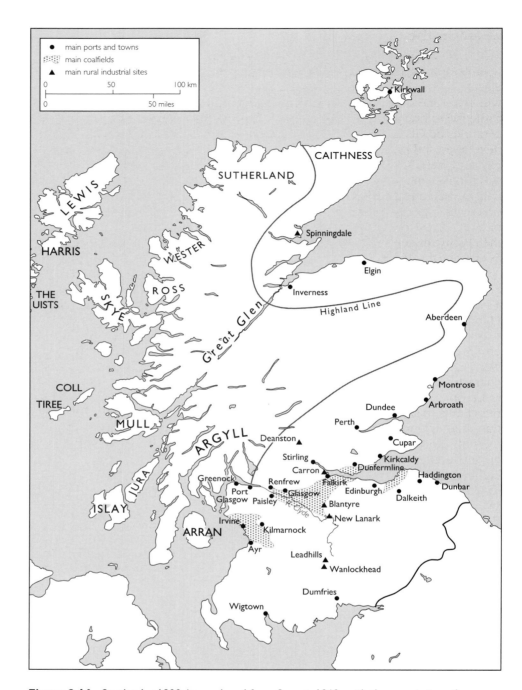

Figure 2.16 Scotland, c.1800 (reproduced from Smout, 1969; with the permission of HarperCollins Publishers)

While these tenements were not part of the English urban fabric, the alleged peculiarity of Scottish municipal control can be overstated: Manchester Corporation's control of its water supply was every bit as strong as Glasgow's. Still, there is truth in this – no English municipality acquired Glasgow Council's powers to carry out night raids on dwellings to check that its rulings on overcrowding were being observed; from 1862, metal plates were fixed to 'ticketed houses' specifying the cubic capacity of the rooms and the number of permitted dwellers.

Giant on the Clyde

Medieval Glasgow was a tiny settlement situated at a ford on the Clyde. It was notable only as an ecclesiastical centre. The twelfth-century cathedral stood on elevated ground about one kilometre to the north of the river crossing, for which a stone bridge was built in the fourteenth century. That sloping line from cathedral to bridge, later to be called 'High Street' in its upper part, would for centuries be Glasgow's axis. By the mid-seventeenth century, local trade had developed sufficiently for this north–south axis to be crossed by streets to the east and west. They intersected at Glasgow Cross, forming a primitive grid that still survives in the heart of the city today, the grid's street-names recalling the rising urban trade: Saltmarket, Trongate (a 'tron' was a public weighing machine or a market-place). Some of this seventeenth-century trading ferment was due to banned commerce with England's American colonies. Jamaica's cane-sugar was the raw material from which rum was being distilled in Glasgow's 'sugar-houses' as early as the 1660s.

But it was in the decades after the union of 1707 and the legalizing of Scotland's transatlantic trade that Glasgow's commercial growth took off. By 1745 Scotland was importing six million kilograms of tobacco leaf from Virginia, Maryland and North Carolina; by 1771 the figure was 21 million kilograms. It was nearly all re-exported to Europe; the only Scottish tobacco industry was the erection of a few small snuff mills in Glasgow. But in a broader sense tobacco was a powerful stimulus to Glasgow's technology. For here was the characteristic pattern of colonial trade: exotic raw materials sent in one direction to Europe, manufactures sent in the opposite direction to the colonies. The merchants of sixteenth- and seventeenth-century Seville controlled a lucrative transatlantic trade, receiving shipments of tropical medicinal plants, hides, dyestuffs and silver, and dispatching cargoes of European manufactures to Spanish settlers in Mexico, Peru and the Caribbean. Similarly, the merchants of eighteenth-century Glasgow secured control of tobacco imports from New England and invested in the manufacture of linen cloth and wrought iron to satisfy the demands of English settlers in Virginia and Maryland. And Glasgow's rich mineral resources facilitated the manufacture of iron products for the English colonies. The city was situated on the edge of the central Scottish coalfield (see Figure 2.16), so coal was readily available, some even within the city boundaries. An Italian prelate visiting Glasgow in the fifteenth century was surprised to observe the donation of a black stone to poor parishioners for domestic fuel. And nearby at Coatbridge, in Lanarkshire, there was an abundance of iron ore. Workshops and forges began to sprout in mid-eighteenth-century Glasgow, financed by the 'tobacco lords'. At the same time the opulent tastes of this merchant elite altered the city fabric as they built luxurious mansions. They rose to be city councillors and provosts, dominating urban commerce, industry and government (Devine and Jackson, 1995, pp.141f.).

Comparing eighteenth-century Glasgow and sixteenth-century Seville reveals further similarities, in the location and functioning of the two cities. Both were inland ports, entrepôts for American imports, located on shallow rivers impenetrable to large ships. For both cities, the true gateway to the Atlantic lay further downstream, where the deeper waters of an estuary allowed large vessels to load and unload their transatlantic cargoes. The notorious sandbars and shoals of Seville's River Guadalquivir prevented any but the smallest vessels sailing into the city; instead, the main port arose at Sanlúcar on the Atlantic estuary. Shallows similarly restricted the navigability of the Clyde. The deeper water was downstream at the estuary, and it was here, on the Firth, that Greenock and Port Glasgow grew to serve as Glasgow's outposts. There the American tobacco shipments were received, unloaded into warehouses and sent by cart on muddy roads to Glasgow, for export, via the east coast, to northern Europe. And the same inadequate roads were used to dispatch the growing volume of Glasgow's manufactures to Greenock. But this was about to change. Where Glasgow differed from Seville was in its use of hydraulic technology to convert the Clyde into a navigable highway, which culminated in the city's elevation to the status of a great international port, so inducing a profound change in the urban fabric. This innovative technology is the key to understanding Glasgow's astonishing rise to become the second city of the British Empire.

The first phase of these river improvements, completed by 1775, was the work of an English engineer, John Golborne. He deepened the Clyde by the use of primitive dredging-ploughs, but mainly by the construction of jetties to narrow the river, so increasing the rate of flow and removing material from the bed. But still the greater depth of just over two metres was not sufficient to allow ships larger than 100 tons to sail up as far as Glasgow.

Historians used to say that the American Revolution of 1775 temporarily ended Glasgow's prosperity and industrial growth. The monopoly of the British tobacco merchants was no more, Glasgow's tobacco era was over and the tobacco lords switched their resources to generate a new cotton industry. None of this is now accepted. Trade continued with the USA, and money for the new cotton industry did not come from the tobacco merchants. Instead, Glasgow's cotton industry was financed by the existing linen industry and fed by imports of raw cotton from the West Indies (Devine and Jackson, 1995, pp.170f.). Those imports soared from £150,000 a year in 1774 to £4,800,000 in 1800.

Like Manchester, Glasgow was short of fast-flowing streams to drive the new textile machinery, so the same pattern of development occurred – dispersal of cotton mills to the water power of surrounding areas within a radius of fifty kilometres: Lanarkshire, Renfrewshire, Dunbartonshire and Ayrshire. Also like Manchester, Glasgow was developing into the dominant city of an industrializing region in the grip of its entrepreneurs. In 1784 there was only one cotton-spinning mill in Glasgow itself. But that soon changed through the inventive genius of James Watt, a native of Glasgow. He conceived the idea for his new design of steam-engine, with its separate condenser and rotatory driving-action – one of the key technological innovations of the Industrial Revolution – during a walk on Glasgow Green. The availability of steam power now brought cotton mills into the city.

In early nineteenth-century Glasgow there was a

> peppered insertion of mills, manufacturing sheds, dye-works and warehouses into already overcrowded districts. Factories and shops created a thriving rag-trade district close to Glasgow Cross, the remains of which survive today.
>
> (Devine and Jackson, 1995, p.130)

This industrial 'colonization' of central properties brought a deterioration in the fabric of buildings and in sanitation. The increased demand for labour led to the subdivision of existing housing; slums appeared. There was an influx of workers from the Highlands, and large numbers of immigrant Irish were employed in hand-loom weaving. According to one estimate, by 1840 as many as one in three of all Glasgow's inhabitants were of Irish origin (Devine and Jackson, 1995, p.11). Glasgow's population, some 30,000 in 1750, had by 1821 multiplied to 147,000.

Cotton had become the leading sector of Glasgow's industry; by 1832 there were forty-nine cotton mills operating in the city, employing at least 11,000 workers (Fraser and Maver, 1996, p.102). And while power-looms produced coarser cloth, Glasgow's entrepreneurs in 1820 also controlled 32,000 hand-looms in the neighbouring villages of Bridgeton and Calton (both were absorbed within the city boundaries by 1846), and the outlying townships of Partick and Govan. Weaving would retain its importance in nineteenth-century Glasgow, but cotton-spinning declined from mid-century – according to one interpretation because of an inherent weakness. From the start Glasgow depended on Manchester for its textile machinery and was slow to replace old models (Devine and Jackson, 1995, p.197; Fraser and Maver, 1996, p.105).

In its first phase of expansion Glasgow's textile industry generated, for its finishing process, a bleaching-powder factory, which developed into the world's largest chemical-works. Founded in 1799, the factory was located at St Rollox, little more than a kilometre north of the city centre, on the banks of the Monkland Canal, which had recently been constructed to bring coal from Lanarkshire into the city. Whether calculated or not, the site also had the advantage of protecting the most densely populated urban areas from chemical fumes – the prevailing south-westerly winds prevented pollution. In 1842 the expanding works sprouted a structure so conspicuous that, until its demolition in the 1920s, it was as much the city's landmark 'as St Paul's in London' (Hume, 1966, p.190). This was a huge chimney, 140 metres tall and fifteen metres in diameter at its base, all the more imposing because it rose up from an already elevated site.

But there was growing smoke pollution within the city from other factories, encouraging a trend for the better-off to move to the fresher air west of the city (Devine and Jackson, 1995, p.185; Reed, 1993, p.59). The layout of the city in 1825 (see Figure 2.22, p.56) shows the extensive eastern industrial zone of the densely populated weaving districts of Calton and Bridgeton. To the north, the city's expansion was checked by the quarries of Cowcaddens; beyond that lay Port Dundas with its distilleries and foundries. To the west, Anderston was a village of hand-loom weavers and cloth-finishers. South of the river there was the, as yet, small Govan ironworks, and the new, extensive and select residential areas of Laurieston and Tradeston, laid out as a grid.

This picture of the city was totally changed by technological developments from the 1830s. Glasgow became one of Britain's principal centres of heavy industry and the city acquired a wholly different appearance. In 1828 James Neilson, an engineer at Glasgow Gas Company, devised a modification in the operation of blast-furnaces that greatly reduced fuel costs. The new technique was to heat the blast of air before it entered the furnace, so causing the air to expand; this increased volume raised the force of the blast, with the consequence that the quantity of coke needed within furnaces was almost halved. Neilson's hot blast transformed a Scottish industry that had been uncompetitive because of high fuel costs; now Scottish pig-iron became much more marketable. Output, a modest 38,000 tonnes in 1830 before the adoption of Neilson's heating apparatus, soared to 760,000 tonnes by 1850, by which time every Scottish ironworks had experienced a decade of production with the improved technique. Most of the furnaces were erected a few miles to the east of the city, at Coatbridge on the Monkland Canal, in the heart of Lanarkshire's rich coal and iron-ore deposits (see Figure 2.17 overleaf). But fifteen furnaces also arose on the very outskirts of the city by 1850. The city landscape on the south bank of the Clyde was dominated in the 1860s by Govan's six blast-furnaces, each eighteen metres tall. All of this was financed by Glasgow's manufacturers, following the pattern set by Henry Houldsworth who, in the 1830s, switched from spinning and weaving in Anderston to iron-production in Lanarkshire, at Coltness.

With cheap pig-iron available, large engineering-works sprouted in and around Glasgow. Springburn, two and a half kilometres to the north of the city centre, until 1840 a rural village, became the world's largest locomotive-building works on a site occupying many hectares. At Possil, one and a half kilometres north of the city centre, the firm of Walter Macfarlane built their Saracen Foundry in 1869. Behind its oriental façade 1,200 workers produced iron drainpipes and gutters, bandstands and cast-iron sanitary appliances. And R. Laidlaw and Son, another large company, operating three foundries within the city, manufactured gas-holders, lamp-standards, fire-hydrants and cast-iron pipes, for both the home and export markets (Fraser and Maver, 1996, pp.113–14). Cast iron, along with glass, was used in a bold way for new building types: large shops in the city centre (see Figure 2.18 overleaf).

Marine engineering was another manifestation of Glasgow's booming iron industry. At first, this meant the production of steam-engines of special design for

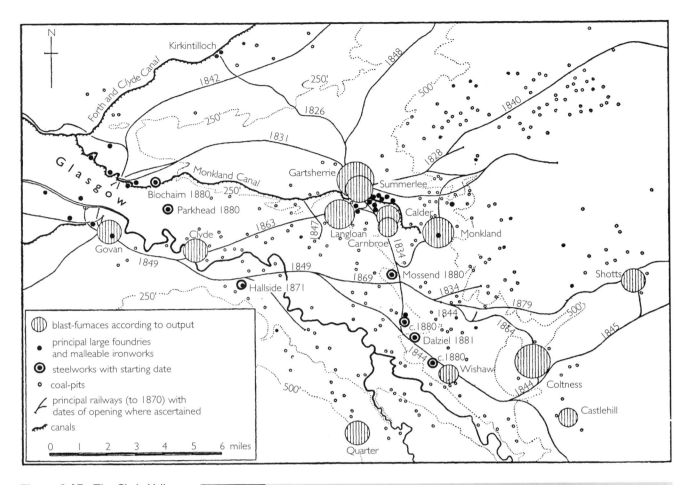

Figure 2.17 The Clyde Valley iron industry, c.1875. Symbols for blast-furnaces are graded according to estimated output at each works. The district of Coatbridge encompassed the furnaces marked at Gartsherrie, Summerlee, Langloan, Carnbroe and Calder (reproduced from Miller and Tivy, 1958, p.171, figure 40)

Figure 2.18 Gardner's store, Jamaica Street – the name commemorates Glasgow's early transatlantic trade (copyright: David W. Wrightson)

the propulsion of paddle-steamers. In 1812 the *Comet*, a steamer powered by an engine built in Glasgow, came into service on the Clyde. This was Europe's first commercially successful steamboat. By 1818, there were seventeen steamers operating on the Clyde from Glasgow to towns on the Firth, and to Liverpool and Belfast. In 1821 their hulls, made of wood still at Greenock and Dumbarton, were being towed upriver to just one kilometre west of Glasgow's central Broomielaw wharf, at Lancefield, where David Napier's works installed the engines. And facing this works on the opposite bank, at Springfield Quay, Thomas Wingate opened a similar engine-works, also in the 1820s (Reed, 1993, pp.43–4). The names of these districts just to the west of Glasgow – Lancefield, Springfield, Fairfield, Meadowside – evoke a pristine rural environment that was soon to vanish beneath heavy industry, as in the area shown in Figure 2.19. For the entire area of the Clyde in and around Glasgow would soon become a zone of hectic iron (and from the 1870s steel) shipbuilding whose mid-Victorian importance is apparent from the records of British output summarized in Figure 2.20 overleaf.

This was an excellent site for shipbuilding yards. The dredging of the 1770s had already established stable riverbanks: the land alongside the river, once subject to flooding, had been secured by depositing on it excavated material from the river-bed. As Figure 2.19 shows, this reclaimed land was flat, free from the steepness that created difficulties for shipbuilding on Tyneside (Reed, 1993, pp.42f.). Glasgow had no shipbuilding firms in 1800; by 1864 there were twenty. Glasgow's share of Clydeside's iron-shipbuilding industry had grown to predominance by 1850 (see Figure 2.21, p.55). Thousands of workers came daily from the city to these yards, travelling by train, and later by tram or steamer.

Figure 2.19 The availability of flat land adjoining the River Clyde allowed the spread of shipyards westwards from Glasgow, as here at Renfrew (courtesy of UCS Collection, Business Records, Glasgow University Archives)

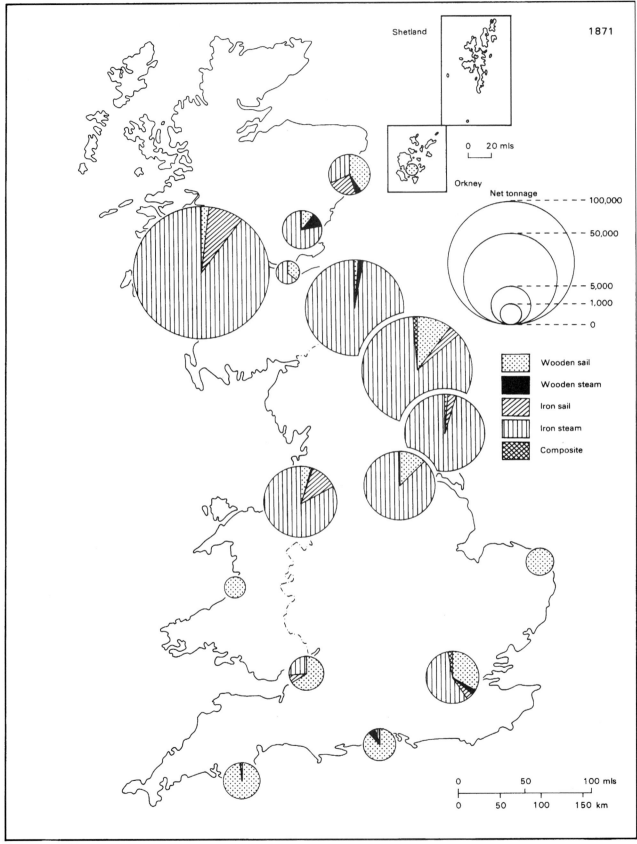

Figure 2.20 Tonnages of ships built in Britain, 1871 (reproduced from Langton and Morris, 1986; with the permission of the publisher)

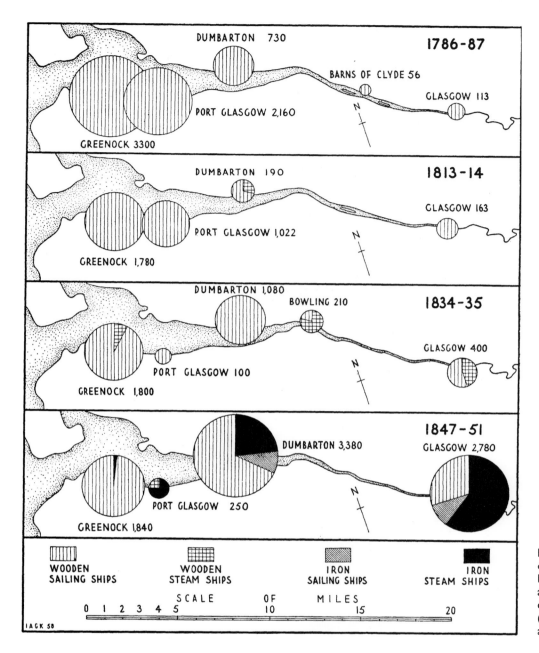

Figure 2.21 Shipbuilding on the Clyde, 1786–1851. Figures are tonnages built annually, to which the circles are proportionate (reproduced from Miller and Tivy, 1958)

As a result of these decades of industrial development, Glasgow in the 1870s had a structure markedly different from the city of fifty years before (see Figure 2.22 overleaf).

The two maps in Figure 2.22 show that there had been a substantial increase in industrial land-use over the period 1825–75. That is particularly evident in the north, where the space between the previously separate centres of Port Dundas and St Rollox has become filled to form an extensive zone of industry and tenements, serviced by railways. In the south too, engineering has become prominent in the Gorbals, and the once select estate of Laurieston has succumbed to industry and railway lines. In all this, Glasgow's earlier mingling of tenements with industry has been reinforced. To the north-west, crescents and a park indicate the development of the elegant, healthier district of the well-to-do. But the trend of Glasgow's western expansion was, above all, due to the new shipbuilding industry on the Clyde. The city's boundaries have greatly extended. Although the

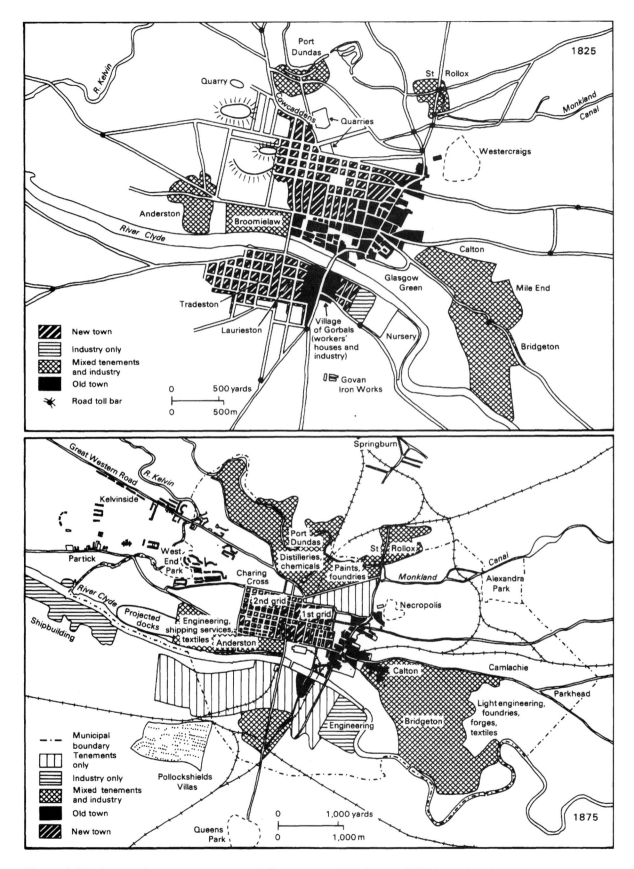

Figure 2.22 Social and economic structure of Glasgow: *top* 1825; *bottom* 1875 (reproduced from Langton and Morris, 1986; with the permission of the publisher)

boundary line is missing from the earlier map, note that neither Calton, Bridgeton, Anderston, Port Dundas nor most of the Gorbals were part of the city in 1825; all had been absorbed by 1875.

Glasgow had spread and would continue to do so. In 1830 the city occupied 754 hectares; by 1912 its area had increased enormously to 7,763 hectares (Fraser and Maver, 1996, p.9). But the incorporation of previously outlying villages was not the cause of Glasgow's striking population growth; instead, natural increase and, probably above all, immigration from Ireland and the Highlands were the reasons. Population growth was explosive; the figures from Glasgow's censuses (Fraser and Maver, 1996, p.142) speak for themselves (see Table 2.1). These censuses also revealed that the inhabitants were employed overwhelmingly in industry: 78 per cent of males and 71 per cent of females in 1851 (Fraser and Maver, 1996, p.167).

By 1871 Glasgow was the UK's second most populous city, and (despite its boundary expansion) the most densely inhabited: 230 people per hectare by 1891 but as high as 405 in the closes off the High Street, according to a survey of 1866 (Reed, 1993, p.84) (see Figure 2.23).

Table 2.1 The growing population of Glasgow

Census year	Population
1801	77,385
1811	100,749
1821	147,043
1831	202,426
1841	274,533
1851	329,097
1861	395,503
1871	477,732
1881	511,415
1891	565,839
1901	761,709

Figure 2.23 The unimproved city, probably the Calton district in the 1860s. Urban by-laws had not yet got rid of the chickens that supplemented the family income (courtesy of Thomas Annan Collection, History and Glasgow Room, Mitchell Library, Glasgow; photograph: Thomas Annan)

The quickening interaction between technological change and Glasgow's form and fabric reached a climax in the resumed hydraulic engineering on the Clyde from the 1840s on. (The following paragraphs draw heavily on Riddell, 1979.[2]) The initiatives of the late eighteenth century had brought improved navigability but, still in the 1830s, the river was too shallow for large ships to sail up to Glasgow, and much silt scoured from around the jetties had settled in other stretches. Now, amidst an industrial boom, there was a pressing need to facilitate exports by developing Glasgow's maritime communications with the world. The Clyde had to be deepened. Accordingly, the trustees of the river – city councillors or their nominees – established an engineering department headed by a succession of civil engineers. They advised a programme of intensive dredging. The device used was the bucket-dredger. The idea of a series of buckets moving on an endless chain, scooping material from a river-bed and then tipping out the deposits on to a container, goes back to the Early Modern Netherlands. But in mid-Victorian Glasgow the technique was developed to its full effect by the use of steam-engines – employed for this purpose before but never on such a scale. Glasgow became the world centre of dredger construction; its Number 8 model (built in 1865) was regarded as the biggest and most powerful in operation

Figure 2.24 Steam-powered bucket-dredger deepening the Clyde, c.1880; a hopper stands alongside (courtesy of Glasgow Museums, Museum of Transport)

anywhere; it could remove 350 tonnes of sand in one hour. Excavated material was loaded on to a hopper barge waiting alongside (see Figure 2.24). Doors in the bottom of the hopper were opened in the deep waters of the Firth to deposit the spoil. By 1845, the steam-powered bucket-dredger had deepened the Clyde to almost six metres at neap tide; for the first time, ships of up to 1,000 tons could sail right into Glasgow. Dredging operations had to be maintained and so, for generations of Glasgow's residents, the large fleet of dredgers and barges, all built in local yards, were a familiar part of the urban environment. Even powerful dredgers were inadequate to deal with layers of hard rock on the river-bed; in 1869 the newly invented dynamite was used to clear these stretches.

Large ocean-liners and steamers, whose function was to maintain industrial connections between Glasgow and Lancashire (through Liverpool), crowded the Clyde's quays, causing delays in unloading cargo and passengers (Fraser and Maver, 1996, p.60). The solution, adopted from the 1860s, would transform the face of Glasgow: extensive wet-docks were constructed to allow the free access and departure of shipping. The capacious tidal basin of Kingston Dock was completed in 1867, a relatively simple operation because the area to be excavated was mostly sand. Its masonry entrance-walls rested on wooden piles (see Figure 2.25). But the much larger Queen's Dock (1872–80) required deep foundations for walls far too massive for wooden piles to support. Instead, a new technique of sinking concrete cylinders was devised, which used a new type of grab-excavator in conjunction with travelling cranes. A succession of large concrete rings, cast on the site, was gradually built up, while material was excavated from their centres. The concrete columns were then driven down by applying several tonnes of cast-iron weights (see Figure 2.26). This was one of the UK's largest docks; it was provided with goods sheds and powered cranes for the loading of coal. Along with the later Prince's Dock (1893–7), it gave Glasgow the new image of a harbour complex (see Figure 2.27). And this made Glasgow a distinctive city: as both a great port and an important centre of heavy industry it combined the functions of Liverpool and Manchester.

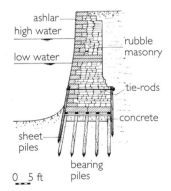

Figure 2.25 The foundations of Kingston Dock (reproduced from Riddell, 1979, p.186; by permission of John Donald Publishers)

---

[2]   Extracts from John F. Riddell's work are reproduced in the Reader associated with this volume (Goodman, 1999).

Figure 2.26 Laying the foundations of Queen's Dock, early 1870s. A crane is about to deposit another concrete ring on the rising column; in the background is a travelling crane, used to excavate spoil from the centre of the hollow columns (courtesy of Strathclyde Regional Archives, Mitchell Library, Glasgow, The Clyde Navigation Collection)

Figure 2.27 Glasgow Harbour, 1923. Queen's Dock and Prince's Dock are in the middle distance. Across Govan Road (bottom right), streets of tenements provide housing for workers in the docks and shipyards (courtesy of History and Glasgow Room, Mitchell Library, Glasgow)

In and out of the city

By the late 1850s more than two million passengers were sailing in or out of Glasgow every year. Some were leaving for America, Australia or New Zealand. For many others it was a local pleasure trip (see Figure 2.28). Sailing remained the preferred option, even though a railway line to Greenock had existed since 1841. By the 1880s, a pattern of city life had developed, worthy of report to MPS at Westminster:

> On the river Clyde, there is a very large traffic indeed by steamers, partly ... for the purposes of pleasure, but also for something that may be called something more than pleasure purposes, because it is for the residential traffic of Glasgow which, as a town, lives out of town in the summer. Every one who can ... either hires apartments, or anything else, gets out of Glasgow in the summer and goes to one or other of the river places, and goes generally as far as Greenock ... or one of those piers – by railway, and then goes on by steamer to one place or another and there stays for the night with his family, and goes back again to earn his bread in the morning.
>
> (quoted in Fraser and Maver, 1996, p.59)

For others still, a river journey meant commuting to work. Some shipyards hired steamers to bring workers from the city to the western shipyards. In 1884 the Clyde Trust introduced a regular river-bus service for this purpose. By the 1890s, twelve of these ferries were carrying three million passengers a year between Glasgow and Whiteinch, the western extremity of Glasgow's shipyards, some six kilometres distant. Ferries also linked the north and south of the city (Reed, 1993, p.52). Much of Glasgow's food supplies, notably Irish pork and beef, came to the city by steamer.

By contrast, one of the effects of Glasgow's railways was to reduce mobility within the city. The multitude of proposed railway projects in the 1840s caused such concern for the fate of the city that Glasgow was the only Victorian city

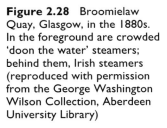

Figure 2.28 Broomielaw Quay, Glasgow, in the 1880s. In the foreground are crowded 'doon the water' steamers; behind them, Irish steamers (reproduced with permission from the George Washington Wilson Collection, Aberdeen University Library)

other than London to bring into existence a Royal Commission to study the problems of the urban railway invasion. But the Commission failed to have any influence (Kellett, 1969, pp.237–9). Instead, rivalry between railway companies for urban space led to duplication of facilities and the lack of any rational master-plan. The southern districts of Laurieston and Tradeston, laid out on a grid as desirable residential areas, had by mid-century been destroyed by the intrusion of the railway, eventually becoming a slum annexe of the Gorbals. Laurieston was just one and a half kilometres from the central business district, yet it was now cut off by railway lines; it was, one of its dejected residents complained, 'as if it were a walled city' (ibid., p.292). Much larger city-barriers developed in the south, through the Caledonian Railway Company's complex of coal depots, marshalling yards and carriage sheds which together covered sixty-five hectares; and, still more extensive – it occupied seventy-seven hectares to the north-east of the city – the concentration of goods yards and engine-works in the area between St Rollox, Springburn and Sighthill (ibid., p.291).

But Glasgow also gained from its railways. Large areas of insanitary, slum property were demolished to make way for railway lines and, at the same time, a new street system – forty new streets and a dozen widened or extended, to provide a rectangular grid of light, space and air in place of what the Victorian housing-reformer Octavia Hill had called 'the perfect honeycomb of slums' (Reed, 1993, pp.86–7). The railway stations, however, so intensified horse-drawn traffic that congestion was an acute problem in the streets of the 1860s. By 1870 900 vehicles an hour were moving through Jamaica Street. Still, the city's inhabitants could always admire the striking iron architecture of one of the new stations (see Figure 2.29).

Figure 2.29 St Enoch Station: the vigorous ironwork of Sir John Fowler and James Blair, 1875–9. The span is sixty-two metres (copyright: David W. Wrightson)

The railways failed to develop an outer ring of dormitory suburbs in Victorian Glasgow. The city remained densely populated and compact. The main middle-class residential area of Kelvinside (to the north-west) and the working-class district of Bridgeton (to the east) were too close – within three kilometres of the centre – for rail services to be attractive. So the well-off used their carriages, the poor walked to work or, from the 1870s, used the municipal trams. Workers' trains were operating in the 1890s, but they did not reduce the city's densely populated areas. Instead there was an outward flow of commuting workers from Glasgow to the Singer sewing-machine works at Clydebank and the steelworks at Coatbridge, augmenting the outward commuting already established by workers travelling by steamer to the shipyards. This pattern of travel to work astonished MPs at Westminster who heard the general manager of Glasgow's Caledonian Railway giving evidence to the Select Committee on Workmen's Trains in 1903:

> It was the reverse of London? You took from your populous centre workmen to work away from the centre and you brought them back at night? – That is so; and that is the rule in Scotland.
>
> (Kellett, 1969, pp.354–5)

Loch Katrine and the momentum of improvement

Glasgow's residents loved their river. The Clyde was the arterial highway that brought them prosperity and employment, as well as their gateway to leisure. Unfortunately, it was also the filthy source of their drinking-water. At the beginning of Victoria's reign, two private companies were pumping up the river's refuse-laden water to provide an intermittent supply for fire-fighting, industrial and domestic use. The city's galloping population growth increasingly revealed the inadequacy of existing facilities. That was brought home by the shock of the cholera epidemic of 1848–9, in which 3,777 people died, including two city councillors. There was a growing feeling that the municipal authorities should take over the task of supplying the city's water, and a conviction that a pure source must be found in the surrounding countryside. But no real progress occurred until 1852, when a letter from John Rankine, engineer and physicist – he was soon to become professor of civil engineering at Glasgow University – to the town council encouraged the council's resolve to municipalize the water supply and to draw on Loch Katrine, fifty-four kilometres to the north of the city, in the Perthshire hills.

There are striking similarities here with what happened in Manchester. In both cases, engineers argued for the advantages of a remote source, unpolluted by agriculture or population. And in both cases an economical engineering solution was sought, which employed gravitation instead of pumping. The same mixture of benefits was promised: progress in fire-fighting, industrial expansion, domestic convenience and improvements in sanitation. Both projects were impelled by a growing sense of urgency over population doubling, and the conviction that private water companies were inadequate for municipal supply. A distinctive feature of Rankine's argument was the rare (perhaps unique) suggestion that high-pressure water supply might provide a non-polluting power source, so ridding the city of its smoking, steam-powered factories (Rankine's letter is reproduced in Anon., 1868, pp.3–8; extracts from it can be found in Goodman, 1999, the Reader associated with this volume).

In October 1852 the town council appointed a committee of its councillors and magistrates, including Robert Stewart, lord provost of the city. Impressed by what had been achieved at Manchester – that would be a continuing source of inspiration – the committee soon appointed Bateman, the director of those engineering works, to report on what should be done in Glasgow. The following March, Bateman's findings declared that Loch Katrine would be the best and cheapest source. Stewart gave his full support, emphasizing its 'absolute necessity' for 'this great and growing city' whose 'immense and rapid increase resembled New York'. He fully expected that, once the Loch's soft water was flowing to the city, bleachers and dyers would flock to Glasgow just as they had done in Manchester. And the city would enjoy 'a perpetual cleansing process' (Anon., 1868, pp.43, 69). But the project was strongly opposed within the council, causing the *Glasgow Herald* to record 'an amount of abuse, vituperation, and insolence, which is unparalleled in the annals of the Town Council' (ibid., p.171). The opposition was led by Councillor Gemmill, who resented any increase in water rates in his Gorbals constituency – a charge to be imposed on landlords, whether they used the new supply or not. And he warned that the huge project would bankrupt the corporation.

Gemmill's voice was soon reinforced by two unexpected difficulties, which halted the progress of the bill in Parliament. There was sudden alarm that the pure Loch water would be turned into poison by the lead pipes conveying it to the city's teeming inhabitants. This was based on chemical tests carried out by Professor Penny of Glasgow's Andersonian Institute. The councillors opposed to the scheme demanded that the town council inspect his water samples. Stewart refused, stating that the matter was already being investigated: approaches were being made to Professors Hofmann, Graham and Faraday, the nation's most eminent chemists. And the provost had sent agents to the USA – to Philadelphia and New York, cities where there had been recent scares over the use of lead water-pipes. The results of these investigations indicated that there was no cause for concern. But Gemmill accused the council of using 'got-up' evidence and rashly promoting a project destined to bring 'excruciating pains and death' to the inhabitants (ibid., p.92). When the opposing party threatened to consult their own chemists, the *Glasgow Herald* commented that 'into these chemical subtleties we will not enter, simply because we do not understand them', and instead maintained its earlier excitement about an engineering project 'worthy to be ranked amongst the greatest works of the ancient Romans' (ibid., pp.75, 170–71).

On top of these difficulties there now came powerful opposition from the Admiralty, which feared that removal of water from Loch Katrine would so reduce the area's water system as to interfere with navigation on the Forth, especially at Stirling. In May 1854 Stewart wrote impassioned letters in an attempt to overcome the resistance. He appealed to the Lords of the Admiralty to consider the importance of the project for the health and comfort, and prosperity of the arts and manufactures, of the second city of the empire. And he sought the intervention of Palmerston, the home secretary, arguing that concern for the river trade of Stirling, and its population of a mere 12,000, must not prevent a project essential to 'the welfare of so large and important a portion of her Majesty's dominions' (ibid., pp.108–11).

In the end, all opposition evaporated through the provost's conciliatory gestures. Stirling settled for £7,000 to purchase a dredging-machine to deepen the Forth; the Admiralty acquiesced. Compensation was paid for losses of salmon-fishing on the Forth. Settlements were agreed with the Duke of Montrose and other owners of land needed for the works. And finally, Gemmill's resistance disappeared with the promise of no increased water rates in the Gorbals.

In May 1855 the bill was passed, giving parliamentary authorization for engineering work to begin, and sanctioning the city's purchase of the private water companies. The Act required the municipality to lay pipes in all streets; but the occupiers or owners of houses were made responsible for installing connecting-pipes to the mains.

Bateman was appointed to supervise the project, whose magnitude is apparent from the statistics on completion: nineteen kilometres of tunnels, with a diameter of two and a half metres, driven through hard rock; sixteen kilometres of bridges, including wrought-iron tubes of four and a half metre span; another six and a half kilometres of cast-iron pipes; a reservoir extending over twenty-five hectares and containing two and a half million cubic metres of water; cast-iron mains from the reservoir to the city eleven kilometres away (see Figure 2.30). In March 1860, after an expenditure of almost one million pounds, Queen Victoria opened the waterworks at a lavish ceremony. And ten years later, a monumental fountain was erected in a city park to commemorate the provost's efforts (see Figure 2.31, p.66). But soon, even engineering on this scale was insufficient for Glasgow's soaring population and, above all, the escalating requirements of industry. To keep pace with demand several extensions to the works were added, culminating in the 1890s in a second aqueduct from the Loch, thirty-seven kilometres long. By the 1880s the city had around 560 kilometres of distribution-pipes and over 4,000 fire-cocks. And, as in Manchester before, there was municipal regulation of plumbing by the city water office (Gale, 1883, pp.180, 185). To preserve the purity of Loch Katrine's water, Glasgow's Water Act of 1892 restricted the number of steamers sailing on the Loch and prohibited all building in the vicinity.

The wider significance of Loch Katrine has been well expressed in an appraisal by Fraser and Maver:

> The success of Loch Katrine cannot be over-estimated in terms of the future progress of sanitary reform in the city, as it dissolved any lingering doubts about municipal interventionism, and demonstrated the enormous public-relations benefits of the civic authority being seen to act for the welfare of the entire community ... Municipal activity in the sphere of sanitary reform progressed even further in 1866, when long-standing plans for a City Improvement Act eventually reached the Statute Book. Blackie ... Lord Provost, was the driving spirit ... Under the terms of the Act, an Improvement Trust was established, which provided for the Lord Provost, magistrates and Council to buy up and clear congested areas that were deemed particularly hazardous to health. An area of approximately eighty-eight acres was involved, centred around Glasgow Cross ... power was given to form new streets and create a public park (Alexandra Park) in the north-east of the city. Finally, the Trustees were authorised to erect housing for the 'mechanics, labourers, and other members of the working and poorer classes'. It was the first time a city in the United Kingdom had taken on such powers of development, although the example of Paris had strongly influenced ... [the] quest to combine sanitary reform with aesthetic improvement ... the reconstruction of the French capital under Baron Haussmann received widespread publicity throughout Europe, and set a precedent for urban renewal which Glasgow hoped to emulate ...
>
> The two gas companies were eventually bought up in 1869 after the failure of private enterprise to light up the wynds and closes. As Lord Provost Samuel Chisholm later pointed out, gas was seen as on a par with water: 'We regard the lighting of the streets of the city and courts, and stairs, and providing a cheap supply of light for the citizens as just about as essential as the supply of clean water' ... The aim was cheapness of supply to make it accessible to all. Within a few years, the price of gas had been halved and by the 1880s no city in the world had so many houses supplied by gas. The prospect of owning the tramways too appealed to sanitary reformers, in the hope that the poor could be carried to their work and away from the crowded urban centre. They insisted that only the city should have control over the streets and, during the 1870s, the Town Council laid the tramlines for lease to the private Tramways Company.

(Fraser and Maver, 1996, pp.410–15)

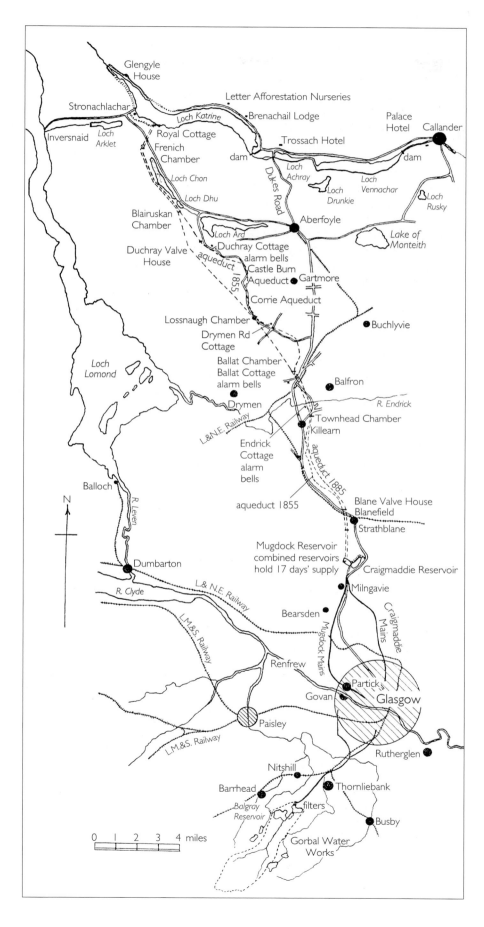

Figure 2.30 The Loch Katrine project: source of supply and line of aqueducts (*Brief account of the origin and development of the works, 1855–1935*, n.d.)

Figure 2.31 The Stewart Memorial Fountain, Kelvingrove Park, commemorates Robert Stewart's efforts as lord provost to secure parliamentary approval in 1855 for the Loch Katrine water-supply project; photograph *c.*1870 (courtesy of History and Glasgow Room, Mitchell Library, Glasgow)

Figure 2.32 Gallowgate, one of Glasgow's central streets, modernized after the reconstruction of 1875. It has been powerfully affected by the new technology of recent decades: note the railway bridge, gas lighting and horse trams. And the façade of the building on the right shows the influence of the new boulevards of Paris (Watson, 1879)

By 1875 the central street of Gallowgate had been transformed beyond recognition (see Figure 2.32). But the marriage of Glasgow to improving technology is perhaps nowhere more powerfully symbolized than in a photograph of the 1890s, showing the provost and town council inspecting another city water source in the Renfrewshire Hills (see Figure 2.33).

Figure 2.33 The town council – with Lord Provost Richmond in top-hat and chain, to the left of the waterfall – inspect the water supply in the Renfrewshire Hills for the Gorbals Waterworks, c.1899 (courtesy of Glasgow Museums, The People's Palace)

2.3   Conclusion

Close parallels are discernible in the industrialization of Glasgow and Manchester. Both cities had long histories: Glasgow originated as a small, medieval, ecclesiastical centre; Manchester's medieval role was similarly small and ecclesiastical, but it was older, with roots in Roman Britain, when it was a fortified hill-town. In the Early Modern period both cities became important manufacturing centres, through the production of textiles. A merchant elite was already visible in seventeenth-century Manchester, and in Glasgow in the following century. Entrepreneurs built their mills far into the rural surroundings of both cities, for there was no adequate water power to be found in either city to drive water-wheels. Manchester and Glasgow were each becoming centres of industrial regions. Then the adoption of Watt's steam-engines led to a spawning of mills within both cities. And along with textile mills, large engineering firms were established. With increasing opportunities for employment, Irish immigrants swelled the population in both cities. While the city centres became overcrowded, air pollution drove the better-off to the surrounding areas. Within both cities, mounting public-health concerns forced the authorities in the mid-nineteenth century to embark on ambitious and expensive projects to bring pure drinking-water from remote hills. The schemes succeeded under the supervision of the same engineer but, in both cases, projected demand was soon overtaken by galloping population increase, so necessitating further extensions to the aqueduct system.

Yet these similarities fall far short of identity. These two case-studies also reveal important differences, which are attributable to the distinctive geography and political history of the individual cities. The terrain of the Clyde, combined with rich resources of both coal and iron-ore on the city's doorstep, favoured the rise of a major shipbuilding industry inconceivable on Manchester's Irwell. Once the Clyde had been made more navigable, Glasgow became a flourishing Atlantic port in a way that Manchester never could. In sharp contrast to Manchester, the local public transport of Glasgow's citizens was mainly on water. Glasgow differed also because of the peculiarities of Scottish law; the system of land-holding helps to explain that characteristic of Glasgow's built environment: the high-rise tenement. And while the same pattern of urban improvements is perceptible in Victorian Glasgow and Manchester, the manner in which they were introduced was different: Glasgow, like other Scottish cities, had a long tradition of strong urban authority, and it was under vigorous municipal direction that the improvements were realized. Manchester had no municipal authority until 1838, and even then it only amounted to the right to supervise a police force; only gradually, in the course of the century, would its powers increase.

All of this tends to show that historical studies of the interaction of cities with technology must consider not only the general effects of technological innovation but also the peculiarities of the individual cities in which the changes occurred.

Extracts

2.1 Faucher, L. (1844) *Manchester in 1844: its present condition and future prospects*, London, Simpkin, Marshall, and Manchester, Abel Heywood, pp.16–19, 90–93

Manchester is an agglomeration, the most extraordinary, the most interesting, and in some respects, the most monstrous, which the progress of society has presented. The first impression is far from favourable. Its position is devoid of picturesque relief, and the horizon of clearness.

Amid the fogs which exhale from this marshy district, and the clouds of smoke vomited forth from the numberless chimneys, Labour presents a mysterious activity, somewhat akin to the subterraneous action of a volcano. There are no great boulevards or heights to aid the eye in measuring the vast extent of surface which it occupies. It is distinguished neither by those contrasting features which mark the cities of the middle ages, nor by that regularity which characterizes the capitals of recent formation. All the houses, all the streets, resemble each other; and yet this uniformity is in the midst of confusion. On closer examination, however, a certain approximation to order is apparent. Manchester is situated at the confluence of a little river, the Irwell, swollen by the waters of the Irk, and of a brook called the Medlock. The Irwell separates Manchester from its principal suburb – the old town which has given its name to the hundred of Salford. On the left bank of the river is another suburb, Chorlton-upon-Medlock, which in 1801, numbered only 675 inhabitants, and which now contains 30,000. The manufactories and machine shops form, as it were, a girdle around the town, and follow the courses of the streams. Factories, seven stories in height, rear their lofty fronts along the banks of the Irwell, and along the borders of the canals, which, penetrating into the town, form an interior navigation. The waters of the Irk, black and fetid as they are, supply numerous tanneries and dye-works; those of the Medlock supply calico-printing establishments, machine shops, and foundries. The banks of the Irwell, which appear to have been the principal seat of this civilization, still form the centre of it. The primitive municipal buildings are scattered along its course. Descending from the hill where the workhouse is situated, you come to the buildings of the College, the Old Church, and the Exchange, and upon the other side, the Court of Sessions, and also the Gaol. From Pendleton to

London Road is one great thoroughfare, crossing the town from east to west. At its extremities are ranged the shops which supply provisions and the necessaries of life; and at the centre, in Market-street, the shops are devoted to luxuries, the libraries and newspaper offices, &c. The aristocratic quarter, called Mosley-street, which joins Market-street at right angles, contains the warehouses of the principal manufacturers; and in the angle of the two streets are concentrated the storehouses for raw material and manufactured goods. The railways, being of recent formation, terminate at the exterior points of the circumference; those of Liverpool, and Bolton to the west, and those of Leeds, Sheffield, and Birmingham, to the east.

From this apparently indifferent combination, there results a great economy both of time and wealth in production. There is perhaps good reason for complaint that too little attention has been paid to the health and convenience of the inhabitants; of the want of public squares, fountains, trees, promenades, and well-ventilated buildings; but it is certain that it would be a difficult task to devise a plan by which the various products of Industry could be more concentrated, or by which the manufactories should be brought nearer to the fuel which feeds them, or more accessible to facilities for disposing of the goods when manufactured. The railways penetrate the town upon immense arcades to the points where it ceases to be inconvenient to load the merchandise upon them, and the canals pass under the streets, and thread their sinuous way in every direction, conveying boat-loads of coal to the doors of the manufactories, and even to the very mouths of the furnaces.

Manchester does not present the bustle either of London or Liverpool. During the greater part of the day, the town is silent and appears almost deserted. The heavily-laden boats glide noiselessly along the canals, not at the feet of palaces, as in Venice, but between rows of immense factories, which divide amongst themselves the air, the water, and the fire. The long trains roll smoothly along the lines of railway, conveying as many multitudes as individuals aforetime. You hear nothing but the breathing of the vast machines, sending forth fire and smoke through their tall chimneys, and offering up to the heavens, as it were in token of homage, the sighs of that Labour which God has imposed upon man.

At certain hours of the day the town appears suddenly animated. The operatives going to, or returning from their work, fill the streets by thousands; or it is perhaps the hour of 'Change [i.e. the Exchange], and you see the chiefs of this immense population gathering to one common centre; but even at those times when the inhabitants relax from their arduous duties and give free course to their feelings, they lose nothing of that serious and angular stiffness, which a too exclusive occupation in industrial pursuits communicates to them …

There are, for nations … periods of transition, which are marked by a large amount of suffering. The manufacturing system in England, and elsewhere, is in this period of trial. The rapidity of its growth; the magnitude of its proportions; everything connected with it … all prove that it is far from a state of maturity …

Amongst the causes which prolong this temporary malady, none act more strongly than the agglomeration of manufactories and operatives in the towns. The centres of industry are seats of corruption … The manufacturing towns are free from one evil which characterizes other large towns, viz. the idleness of the poorer classes. They have, however, a still greater evil in the fermentation engendered by the close contact of all ages, and both sexes, during the long hours of labour …

… urban industry, whatever may be its extent, whether it embrace a population of 30,000 or of 300,000, presents equally unfavourable conditions, both for the morality and health of its operatives. Reformation therefore requires that the close contact of the operatives in the manufactories should be diminished, and that manufactures injurious by their too close proximity, should be diffused over a greater space. The factories with their seven stories of windows, recall to the mind the buildings of ancient Rome, which were compared to islands (*insulæ*), doubtless to indicate the necessity of isolating them, and to afford a greater amount of space and fresh air around them …

In the natural order of society, towns are destined for the concentration of commerce and luxury, literature, science, and the fine arts. It is there where the merchandise of society is accumulated and exchanged, but it is not there that productive industry should locate itself. Towns were originally commercial marts, and their origin clearly bespeaks this to be their final destiny. To the towns belong the stores, warehouses, countinghouses, banks, museums, libraries, colleges, clubs, academies, and mechanical and liberal arts; their function is sufficiently comprehensive, without adding to it that of Industry …

The increased facilities of communication now afforded by roads, by canals, and by railways, render possible the *decentralization* of manufactures, a result much to be desired. A factory may now be established close to a coal mine, or by a canal, which shall convey to it fuel, or by a waterfall, without losing the advantages of proximity to a great market. The spinners of Hyde or of Turton, can come in less than an hour to the Manchester Exchange as easily as though they were located in Little Ireland or on the banks of the Irk. Distance is daily becoming cancelled, and the economy of time is everywhere facilitated. There is therefore no longer any reason for struggling for a few feet of land in the midst of some filthy purlieu, and at the risk of the general health.

The superiority of rural over urban manufacture is a conclusion not only deduced from reasoning but, if I mistake not, a matter of experience also.

2.2   Kellett, J.R. (1969) *The Impact of Railways on Victorian Cities*, London, Routledge and Kegan Paul, pp.171–4

The distances which separated Manchester from the specialised manufacturing villages and towns which surrounded her were not too short for effective and profitable railway service, nor so great that the linkage became cumbersome and expensive. A pair of circles drawn at intervals of ten and thirty miles [sixteen and forty-eight kilometres] from Manchester would include most of the satellite manufacturing and processing centres …

Railways effectively created in Lancashire, even within two decades, a system of cities which was dominated operationally by Manchester …

The *Preston Chronicle* described rapturously the extraordinary speed with which the thirty-mile [forty-eight kilometre] Preston linkage worked as early as 1839. Brokers despatched the cotton yarn from Manchester at 3 a.m., arriving (the newspaper stated precisely) at 9.08 a.m., being converted to cloth by 11.30 a.m., sent back as shirting material on the 4.20 p.m. train, arriving at 7 p.m. and being put on sale the next morning – 'the very millennium of railway velocity'. The broadside *Reasons in Favour of a Direct Line of Railroad from London to MANCHESTER* (1846) spoke of Manchester's special claim to speedy rail communication, since 'the productions of its looms and mills form indisputably the largest and most important branch of our national industry', and also stressed the unexpected importance to industry itself of the rapid flow of passenger traffic, small masters, and salesmen by train, causing 'changes in all departments of trade far beyond the contemplation and conception of those who originated these schemes' … Neither Liverpool, Glasgow nor London employed its rail linkages in quite the same way – virtually as production lines in a system of industrially inter-related cities …

Birmingham was more akin to Manchester in the functional role it fulfilled for a hinterland of smaller manufacturing towns; but its staple manufactures were neither so perishable, so bulky to store, nor so subject to the whims of season and fashion. In Manchester, the Board of Trade reported, the effect of the railways had been to break down, even by 1845, the old system of manufacturing and of forwarding large quantities of goods 'in anticipation of the probable demand of particular seasons and markets'. Instead there was a considerable saving of capital tied up in stocks, and a reduction to reasonable physical compass of the urban space demanded by the warehousing associated with an annual output of a million yards [914,000 metres] of material and 140 million lb. [64 million kilograms] of twist and yarn. Also, and equally important, there was a reduction in the risk of fluctuation and loss arising from miscalculations as to the probable nature and extent of demand in distant markets …

The dependence of Manchester upon a ring of specialised towns admirably situated for a fixed-route transport system to operate most efficiently, and the peculiar force, in Manchester's staple industry, of the inventory factors mentioned in the Board of Trade report, help to explain why the city became an arena for sharp contest between so many railway companies. They also help to explain the Corporation's attitude towards the railway companies.

At first the Corporation could not have been more complaisant. Its Improvement Committee reported, on 22 September 1845, that the land required to improve the approach to London Road station had been purchased and laid to the street, adding, almost gratefully, 'The Manchester and Birmingham Railway company have agreed to pay half the cost of this improvement.' The Corporation, of course, paid the other half. At the other end of the town a mere intimation from the Engineer of the Manchester and Leeds railway company, on 21 June 1846, that the land bounded by the Irk might be required for railway purposes, and requesting the re-direction of an intended new road, was enough to cause the Corporation to 'examine the locality and alter the plans'; and subsequently to arrange to sweep away the old Apple market, and straighten or eliminate several small streets, to secure a more commodious approach to the station. The Committee for General Purposes, having considered a letter from the Lancashire and Yorkshire railway company on 1 December 1847, recommended that all clocks under the control of the Corporation, and that of the church wardens, should be adjusted by the necessary nine minutes, to conform with railway time; a symbolic act of homage by a mercantile and industrial community to the new masters of its traffic.

By the end of the following decade the Corporation's courtship of the railway companies was over. The Corporation itself had been split, and unscrupulously manipulated, and the poor amenities, together with the 'insufficient and unnecessarily expensive' services had led to 'a sense of bitter disappointment.' Thereafter, the Corporation adopted a far more suspicious and legalistic attitude towards the private companies providing the city's rail transport.

References

ANON. (1868) *Water Supply from Loch Katrine: the discussions in the Town Council,* published for private circulation, conserved in the Mitchell Library, Glasgow.

BATEMAN, J.F. (1884) *History and Description of the Manchester Waterworks,* Manchester, T.J. Day.

BRIGGS, A. (1968) *Victorian Cities,* Harmondsworth, Penguin Books.

CARTER, C.F. (ed.) (1962) *Manchester and its Region,* Manchester, Manchester University Press.

CHALLONER, W.H. (1959–60) 'Manchester in the latter half of the eighteenth century', *Bulletin of the John Rylands Library,* vol.42, pp.40–60.

DEFOE, D. (1962) *A Tour Through the Whole Island of Great Britain,* London, J.M. Dent, vol.2.

DEVINE, T.M. and JACKSON, G. (eds) (1995) *Glasgow,* Manchester, Manchester University Press, vol.1.

FAUCHER, L. (1844) *Manchester in 1844: its present condition and future prospects,* London, Simpkin, Marshall, and Manchester, Abel Heywood.

FLINN, M.W. (ed.) (1965) *Chadwick's 'Report on the Sanitary Condition of the Labouring Population of Great Britain',* Edinburgh, Edinburgh University Press, pp.1–73.

FRASER, W.H. and MAVER, I. (eds) (1996) *Glasgow,* Manchester, Manchester University Press, vol.2.

GALE, J. (1883) 'On the latest additions to the Loch Katrine water works', *Transactions of the Institution of Engineers and Shipbuilders in Scotland,* vol.26, pp.151–94.

GILES, C. (1993) 'Housing the loom, 1790–1850: a study of industrial building and mechanisation in a transitional period', *Industrial Archaeology,* vol.16, pp.27–37.

GOMME, A. and WALKER, D. (1968) *Architecture of Glasgow,* London, Lund Humphries.

GOODMAN, D. (ed.) (1999) *The European Cities and Technology Reader: industrial to post-industrial city,* London, Routledge, in association with The Open University.

HASSAN, J.A. (1984) 'The impact and development of the water supply in Manchester, 1568–1882', *Transactions of the Historic Society of Lancashire and Cheshire,* vol.133, pp.25–45.

HUME, J. (1966) 'The St Rollox chemical works 1799–1964', *Industrial Archaeology*, vol.3, pp.185–92.

KAY, J.P. (1833, 2nd edn) *The Moral and Physical Condition of the Working Classes Employed in the Cotton Manufacture in Manchester*, London, James Ridgway.

KELLETT, J.R. (1969) *The Impact of Railways on Victorian Cities*, London, Routledge and Kegan Paul.

LANGTON, J. and MORRIS, R.J. (eds) (1986) *Atlas of Industrializing Britain, 1780–1914*, London and New York, Methuen.

LLOYD-JONES, R. and LEWIS, M.J. (1988) *Manchester and the Age of the Factory*, London and New York, Croom Helm.

LOVE, B. (1842) *Hand-book of Manchester*, Manchester, Love and Barton.

MILLER, R. and TIVY, J. (1958) *The Glasgow Region: a general survey*, Glasgow, British Association for the Advancement of Science.

MORRIS, R.J. (1990) 'Urbanisation and Scotland' in W.H. Fraser and R.J. Morris (eds) *People and Society in Scotland*, Edinburgh, John Donald, vol.2, pp.73–102.

POOLEY, M.E. and POOLEY, C.G. (1984) 'Health, society and environment in Victorian Manchester' in R. Woods and J. Woodward (eds) *Urban Disease and Mortality in Nineteenth-century England*, London and New York, Batsford and St Martin's Press, pp.148–75.

REED, P. (ed.) (1993) *Glasgow: the forming of the city*, Edinburgh, Edinburgh University Press.

RIDDELL, J.F. (1979) *Clyde Navigation: a history of the development and deepening of the River Clyde*, Edinburgh, John Donald.

SCOLA, R. (1992) *Feeding the Victorian City: the food supply of Manchester 1770–1870*, Manchester, Manchester University Press.

SIMMONS, J. (1986) *The Railway in Town and Country*, Newton Abbot, David and Charles.

SISSONS, M. (1994) 'Housing the loom: a comment', *Industrial Archaeology*, vol.17, pp.86–8.

SMITH, R. (1953–4) 'Manchester as a centre for the manufacture and merchanting of cotton goods, 1820–30', *University of Birmingham Historical Journal*, vol.4, pp.47–65.

SMOUT, T.C. (1969) *A History of the Scottish People 1560–1830*, London, Collins-Fontana.

URE, A. (1836) *The Cotton Manufacture of Great Britain*, London, W. Knight.

VIGIER, F. (1970) *Change and Apathy: Liverpool and Manchester during the Industrial Revolution*, Cambridge, Mass., MIT Press.

WALTON, J.K. (1990) 'The north-west' in F.M.L. Thompson (ed.) *Cambridge Social History of Britain 1750–1950*, Cambridge, Cambridge University Press, vol.1, pp.355–414.

WATSON, J. (1879) 'Improvements in Glasgow and the City Improvements Acts: origins of the Artisans Dwellings Act', *Royal Institute of British Architects: Transactions*, 1st series, vol.29, pp.153–61.

WHITWORTH, J. (1847) 'On the advantages and economy of maintaining a high degree of cleanliness in streets and roads, with an account of the construction and operation of the street sweeping machine', *Minutes of the Proceedings of the Institution of Civil Engineers*, vol.6, pp.431–65.

WILKINSON, S. (1982) *Manchester's Warehouses: their history and architecture*, Swinton, Neil Richardson.

Chapter 3: TWO CAPITALS: LONDON AND PARIS

by David Goodman

3.1 London: the unique city

London and the Industrial Revolution

Britain's Industrial Revolution was a storm that passed over London and broke elsewhere. While Manchester, Glasgow and other cities in the North and the Midlands were caught up in the whirlwind of industrialization, London watched from the sidelines. With no coal deposits or iron ore in its vicinity, London was ill-equipped for the new heavy industry; nor was there any sprouting of the factory system in the metropolis. While the industrial regions toiled and created prosperity, London made no important manufacturing contribution; yet, because of its dominance over the rest of the kingdom, it tapped the wealth generated in the provinces. In short, London, like other European capitals, was a parasite.

That was how historians used to interpret the role of London in the Industrial Revolution. J.L. Hammond, for example, a much-read liberal historian, famous for his biting criticism of nineteenth-century Britain's ruling class and pessimistic portrayals of the social consequences of urban industrialization, expressed such views in the 1920s (in a book review in the *New Statesman*, 21 March 1925). And, still in the 1960s, the unproductive, parasitic role of London (and Paris) in the Industrial Revolution was being forcibly expressed by the most influential of French historians, Fernand Braudel (1974, p.440). These images of London are now largely discredited and superseded. Instead, much more positive assessments of London's contribution prevail.[1]

Far from being a bystander in the Industrial Revolution, London was a flourishing centre of manufactures and a source of innovative technology. In fact it was the kingdom's principal manufacturing centre, and had been so ever since the later Tudors, when metalworking, the manufacture of clothing and leather shoes, and numerous other crafts had assisted London's rise to economic dominance (Barker, 1989). The strength of this tradition was evident in the census of 1851, which showed that more Londoners were occupied in manufacturing than in any other activity: no less than one-third of the labour force was so employed, some 373,000 workers, making the capital the kingdom's largest manufacturing city by far (Schwarz, 1992, pp.11–13; see Figure 3.1 overleaf).

There were, it is now clear, three reasons for this outstanding manufacturing role. The first was the capital's population growth, astonishing even in the context of the demographic revolution occurring in much of Europe. There was nowhere approaching London's size. From an estimated 200,000 in 1600, its population was already doubling under the Stuarts, reaching nearly 400,000 by 1650. The best estimate for 1750 is 675,000. The first census of 1801 recorded 900,000, making London the world's largest city. Subsequent censuses showed that it had become the greatest city ever: 3.9 million in 1870; five and a half million by 1891, when Sidney Webb, the socialist historian, declared that

---

[1] For an illustration of this reinterpretation, see the extracts from Roy Porter's synthesis reproduced in the Reader associated with this volume (Goodman, 1999).

Figure 3.1 Occupations in London (both sexes); figures taken from the 1851 census (Schwarz, 1992; by permission of Cambridge University Press)

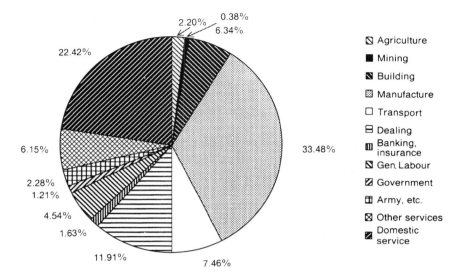

'London is more than a city: it is a whole kingdom in itself', with 'revenues exceeding mighty principalities', and 'with its suburbs it exceeds all Ireland' (Porter, 1994, p.207) (see Figure 3.2). By the end of the century, one in five of the population of England and Wales lived in the capital. To use the language of geographers, London was a 'primate city': it contained a high proportion of the national population and was of an order of magnitude beyond that of the next largest provincial cities. Such was the population and extent of London in mid-century that the Royal Commission, appointed in 1854 to investigate the desirability of putting the monster city under a single municipal authority, concluded that it was wholly inappropriate because London was

> a province covered with houses ... [whose] diameter from north to south and from east to west is so great that the persons living at its furthest extremities have few interests in common; its area is so large that each inhabitant is in general acquainted only with his own quarter, and has no minute knowledge of other parts of the town.
>
> (quoted in Young and Handcock, 1956, p.659)

Not until 1889 would sprawling London be granted a unified representative government, the London County Council.

This unequalled concentration of population supplied a plentiful reservoir of labour, skilled and unskilled, a basic condition for manufacturing. And the presence within the city of much surplus wealth constituted a powerful, local consumer-market for manufactured products. The basic needs of the huge population stimulated technologies old and new. The enormous food demands of the capital were felt as early as the sixteenth century in the neighbouring counties of Middlesex and Kent, where agriculture was, in part, conditioned to meet London's needs. This dependence on the countryside had so intensified by the nineteenth century that farmers, market-gardeners and livestock-breeders throughout the length and breadth of Britain were in some way employed in the service of Londoners' diet. And that stimulated the manufacture in London of a variety of forms of wagon for supply and distribution. Similarly, distant links with the capital were forged to satisfy that other basic requirement: fuel. From the seventeenth century, the mines of Northumberland and Durham supplied London with rapidly increasing quantities of 'sea-coal', as it was then called, because it was shipped from Newcastle. By 1750 coal cargoes totalling 640,000 tonnes were delivered for London's domestic and industrial uses. Here there were powerful incentives for technological innovation. As the northern mines were dug deeper, so water was encountered. The 'miner's friend', a steam-driven

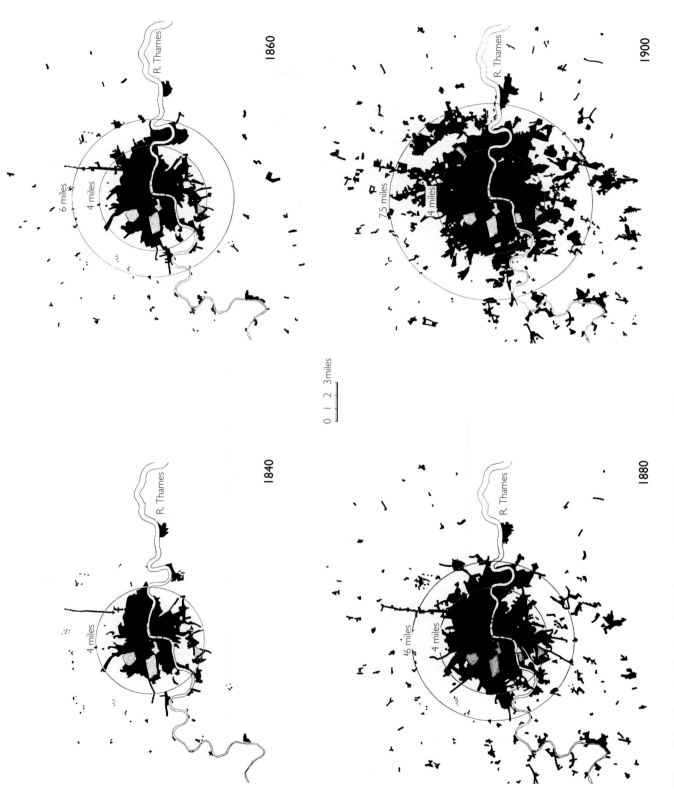

Figure 3.2 Growth of London: the built-up area, 1840–1900 (Rasmussen, 1982; by permission of Jonathan Cape)

suction machine for pumping water out of coal-mines and the first operational steam-engine, was invented by Thomas Savery in 1698, and manufactured in his workshop off London's Fleet Street; it proved inadequate, as the water could only be lifted to a small height, but it was used, around 1712, to pump water from the Thames at the Strand works of one of London's water-supply companies. As Londoners burned coal in greater quantities, so the capital became notorious for its fogs. By the 1790s, when Benjamin Thompson (Count Rumford), a wandering New Englander, scientist and philanthropist, settled in London, he was appalled by the dark cloud of smoke that hung over the capital. Anxious to remove the cause, which he attributed to an inefficient burning of coal, he busied himself with improving the design of fireplaces.

A second reason for London's manufacturing prominence was its function as a communications centre. That role went all the way back to the Roman conquerors, who had selected the site for Londinium at a point where the Thames was fordable, close to its estuary – a safe anchorage with easy navigation to the open sea. And the Roman roads throughout Britannia were built to converge on this provincial capital. In the Industrial Revolution, London's communications were further enhanced, and metropolitan manufacturers enjoyed ready access to markets domestic and overseas.

Third, the gradual acquisition, from the mid-eighteenth century, of a vast empire was a ferment for London's trade and industries. Fortunes were made in overseas trade. The square mile of the City (the fortified area of Roman London) became a place for merchant bankers and insurance offices – these organized London's first fire-brigades. The new wealth created a demand for luxury products. Ships sailing in growing number to the distant colonies stimulated the manufacture of scientific instruments – telescopes, barometers, terrestrial globes and, eventually, marine chronometers – all made in the capital's specialist workshops. Exotic imports from the empire, such as American tobacco and West Indian molasses – converted by London's distilleries to liquor for a wide market – were processed in London and then re-exported. Jamaican mahogany generated a new era of cabinet-making in eighteenth-century London, as in Thomas Chippendale's famous workshop in Long Acre. As a now great international entrepôt, London's port was a hive of activity. And the rising commerce had its effect on the fabric of the medieval city: the city gates were taken down in 1760 to facilitate traffic.

The river was the key to London's prosperity. Along its banks arose a cluster of industrial plants, intimately associated with the port and shipping, or located there for the convenience and economic advantage of receiving coal-fuel by waterways before the railway era. It is thought that 'a very large proportion of Londoners must, at some time during the year, have worked on the riverside' (Schwarz, 1992, p.9). The waterfront at Lambeth had been a site for potteries since the seventeenth century, and a photograph of the 1860s still shows the conspicuous kilns and tall chimneys of the Royal Doulton Pottery, close to this stretch of the river (see Figure 3.3).

During the nineteenth century, more than a dozen gas-works were established along the Thames, as at Beckton, where coal, landed directly from barges, was heated in large retorts to produce the gas. London was notably precocious in implementing the new gas lighting. By the 1810s the metropolis had forty-four kilometres of cast-iron gas mains for street lighting; in the 1820s the roads were dug up and narrow distribution-pipes to light individual houses were laid, as was witnessed by the immigrant Bavarian artist George Scharf (see Figure 3.4).

And after the coal-tar residues of gas-production were gradually discovered to be a valuable source of creosote, naphtha and aniline dyes, a large chemical-works to manufacture these materials was established on the Thames at Silvertown, on the eastern edge of the metropolis.

Figure 3.3 Industrial aspect of the Thames at Lambeth, c.1866 (photograph: V&A Picture Library)

Figure 3.4 Installing London's gas mains, as depicted in a painting by George Scharf, 1825 (© The British Museum)

But the most labour-intensive industry on London's waterfront was shipbuilding. This had been of national importance ever since the sixteenth century, when Henry VIII established great yards at Deptford to build large wooden ships for commerce and war. Still in the mid-nineteenth century, great ships were rising on the stocks around the Isle of Dogs at Millwall and Blackwall (see Figure 3.5 overleaf). London was at the forefront of the latest technology, soon participating in that nineteenth-century twin revolution in traditional shipbuilding: the substitution of iron for timber, of steam power for sail. The world's first iron paddle-steamer, the *Aaron Manby*, was assembled in 1822 at London's Rotherhithe shipyard from wrought-iron sections transported from an ironworks in Tipton in the Black Country (there was no such iron to be had in London). New designs of ship machinery came out of the engineering-works established on the river, near the yards, such as that of John Penn and Sons of Greenwich, whose marine water-tube boilers were photographed under construction around 1860 (see Figure 3.6). London's ropewalks and coopers' yards were the scenes of hectic craft activity associated with shipbuilding. And nearby were innovative food-preservation plants, such as Donkin and Hall's factory at Bermondsey where, in 1814, there was the first commercial production of tin cans for food – tinned soups and meat for the Royal Navy – and the later, large cannery at Houndsditch.

Figure 3.5   A London shipyard, 1865 (photograph: Hulton Getty)

Figure 3.6   Penn's engineering-works, Deptford, 1865 (photograph: Somervell, 1951)

It is estimated that as many as 27,000 were employed in shipbuilding and marine engineering in London in 1865. But after that, the industry rapidly collapsed, a victim of the capital's distance from iron ore, its high wages and strong unions (Pollard, 1950–51).

Economically, London was strongly polarized, not only between a rich West End and an impoverished East End. Land was much cheaper south of the river, where there were marshes and, until the 1820s, the population was much less dense. There was still only one bridge in the mid-eighteenth century, and only when another two were built across the wide river (Westminster 1750, Blackfriars 1760) did London's two riverbanks begin to be united. Historians are not yet sure whether these bridges were built to provide access to more land for housing or to improve communications for trade and industry. But it is clear that the attraction of cheaper land induced Henry Maudslay, the inventor of a screw-cutting lathe, to move his works across the river to Lambeth around 1820. There, his firm of Maudslay, Sons and Field expanded to employ 1,500 workers; it became a nursery for the development of British precision machine-tools. This was London's largest engineering enterprise and exceptional. The 1861 census may show that there were 13,000 machine- and toolmakers, and 35,000 metalworkers in London, but they were scattered over numerous small workshops north of the river – only twenty-six of the capital's engineering firms employed more than twenty people.

This form of production in workshops, rather than large factories, remained the most characteristic feature of Victorian London's industries. Lack of space and high land-costs on the crowded north bank were the reasons. The long-established watchmakers were concentrated in Clerkenwell; the multitude of parts of which a watch consists were manufactured in the congeries of workshops in this district, a type of assembly-line dispersed through its streets. The typically cramped London workshop of the later eighteenth century was depicted by Johann Zoffany in his portrait of an optician (see Figure 3.7 overleaf). There were no Manchester-style multi-storey textile factories; instead, a residual manufacture of silk was carried out in Bethnal Green's Victorian hovels – the bulk of the industry, planted by seventeenth-century Huguenot refugees settling in Spitalfields, had long since been transferred to Macclesfield (in Cheshire) where there was cheaper labour. And the use of sweated labour together with the introduction of the sewing-machine, after 1850, permitted the manufacture of ready-made clothing to flourish in the poky workshops and dwellings of the East End. Luxury clothing for the West End's bespoke tailoring and fashionable dresses were made in the workshops of Savile Row and Great Portland Street. But even small workshops could be uneconomical. Some Victorian printers' workshops were forced out of central London by the high cost of land. But, at the same time, Fleet Street's newspaper industry was flourishing as a result of new technology, notably at *The Times*. With rising literacy and the repeal, in the 1850s, of taxes on newspapers and advertisements, conditions favoured mass production of newspapers for a potentially vast market within the capital and distribution throughout the country by the railway network. So, in 1857, *The Times* invested in the latest American rotary-press technology and was printing at the rate of 20,000 impressions an hour, twenty times faster than the first steam-powered press of 1814, eighty times faster than the hand-press. *The Times* needed this advanced technology to compete with the circulation of the new *Daily Telegraph* (see Figure 3.8 overleaf).

It is a travesty, then, to portray London as idling during the Industrial Revolution. A travesty also to deny the presence of heavy engineering, though large industrial plants were few. London was a hive of small workshops and a fertile soil for technological innovation.

Figure 3.7 A London optician in his small workshop: *John Cuff and His Assistant*, painted by Johann Zoffany, 1772 (The Royal Collection copyright 1998 Her Majesty Queen Elizabeth II)

Figure 3.8 The Hoe ten-feeder, sheet-fed rotary press, which was capable of 20,000 impressions an hour. Hoe machines with four, six, eight or ten feeder stations were used by many newspapers during the 1860s and 1870s (courtesy of Science and Society Picture Library/Science Museum)

New technology at the docks

London's river, by the end of the eighteenth century, had become intolerably
congested. This was partly the result of expanding overseas trade, with
incoming ships queuing up to sixteen abreast. It was also the consequence of
the law: all imports and exports were subject to duty imposed by Customs and
Excise. To enforce the regulations, ships were restricted as to where they could
offload their cargoes; they could do so only along the 'legal quays' around the

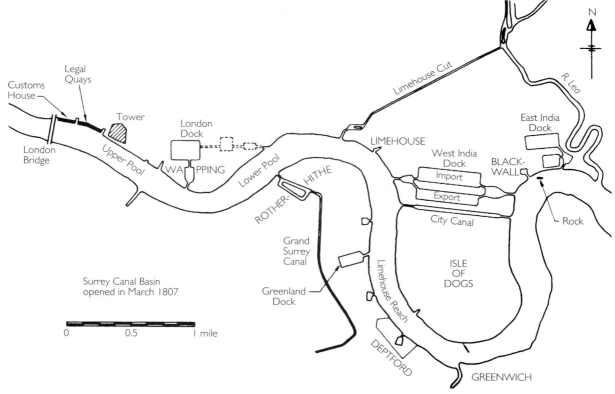

Figure 3.9 The port of London in 1806 (adapted from Skempton, 1978–9, p.89)

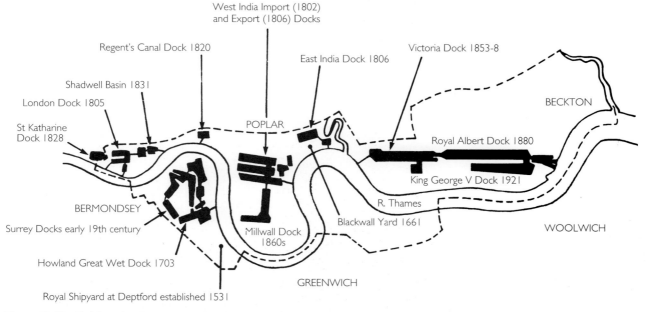

Figure 3.10 Further development of the port of London (adapted from Al Naib, 1998, with the permission of the author)

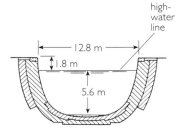

Figure 3.11 Greenland Dock, London, section of entrance lock, 1809 (adapted with permission from Skempton, 1981–2, p.75)

Customs House, which extended for a mere 464 metres, or, after additional payment, along the 'sufferance quays', a narrow zone on the opposite bank. These restrictions, the frustrating delays, the pilferage of valuable merchandise, brought petitions for change in the 1790s from merchants of the private trading companies. And recognition by Parliament of the merits of the merchants' case soon led to Acts relaxing the regulations and authorizing the construction of wet docks along the river. This was the origin of London's series of commercial docks (see Figures 3.9 and 3.10); in their construction, on an unprecedented scale by expert engineers, difficulties were overcome by technology both traditional and innovative. All the designs had to take account of the unusually large variation of the Thames tide; lock gates would be needed to maintain a constant water-level in any dock.

The import dock of the West India Company, the first to be built, was completed in 1802 on that great loop of the Thames at the Isle of Dogs. This was desolate marshland and the site had to be protected from thieves. The docks were therefore surrounded by high-security walls, and the adjacent warehouses had cast-iron window-frames with deterring spikes. The building materials (bricks, limestone, sand, timber) were brought in by water to jetties and then transported on wooden railways to the site, which was excavated by steam power. The dock walls were of brick, 1.8 metres thick, and set in hydraulic lime-mortar. The walls were given greater stability by their concave shape, a form the chief engineer, William Jessop, had earlier introduced at Dublin docks and which would become standard (see Figure 3.11). After nearly 200 years, Jessop's walls remain in good condition. Puddle clay at the dock bottom reduced seepage (Skempton, 1978–9, pp.90–94). Covering twenty-six hectares, this dock would long remain the world's largest.

The competing London Dock Company received parliamentary authorization to build at Wapping. Here the Scottish engineer John Rennie was in charge. For the most part he followed Jessop's designs. But a notable innovation was the first use of iron railways on a construction site. Moving along the rails, horse-drawn wagons carried away huge quantities of excavated earth (Skempton, 1978–9, pp.96–7). By 1806 London Dock, with its quay warehouses, was complete (see Figure 3.12).

Figure 3.12 London Dock, from an engraving by William Daniell, 1808 (reproduced from Broodbank, 1921)

Figure 3.13 Impenetrable customs wall, St Katharine Dock, 1852; now demolished
(photograph: Malcolm T. Tucker)

Massive dock warehouses became increasingly conspicuous along London's nineteenth-century waterfront. They had to be structurally strong and protected from the weather, fire and theft – hence their forbidding exterior (see Figure 3.13). The high compressive strength of cast iron made it the ideal material for columns supporting the heavily laden floors of warehouses. Iron columns were already in use in 1810 at London Dock. And soon after this, in a tobacco warehouse at the same dock, a forest of tree-like, branching cast-iron columns was used to support the roof-trusses with spans of sixteen and a half metres (see Figure 3.14 overleaf). There were no bolts or rivets, just joints that slotted together. The area of this single-storey warehouse – 19,580 square metres – led a contemporary to declare that it covered 'more ground than any public building except the pyramids of Egypt' (Fox, 1992, p.233). Still larger were the famous wine vaults at London Dock; the stone structure covered 8.9 hectares below quay level. Rennie designed a special shed in 1817 to store the precious mahogany imports of the West India Company. The innovation here was a crane travelling on rails above the roof; five-tonne mahogany logs were stacked by this means. The most spacious of all London's warehouses were those of St Katharine Dock, completed in 1828. With storage space of 120,000 square

Figure 3.14 New Tobacco Warehouse, 1813, London Dock (photograph: Malcolm T. Tucker)

metres, they were built right up to the water's edge, so permitting the direct hoisting by wall cranes of cargoes from ship to warehouse (see Figure 3.15). The floors were supported by cast-iron columns and some wrought-iron girders to provide strength in tension.

St Katharine Dock was a massive undertaking supervised by the famous civil engineer Thomas Telford. A total of 1,250 houses were demolished, displacing 11,000 people. The scale of the operation was captured in a contemporary engraving (see Figure 3.16). Horse-drawn wagons on three railways carried off the spoil. An engine drove piles, thirteen and a half metres long, three metres down into London's clay to form a coffer-dam from which water could be pumped out and stone wall-foundations laid. The entrance lock was built of hard sandstone shipped from Yorkshire; the masonry here

Figure 3.15 D Warehouse, St Katharine Dock, designed by Philip Hardwick and begun in 1827 (photograph: © Crown copyright.RCHME)

Figure 3.16 Building St Katharine Dock; *The works in their progress as they appeared in the month of January,* engraving by J. Phelps (after W. Ranwell) (courtesy of The Museum in Docklands/PLA Collection)

was superb, requiring no maintenance for over a century (see Figure 3.17). To maintain the level of the lock at low tide, two sixty-kilowatt steam-engines were used to pump water from the river; it was done in minutes.

The first ship entered St Katharine Dock in 1828: it was still a sailing-ship. The grave weakness of this and the other docks – by now they covered seventy hectares and had cost over seven million pounds – was the assumption that the size of ships would not change. Unfortunately, the new steamships required much more than the 13.75 metres width of St Katharine Dock's entrance, which soon made it obsolete. Other, wider docks were now essential. Victoria Dock, built in 1853–8, was designed for a new era: twenty-five metres wide to receive the largest steamships, it was equipped with hydraulic power to open lock gates in a minute and a half, hydraulic capstans to replace thirty men operating a windlass, hydraulic cranes on the quayside, and connections to the railway network (Greeves, 1982, pp.9–10).

The docks stimulated road-building to the City – Commercial Road, for example – and housing for the growing army of dockworkers. These effects are discernible at Tilbury, forty-two kilometres downriver on the deserted Essex marshes, where the West India Company, facing ruin, planned a new dock to attract shipping away from its competitors. Rapid transit of goods by rail was promised, and savings in pilot charges. The work required extensive drainage operations using sumps up to twenty-one and a half metres deep. A first-class hotel was built for ships' passengers, houses for staff and tenement blocks for the dockers. All was completed by 1886, but the huge expenditure and competitors' tactics soon forced the West India Company into liquidation. Subsequently amalgamations occurred, and eventually the formation in 1908 of the Port of London Authority (Greeves, 1982, pp.13–15). Just like many other stretches of London's river, this area of Essex had been changed beyond recognition by the demands of trade and the tools of the latest technology.

Figure 3.17 Detail of the entrance to St Katharine Dock (photograph: J. Flowerday; courtesy of The Museum in Docklands)

Change in the building industry

The river was important for the metropolis's expanding building industry – so important that Thomas Cubitt, the most enterprising and successful of London's builders, declared that without 'ample means of communication' between his extensive works and the Thames, 'I might as well abandon my property' (Hobhouse, 1971, p.291). Building materials are heavy, and difficult and expensive to transport. Before the railway age, waterways provided the best means, and London was particularly well served by navigable rivers and canals.

In the early nineteenth century, the capital's explosive population growth and public works generated a huge demand for building. In 1825, 232 million bricks were manufactured in London. Building held out the prospect of wealth, though it was speculative and risky. Cubitt, the son of a Norfolk carpenter, settled in London in 1809 after serving as a ship's carpenter on a voyage to India. At his death in 1855 he left over a million pounds. From master carpenter he had greatly expanded his activities to emerge as the first master builder. That represented a pioneering reorganization of the building industry: previously building was done by master craftsmen who only employed workers in their own particular trade; now Cubitt, from 1820, employed a whole body of workers skilled in all the building crafts. It was a change that has been interpreted by one historian as a response to London's demand and the need for firms to expand to meet risks; he denies that this restructuring of the building industry was in any way caused by changing technology (Cooney, 1955–6), though that is unsustainable. Cubitt's example was soon imitated: the 1851 census showed fifty-seven London building firms with at least fifty workers; nine had over 200.

Cubitt built much of Bloomsbury, Pimlico and Belgravia. His reputation brought royal commissions to extend Buckingham Palace, install a hot-water system at Balmoral, and build Osborne House for Queen Victoria on the Isle of Wight – the cement for this was delivered entirely by waterways from Cubitt's extensive workshops on the north bank of the Thames, bordering Chelsea.

There is no doubting Cubitt's interest in new technology or its prominent role, once he had established himself as a master builder. Cubitt was a member of London's Institution of Civil Engineers and communicated the results of experiments on the strength of iron girders and bricks. The shaft of the brick chimney of his works – thirty-two metres tall, and a landmark in Victorian London – was used to investigate the expansion of brickwork. His extensive riverside works (see Figure 3.18) were fully operational by 1842. Cubitt's biographer presents a vivid account of the multiple activities within:

> The size of the works made it possible to break down the traditional processes so that the fullest possible mechanisation could be employed. Thus there was a large carpenters' shop where the rougher structural timber could be prepared, carefully equipped with water tanks round the walls in case of fire. There were sawing and cutting-out shops serving both carpenters and joiners, where the cruder wood-working could be done rapidly by machinery. Much of this was

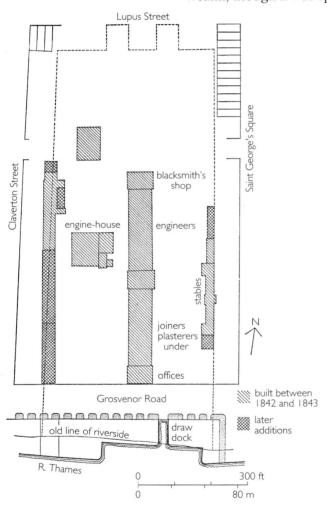

Figure 3.18 Cubitt's workshops at Thames Bank, 1842–56 (reproduced from Hobhouse, 1971)

made by Cubitt himself, and included vertical frame-saws for cutting out deals or floorboards which replaced the traditional manual saw-pits. He also made circular saws, rather more sophisticated in application, which were used to prepare timber for finishing by hand in the joinery shops. One saw-bench at least was equipped for cutting mouldings and architraves and skirtings. The joiners' shop was some two hundred feet [sixty metres] long, equipped with some thirteen to eighteen benches each worth £300, and so spacious and lofty that contemporaries were amazed. Cubitt was pioneering a system by which most of the woodwork was prepared at his works instead of on the premises, and this was so unusual that he had to explain it to the London Assurance Corporation, so that they would not demand the extra premium usual for houses 'in process of finishing': 'It should be observed that my houses are not subject to the risk of the usual mode of Building as the work is all prepared at Thames Bank in the Shops.' This eliminated the very serious fire hazard in an unfinished house created by loose shavings lying around where the men were working by candlelight, as well as providing better working conditions and opportunities for mechanisation.

Next to the joinery works was a four-storey block, about eighty feet [twenty-four metres] wide, where marble was worked. Marble slabs for staircases and sideboards, and marble fireplaces were made on a large scale … The marble was ordered direct from Carrara and Paris, and not only through London agents of the quarry-owners …

There was also a large and well-equipped engineering works, which contained a 'stocking machine' for cutting stocks for wheels, a machine for cutting timber-moulding, lathes and drilling-machines, and also machinery for working iron and steel, both for planing it and drilling it.

Some of the pig-iron seems to have been smelted on the premises, and Cubitt's experiments with iron girders were designed to help his ironfounders in improving their product … Cubitt's works produced a good deal of machinery, which would not today be regarded as part of the business of a builder. In 1844, he supplied crab engines and hoists for the building of the Embassy at Constantinople. He took a great interest in the development of machinery for brickmaking, and some of this was made or adapted in his own works …

Wrought and cast iron were used to a very large extent in contemporary buildings for fireplaces, stoves, balustrades for staircases and for external balconies, and for gas-fittings, as well as for stable mangers and other fittings, and in cast-iron gratings for drains and sewers.

In addition, he seems to have made his own baths, cisterns and other plumbing equipment … [and] water closets, which Cubitt claimed he had installed in all his buildings …

(Hobhouse, 1971, pp.286–8)

From the Thames, a pipe brought in water for the works' steam-engines. To the Thameside wharf, tonnes of hay and oats were brought by barge to feed horses kept in the works' stables. The horses were needed to pull barges, built by Cubitt and laden with building materials arriving from all over the country – stone paving from Yorkshire, Purbeck stone for staircases and sinks, roofing slates from North Wales quarries, and Medway cement by the tonne.

But brick was London's staple building material – not just because building regulations after the Great Fire had encouraged its use:

London is, of course, a brick city not only because of the absence of local stone, but because of the easily accessible beds of brick earth in most localities. The traditional London method of brickmaking, where hand-made bricks were burnt in clamps, required so little fixed capital equipment that brickfields were opened up throughout the metropolitan area, not only in suburban areas like … Islington and the notorious Notting Hill Potteries, but often on building sites themselves …

In the early nineteenth century … bricks were made in a number of different ways – by house builders on the site on which they intended to use the bricks, by large contractors for brickwork, who would open up a leased brickfield in the locality to provide bricks specifically for that contract, and by brickmakers, basically suppliers to other builders …

London brickmaking practice at this time differed from that of the [Staffordshire] Potteries and other brickmaking areas in that the bricks were clamp burnt and not baked in a kiln. There were various other differences between London traditions and those of the rest of England, partly due to its metropolitan position near a large centre of population and partly to the nature of the local brick earth.

There were three qualities of brick earth available in the London area: strong clay, producing very hard and sound bricks, but difficult to handle because bricks made from it were prone to shrink and crack in drying and warp in firing; loams, containing a higher proportion of sand, but so full of stones that they usually required washing; and malm, an almost perfect brick earth, containing a proportion of chalk, requiring no additions, from which the best quality of bricks could be made, but of which the supply was running short. Most London bricks were made of loams treated with natural or artificial malm, though some expensive bricks were made out of pure malm.

(Hobhouse, 1971, pp.302–3)

Figure 3.19 The main centres of stock brick-making along the River Thames in Kent and Essex during the later nineteenth and early twentieth centuries (Hobhouse and Saunders, 1989; with the permission of the London Topographical Society)

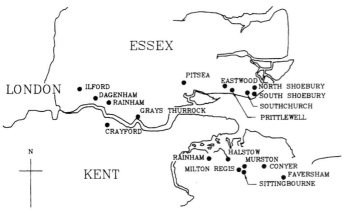

By the nineteenth century, metropolitan demand for bricks exceeded what city resources could supply, and there was increasing resort to surrounding counties, to brickworks dotted along the Essex and Kent banks of the Thames and its tributaries (see Figure 3.19). And in 1852 Cubitt purchased twenty-nine hectares of land containing brick earth at Burham (Kent), close to the Medway for water transport. When brick-manufacture began here, Cubitt adopted all the latest technology which in the past decade had been transforming long-established practice.

Making bricks was a slow, laborious, seasonal occupation. London clay was mixed with ground chalk and water, and left for months to weather. Then, after being worked in a pug-mill ('pugging' was the process of breaking down the clay into particles by mixing and kneading), the clay was taken to a moulding-stool (see Figure 3.20), where a gang, often a family, worked together to produce some 5,000 bricks in a day: a man delivered the clay, a woman 'clot-moulder' formed the clay into lumps for the male moulder to shape with a mould (by the 1850s iron had replaced the wooden box); a 'taking-off boy' removed the finished bricks on to a long barrow for men to wheel away to the 'drying hacks', where they were left in rows to harden. Finally, the bricks were stacked in long lines in a clamp, with a mixture of wood, coal and ashes placed in layers between each course. This mixture was ignited to bake the bricks.

Figure 3.20 Moulding-stool of 1850, where bricks were made by hand by a brick-making team

Key a: lump of brick earth; b, c: sand for preventing the brick sticking to the mould; d: bottom of the mould, called the 'stock-board'; e: water-tub; g: pile of pallets on which finished bricks are placed; h: finished bricks on pallet, ready for 'taking-off boy' to put on his cart. The clot-moulder stood at m, the moulder at k, and the 'taking-off boy' at n; o was the cuckhold, a concave shovel used to bring the clay from the pug-mill to the moulding-stool (reproduced from Hobhouse, 1971)

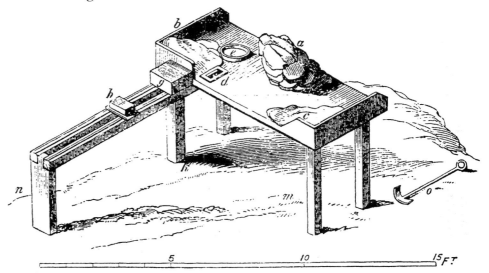

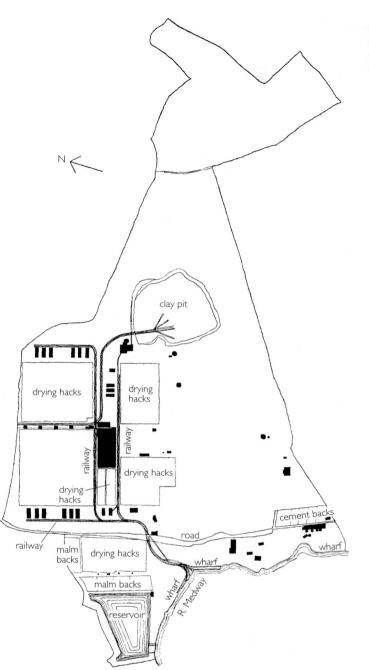

Figure 3.21 Plan of Burham Works, c.1856 (reproduced from Hobhouse, 1971)

At Cubitt's new Burham Works (see Figure 3.21) much of this manual labour was eliminated:

> Cubitt developed a large modern brickfield where clay pit, malm backs, machinery sheds, drying hacks and kilns were all connected by rails, cutting out much of the back-breaking labour of moving wet clay and clay goods. Another set of rails connected the kilns with the wharf, where Cubitt had built a massive river wall, from whence the bricks were loaded on to barges. A large reservoir supplied piped water to the works, and steam wash-mills and a portable steam-engine for pugging further reduced the manual labour of the works.
>
> Such mechanisation of the brick industry had been welcomed by the *Builder* as putting an end to the 'degrading labour to which the brickfield has subjected our species, and most revolting of all to see women put to the drudgery of horses and engines; little children too, who in a country like this should be at school, disguised past recognition in the mixed sweat and plasterings of clay and mud which encumbered their attenuated frames.'

> Mechanisation of the brickmaking process had begun with the various mills which prepared the raw materials, formerly mixed manually or operated by horse power … The new departure was the replacement of the traditional 'stool' system with its family structure of skilled brickmoulder, female 'clot-moulder' (who moulded the clay for the brick very roughly), and boys moving the bricks themselves … the innovation at Burham was the employment of unskilled labour to work the machinery …

The *Builder*, as usual well-supplied with information about Cubitt's activities, had already reported to its readers, in the course of a leading article on brick-making:

> Amongst those who are entering largely into the manufacture of bricks is Mr. Thos. Cubitt, who has opened extensive grounds on the Medway, set up steam engines with lofty furnace shafts, and is otherwise preparing for large operations in machine-made bricks. The arrangement he has in use at Thames-bank is that known as the Ainslie machine, with some improvements. One of these, attended by three boys, turns out 1,000 bricks an hour, a limit fixed, not by the machine, but by the ability of the attendants to remove those made. The clay passes through two rollers out of the pug-mill, by which means the air is driven out … Oil runs in behind the die, to facilitate the passage of the clay through it, and this assists in giving a smooth face and ends to the brick, while the wire which cuts each off leaves a rough top and bottom for the mortar.

(Hobhouse, 1971, pp.310–12)

The Ainslie machine mentioned in this report relates to John Ainslie, a Scottish farmer who had patented a series of inventions in the 1840s to mechanize production of bricks and tiles at his plant in Alperton, rural Middlesex.

From Cubitt's Medway site, a large Thames barge could carry up to 50,000 bricks. But it was the railways that opened a new phase in brick-manufacture and supply, facilitating the rapid, cheap transport of bricks made from new sources of clay: special red bricks from Fareham (Hampshire) were brought by rail to build the Albert Hall (1864–71) and South Kensington's Natural History Museum (1873–81); other red bricks, from Nottingham, were insisted upon by the architect of St Pancras Station. And at Arlesey in Bedfordshire, the new railway link to London induced Robert Beart to begin production for the London market. He harnessed steam power to force clay into brick-making machines, the emerging bricks cured within twenty-four hours in steam-drying stoves. His huge works had a siding off the main York–London line. And, indispensable for his business with the metropolis, he had an office and depot at King's Cross Station. By 1858 his annual production was eight million bricks; yellow or white, some of them attracted interest in fashionable South Kensington estates (Hobhouse and Saunders, 1989, pp.13–14).

London transport

The capital dominated the nation. That was ensured by its function as the seat of government, as the booming centre of overseas trade, and through the sheer magnitude of its population. And London's influence and control were further enhanced by its position at the hub of the national communications system. Transport technology strengthened the power of the capital.

For centuries London had dominated the overland carrying trade; its hauliers operated large wagons pulled by teams of horses over a road system largely created by the Romans. Two notable new trends are evident in the first phase of the Industrial Revolution. From the mid-eighteenth century, there was a conspicuous increase in passenger traffic, attributable to rising wages. Over the period 1750–1830, journey times were greatly reduced, especially after highways were greatly improved by John McAdam (1756–1836), who introduced layered roads of small interlocking stones to provide a road surface ('macadam') impervious to water, so eliminating former quagmires, and Thomas Telford (1757–1834), who concentrated on improved stone foundations. With proper drainage and reduced camber, fewer coaches overturned, and this greater safety permitted the use of lighter, and therefore speedier, coaches. Already in 1750, passengers travelling between the capital and the provinces had a choice of vehicle: a fast coach, a slow coach, and a cheaper but rougher ride on a goods wagon. Greater speeds had become possible not only as a result of better roads but also from the introduction of new types of horse: now bred for speed instead of strength, these horses were grouped in teams that were replaced by fresh animals at designated stops on the journey. The fastest coach to Edinburgh in 1750 took ten days; this was cut, in the 1760s, to four days by 'flying coaches', and to only two by 1830. Similarly, Manchester, a three-days' coach journey from London in 1750, could be reached in thirty hours in 1820. And journeys became more comfortable after Obadiah Elliot's invention in 1805 of an improved spring suspension.

The long coach journeys were performed in stages, the halts at coaching inns providing rest and refreshment for horses and travellers. Inns became a familiar part of the London scene during the coaching era. But there was nothing new about the building form – a medieval type with courtyards, stables and galleries leading to guests' rooms (see Figure 3.22).

By the 1820s, London had an established short-stage coach service from the City and West End to surrounding villages such as Paddington and, after three new bridges had been built across the Thames in 1816–19, to the rapidly growing Clapham and Camberwell (see Figure 3.23). Fares were high, journeys slowed by frequent stops at picking-up points (public houses), and capacity was limited to a maximum of twelve

Figure 3.22 Courtyard of a coaching inn, Bishopsgate, 1865 (photograph: V&A Picture Library)

Figure 3.23 Short-stage coach traffic in 1825 (reproduced from Barker and Robbins, 1975, p.15)

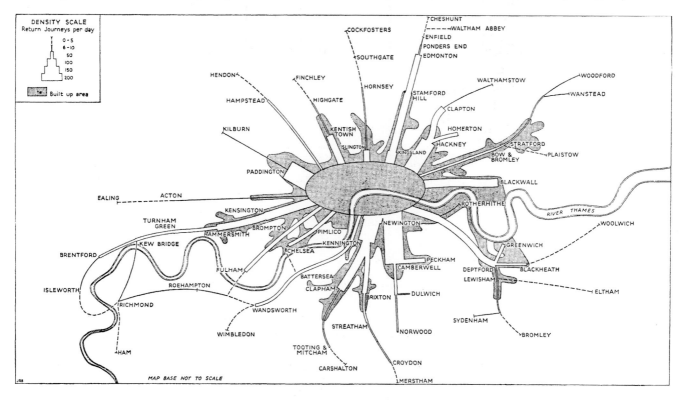

Figure 3.24 Shillibeer's omnibus, 1829 (London Transport Museum; copyright London Transport)

people on the larger four-horse coaches. Competition between private companies is a dominant theme in the history of British transport, and London's short-stage coach operators soon felt the pinch. It came first from the new Thames steamships, which from 1815 took increasing passenger traffic to Twickenham, Richmond, Greenwich and Margate. Their advantage over coaches was their capacity to carry hundreds of passengers at a time; but bad weather could make the river journey a trial for civil servants commuting from Gravesend or Greenwich by season ticket (six guineas a year from Gravesend in 1830). By 1837 a stagecoach operator in Greenwich was complaining that his business had been seriously reduced by the competition of steamship companies (Aldcroft and Freeman, 1983, p.172). An inquiry of 1854 found that 15,000 travelled to the City daily by steamer.

Figure 3.25 Saloon bus, c.1858 (photograph: Hulton Getty)

The short-stage coach gave way to London's horse-drawn omnibus, more the result of an enlarged seating capacity (interior, longitudinal benches) than any important innovation in coach design. It came from Paris, where the concentration of population in multi-storey buildings favoured a short-distance urban service, which was inaugurated in 1828. Its imitation in London the following year was explicit in the first omnibus service introduced by George Shillibeer, a coach-builder. He announced transport 'in the Parisian mode' from Paddington to the Bank for a one shilling fare. His three-horse coaches, seating twenty, all within, ran five times a day; the eight-kilometre journey could be done in forty minutes (see Figure 3.24). Speed and punctuality were his strategy for success: 'Other coaches hang about at

Figure 3.26 A London horse-bus stable, c.1885 (photograph: London Transport Museum; copyright London Transport)

the public houses … but I go right away whether I have passengers or not.' *The Times* reported its initial flourishing: '15 ladies and gentlemen were frequently to be seen running after it when it was completely full' (Barker and Robbins, 1975, pp.17–21). Shillibeer's success was short-lived; his omnibuses proved no more economical than the short-stages, and he went bankrupt. Other operators introduced two-horse versions of omnibuses with outside seats on the roof, a total passenger capacity of twenty-four. The London omnibus became established as the main form of public transport in the centre (see Figure 3.25). In 1846 reduced fares of one penny for the shortest distance made it widely affordable, though not for the working classes. By the 1860s the London General Omnibus Company, with a stock of 500 buses and 6,000 horses, was providing a service for 40 million passenger-journeys a year. By then, bus roofs had been raised to permit standing-room, ventilation had been increased, and handrails were provided to help the climb to the roof seats. And London had to accommodate large horse-bus stables (see Figure 3.26).

Figure 3.27 Paving Oxford Street with wooden blocks, a scene captured by George Scharf's painting of 1838. The perceived success of this experimental road-surface soon led to the use of wood paving all over London (© The British Museum)

Figure 3.28 St Pancras Station from the air. The Euston Road runs from left to right across the picture, about two-thirds of the way down. Somers Town goods station is on the left, separated from the passenger station by Midland Road. To the right of the station is Pancras Road, with the curved Great Northern Hotel backing on to it; and, further still to the right, King's Cross Station. The narrow waist of the railway entrance into St Pancras is clear at the top of the picture, hemmed in by the gas-works on the right of the line and old St Pancras churchyard, with its trees, opposite. The Regent's Canal curves round at the top right-hand corner of the picture (photograph: Aerofilms)

London's increasing volume of horse-drawn traffic stimulated experiments with various materials for the street surface. The deadening of noise was a prime consideration, reducing the wear and tear on streets, horses and vehicles another. Long before asphalt or tarmac, Purbeck stone and Aberdeen granite were tried. And in the 1830s and 1840s the widespread use of wood for paving Oxford Street and other central streets was recorded in the paintings of George Scharf (see Figure 3.27). The *Gentleman's Magazine* in February 1839 reported that vehicles passed over the wooden surface as quietly as if rolling over a carpet.

While the omnibus prospered in the centre, the short-stage coach services to London's suburbs collapsed in the wake of the railways. The horse-drawn coach could not match the capacity or speed of the steam train. London's first railway, from Deptford to Bermondsey, began operating in 1836. Many other lines soon followed. From all directions the railway companies converged on London. Their approach routes were largely determined by the compliance of large landowners. The Duke of Bedford resisted any railway invasion of his fashionable Bloomsbury estates; he even erected gates to keep out the heavy road traffic bound for London's railway termini. By contrast, the estates of Earl Camden and Baron Somers, the Brewers' Company property around St Pancras, and bishops' land south of the river were already dilapidated areas, and the owners readily sold out to the railways (Kellett, 1969, pp.244–62). Camden Town consequently became a zone of railway lines, and St Pancras was similarly consumed (see Figure 3.28). When the railway promoters projected a spate of schemes which seemed to threaten the entire fabric of London's historic centre, Parliament stepped in to protect the capital and seat of government. In 1846 Parliament asked the queen to appoint a Royal Commission on metropolitan railway termini to consider whether it would be in the public interest to permit railways to enter the centre of London. Witnesses were soon giving evidence. Some spoke of the benefits of letting in light and air to the city by demolishing slums for railway building, others warned of greater traffic congestion, an increase in smoke and noise, and the damaging interference of railway vibrations in the precision work of Clerkenwell's watchmakers. Within weeks the Commission presented its recommendations: no railway station to be built within a specified central zone, except for Waterloo on the south bank. The designated exclusion zone would subsequently be breached at Charing Cross and Liverpool Street; but otherwise it held sufficiently for the sociologist of London, Charles Booth, to declare in 1890 that central London was a 'charmed circle', impervious to railways (Kellett, 1969, p.282).

London would therefore have no grand, central station typical of the largest of American and European cities. Instead, by the end of the century, London was surrounded by a ring of fourteen termini, and journeys had to be broken to reach the centre or transfer to another railway line. But London was the hub of the national rail network (see Figure 3.29). Inter-station traffic and the movement of carts from railway yards to central markets and warehouses generated huge traffic jams. Around 1850, a commuter from Brighton complained that it was quicker to walk from Victoria, his arrival point, to his office in Trafalgar Square than to take an omnibus or cab. At the same time, a shopkeeper on Ludgate Hill, fed up with the traffic and parking problems that were taking customers away, moved elsewhere (Barker and Robbins, 1975, p.65). And inter-station traffic so congested London Bridge in 1860 that at peak

Figure 3.29 Britain's railway network, 1872 (reproduced from Langton and Morris, 1986)

Figure 3.30 Heavy traffic on London Bridge, 1892. The great variety of vehicles includes carts, wagons, hansom cabs and horse-buses. Congestion was attributed to the railway station nearby (photograph: Hulton Getty)

times 1,700 horse-drawn vehicles were crossing it hourly, a ten-fold increase on 1850; this would remain a traffic blackspot in 1900 (see Figure 3.30).

The capital's worsening traffic problems led, around 1850, to the then startling proposal of an underground railway. One of the principal promoters of the scheme was Charles Pearson, City solicitor and, in 1847–50, MP for Lambeth. Another was William Malins, a shadowy figure – still, very little is known about him – who temporarily became chairman of a company, soon to be called 'The Metropolitan', created to secure approval for the project and to operate the line. Malins spoke forcibly on the subject:

> It must be obvious that the constantly accumulating number of omnibuses, wagons and conveyances of all sorts would, if it continued two or three years longer, render London almost insupportable for purposes of business, recreation and all ordinary transit from place to place … doubling the thoroughfares by means of *sub via* [subway] railways [is] the only mode of accommodating the increased traffic of London.
>
> (Barker and Robbins, 1975, p.106)

But support for the underground line was also motivated by interests other than the relief of London's traffic congestion. The large railway companies saw the scheme as a means of connecting their lines to the centre of the city, the cherished goal they had long been denied. So, in 1853, the Great Western Company committed £185,000 for shares in the underground company. In return, the project was modified so as to extend the proposed line to a station directly opposite the Great Western's main-line terminus; another branch-line connected to the Great Western's overground line.

Legal constraints influenced the detailed planning. The law required anyone digging beneath a building to purchase it, even if only part of the foundations was undermined. An underground railway in London would therefore be economically feasible only if it passed beneath open spaces and roads. These considerations favoured a roughly west–east route from Paddington towards the City: a large thoroughfare, the New Road, had recently been built in this area, and tunnelling beneath it would not entail damage to houses (Barker and Robbins, 1975, p.104).

Parliamentary approval was eventually secured, with the help of Malin's rhetoric – he spoke of the benefits to humanity and to the Post Office (proposing a link to the central sorting office to speed delivery of the capital's rapidly growing volume of letters). But formidable difficulties remained. How could the company raise the necessary one million pounds to fund the vast engineering project? The 1850s were a time of economic difficulty, and there was a general shortage of capital for investment. Lack of progress forced Malin's resignation as company chairman. Only when the bank rate fell in 1859 did it become easier to borrow money and to find sufficient purchasers of shares. But investment and general public support were also hindered by psychological barriers, as *The Times* explained in November 1861, when it considered the prospect of Londoners venturing into 'dark, noisome tunnels, buried many fathoms deep, beyond the reach of light or life; passages inhabited by rats, soaked with sewer drippings, and poisoned by the escape of gas mains'. It all seemed 'an insult to common sense to suppose that people who could travel as cheaply to the city on the outside of a Paddington bus would ever prefer, as a merely quicker medium, to be driven amid palpable darkness through the foul subsoil of London' (Barker and Robbins, 1975, p.118).

According to the standard history of London transport (Barker and Robbins, 1975, p.118), all fear of underground transport would be dispelled by the proven safety of the operating railway, the result of the introduction of a new method of visual signalling (the 'block system'): signalmen stationed in boxes used a disc instrument to indicate that the line was clear.

The building of the tunnels was made easier by the soft clay beneath London. John Fowler, a railway engineer, supervised construction. Avoiding sewers was a source of concern, and when the Fleet sewer burst into the excavations at King's Cross in June 1862, work had to be suspended for months. Meanwhile various designs of steam locomotive had been tested. Special features were needed – above all, high acceleration and the removal of smoke. Eventually Daniel Gooch's coke-burning locomotive was selected; the exhaust steam was passed into tanks beneath the boiler, a supposed solution for the removal of fumes in the tunnels. In January 1863 everything was ready for the opening of the Metropolitan, the world's first underground railway, a six-kilometre stretch from Paddington to Farringdon Street (see Figure 3.31 overleaf). Passengers waited on platforms lit by gas lamps; some stations had roofs of wrought iron and glass. They entered the ten-seater carriages which, at first, were open. By 1865 the line had been extended to its terminus at Moorgate, within walking distance of the Bank of England and the City's bustling financial centre. By the 1870s the Metropolitan was flourishing, with 40 million passenger-journeys a year.

Subsequent developments would, by 1884, make the line part of a circle linking London's railway termini. But the engineers had been optimistic about improving the condition of the air in the tunnels. Ventilator shafts were tried, a series of 'blow-holes' cut from the crown of the tunnel to the middle of the road above, between Edgware Road and King's Cross – iron gratings in the road are still visible today (Barker and Robbins, 1975, p.125). Passengers' complaints persisted. The fume nuisance would only be solved when steam locomotives were eliminated as a result of electrification (for this, see Chapter 4).

Figure 3.31 The world's first underground railway: Baker Street Station, Metropolitan Railway, 1863 (London Transport Museum; copyright London Transport)

Finally, what were the consequences of the overground railways for London? An estimated 75,000 lost their homes through railway building in 1850–1900; the displaced population worsened existing slums. There was some environmental damage, such as the railway bridge over Ludgate Hill (see p.29). Social zoning resulted from cheap fares on the Great Eastern Company railway; from the 1860s that led to a marked increase in the working-class populations of Walthamstow and Tottenham. Croydon and Bromley also grew in the 1880s; only the speed of the railways could have made commuting possible at those distances. By 1851 some could live by the seaside at Brighton and take the seventy-five-minute express to and from their work in London. The capital's fuel supplies, long brought by sea from the north-east, began to come by rail – by 1867 more than by sea. The advantage here was that supplies were less subject to stoppages through stormy sea-crossings. Londoners now had a wider choice of better coal: Swansea anthracite and Derby coal. And huge quantities of perishable foodstuffs were hurtled by train from the countryside to boost Londoners' diet. By 1880, 93 million litres of country milk a year arrived in the capital from dairy farmers as far away as Bridgwater in Somerset, 240 kilometres along the Great Western Railway from Paddington.[2] In these and many other ways, the construction of the railway network and its unprecedented heights of travel speed permanently changed the fabric of London and the lives of its citizens.

[2]   For the controversy over the importance of the railways for London's milk supplies, see the extracts from E.H. Whetham, David Taylor and P.J. Atkins in the Reader associated with this volume (Goodman, 1999).

Engineering for metropolitan health

In 1847 William Farr, statistician, announced his gloomy findings on London's mortality rate: in the metropolis thirty-eight more people died every day compared with the proportion in the immediate surroundings. Farr was employed in the office of the registrar-general to compile figures and annual reports on the causes of death in England. His explanation for London's high mortality was the poisonous vapour, a 'disease-mist', hovering over the capital 'like an angel of death'. Generated by open sewers and cesspools, graves and slaughterhouses, this deadly cloud was responsible for outbreaks of cholera, smallpox, measles, influenza and a variety of fevers (Flinn, 1965, pp.28–9).

This general and imprecise diagnosis of epidemic disease was a commonly held opinion in early Victorian medicine – there was no bacteriology until the last decades of the nineteenth century. Physicians referred to a poison in the air, 'malaria' (from an Italian combination of two words, *mal*, meaning 'bad', and *aria*, 'air'). They saw no real difference between the malaria of the Indian jungle and the malaria of crowded cities, produced from their pestilential smells of accumulated excrement and gas-works. Was there not compelling evidence from the case of the seamen struck down with fatal cholera when their ship was eight kilometres from shore, the very point where the land-smell first became perceptible (Smith, 1838)? These beliefs led appointed physicians to recommend the following measures for removing the danger to life in crowded cities – above all, in London: an efficient system of sewers to remove refuse; separation from densely populated areas of malaria-producing slaughterhouses, gas-works, cow-sheds and cemeteries; ventilation, by wider streets and better-built houses, to dilute and carry away the aerial poison; and a plentiful water supply to wash the poison from streets, houses, clothing and persons (Arnot and Kay, 1838). Such ideas, by the 1850s, provided the momentum for a gigantic engineering project, the drainage of the metropolis, which has ever since protected the sanitary environment of Londoners.

By mid-century, attention had become increasingly focused on the river. As early as 1825 there were clear signs that the river was unhealthy: fish, once plentiful enough for abundant catches between Putney and Woolwich, had moved downstream (Luckin, 1986, p.12). Industrial pollution was to blame, and population growth – the increasing adoption of the water-closet sent large quantities of excrement into the Thames instead of to the cesspools, which were now closed. And water from the Thames, supplied by private companies, was London's principal source for washing, cooking and drinking. A terrible cholera epidemic struck London in 1849, killing 18,000. The General Board of Health was impelled to issue a report calling for a gigantic scheme of public works to provide London with pure water and sewers, both to be under the management of the state, just as ancient Rome had given extraordinary powers to its aediles, state officials who strictly supervised the water supply (General Board of Health, 1850, pp.286–8). But such centralization was against vested interests and the *laissez-faire* political tradition. Who else could intervene? London had no municipal authority; in March 1855 *The Times* declared:

> We may really say that there is no such place as London at all, the huge city passing under this title being rent into an infinity of divisions, districts, and areas ... Within the metropolitan limits the local administration is carried on by no fewer than 300 different bodies, deriving powers from about 250 different local Acts.

(quoted in Young and Handcock, 1956, p.667)

One reason for the 1854 Royal Commission's opposition to a single municipal authority for London was its 'bisection by the Thames', a 'natural circumstance' that prevented roads, sewers, gas- and water-pipes running continuously from north to south. The normal municipal control of these utilities was impossible in London; a metropolitan municipality was therefore impracticable (Young and Handcock, 1956, p.659). But something had to be done. The smell of the river

Figure 3.32 Faraday gives his card to Father Thames (*Punch*, 21 July 1855)

gave rise to a cartoon in the satirical magazine *Punch*, which showed a famous scientist, Michael Faraday, holding his nose and handing his card to a filthy personification of the Thames; the caption reads: 'And we hope the Dirty Fellow will consult the learned Professor' (see Figure 3.32). It took another cholera outbreak in 1854 to stir the government into action. An Act of Parliament of 1855 created the Metropolitan Board of Works specifically to provide London with an efficient sewage system. Consisting of a chairman and delegates from London's numerous parishes and vestries, it was 'a semi-representative body' controlled by the government (Owen, 1982, p.47). Its jurisdiction 'was determined neither by the facts of civic history nor by human geography but by the network of drains and sewers' (Briggs, 1968, p.322) – in other words, it was a metropolitan authority defined by technology. The crucial appointment to the board was the chief engineer, Joseph Bazalgette. He saw his task in terms of preventing 'the fearful destruction of life from malaria', a scourge that threatened wherever 'large numbers of the human species congregate upon a limited space, without provision being made for the rapid removal of the refuse thereby produced' (Bazalgette, 1864–5, p.280). In London the sewers were of defective design, even open and offensive. They ran at right angles to the Thames. They would have to be rebuilt and their contents collected by large, intercepting pipes, placed below them and running parallel to the river, discharging the foul contents at a safe distance from the capital. He ruled out separate drainage systems for rainwater and sewage, a doubling of drains that would be prohibitively expensive.

For too many months, Bazalgette's plans were discussed. The inaction was abruptly ended by 'the great stink' of the hot summer of 1858, when the unbearable smell of the Thames forced the House of Commons to adjourn. Parliament increased the board's powers, including authorization to borrow the necessary funds – up to three million pounds, to be repaid by a sewer rate. Work began in January 1859 on Hackney Common; it would not be completed until 1874 because of unforeseen obstacles, as well as the sheer magnitude of the project (see Figure 3.33), evident in Bazalgette's later summary:

> A primary object sought to be attained in this scheme was the removal of as much of the sewage as practicable by gravitation, so as to reduce the amount of pumping to a minimum. To effect this, three lines of sewers have been constructed on each side of the river, termed respectively the High Level, the Middle Level, and the Low Level. The High and the Middle Level Sewers discharge by gravitation, and the Low Level Sewers discharge only by the aid of pumping. The three lines of sewers north of the Thames converge and unite at Abbey Mills, east of London, where the contents of the Low Level will be pumped into the Upper Level Sewer, and their aggregate stream will flow through the Northern Outfall Sewer, which is carried on a concrete embankment across the marshes to Barking Creek, and there discharges into the river by gravitation.
>
> On the South side, the three intercepting lines unite at Deptford Creek, and the contents of the Low Level Sewer are there pumped to the Upper Level, and the united streams of all three flow in one channel through Woolwich to Crossness Point in Erith Marshes. Here the full volume of sewage can flow into the Thames at low water, but will ordinarily be raised by pumping into the reservoir.

(Bazalgette, 1864–5, p.294)

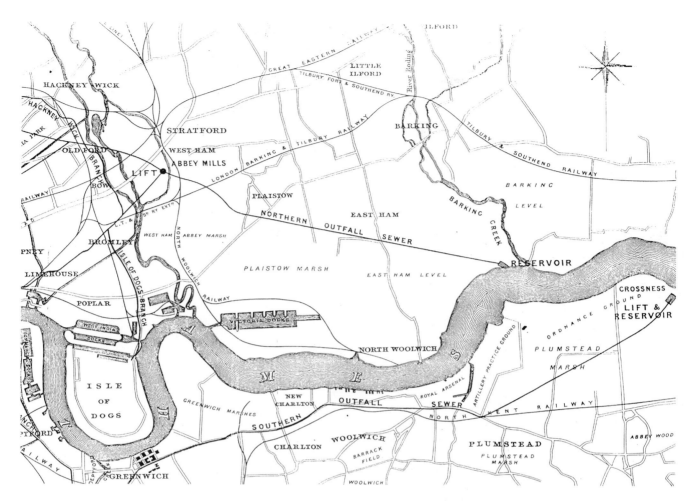

The main sewers, up to 3.7 metres in diameter, were made of brick and given an oval shape for greater strength and to maximize the flow while minimizing the brickwork and cost. But laying the sewers was fraught with difficulties arising from the nature of London's subsoil and the existence of concentrations of other public works. Water flooded in when they tunnelled under the Regent's Canal. The Middle Level sewer had to be carried over the Metropolitan Railway by a wrought-iron aqueduct of forty-six metres' span; but the railway traffic could not be stopped during construction, so the aqueduct was built on a stage one and a half metres above the intended level, and then lowered by hydraulic rams to a few centimetres above the locomotives' chimneys. Elsewhere, especially on the marshy south bank, tunnellers encountered quicksand. A special technique was developed to pump the water out, using large, sunken, iron cylinders with attached pipes, until the subsoil was rendered firm enough to build sewers on. At Barking, the ground would not sustain a reservoir, so concrete foundations were laid over six metres deep. Where sewers were forced overground by the obstacles of rivers and railways, strange shapes, resembling railway embankments, appeared on London's outskirts (see Figure 3.34 overleaf). And to avoid disrupting Fleet Street and the Strand, an embankment was projected which would also reclaim fifteen hectares of smelly mudbanks. Here there was long delay owing to opposition from wharf owners and other landowners. Completed in 1874, the embankment housed a sewer, and a subway for gas-pipes, water-pipes and telegraph-wires (see Figure 3.35). At Abbey Mills the world's largest pumping station was erected, with powerful steam-engines capable of lifting 430 cubic metres of sewage per minute to the required height of eleven metres.

Figure 3.33 Section of Bazalgette's great drainage project for the metropolis, showing the transfer of London's sewage to two outfalls to the east of the capital, at Barking and Crossness. The plan also helps to make clear how frequent obstacles to this engineering work arose from the presence of marshy land and functioning railways (Institution of Civil Engineers, 1864–5, plate 14)

Figure 3.34 Northern outfall sewer: section of embankment, culverts and substructure (Institution of Civil Engineers, 1864–5, plate 18)

Figure 3.35 Section of the Thames embankment, showing (1) the subway, (2) the low-level sewer, (3) the Metropolitan railway and (4) the pneumatic railway (*Illustrated London News*, 22 June 1867, p.632)

The scale of the project can be grasped from some figures: 130 kilometres of main sewers; 318 million bricks; 2.7 million cubic metres of excavated earth; 670,000 cubic metres of concrete. The final cost, £4.6 million, was swelled by increasing labour costs and a 20 per cent rise in the price of bricks, the result of soaring demand by railway companies and other public works. Bazalgette received a knighthood. But further expenditure was soon necessary to solve a problem he had treated optimistically. His observations on the tide at the Barking outfall – he watched the movement of floated objects – had persuaded him that the discharged sewage was too far away (twenty-two kilometres) from central London to be brought back by the tide. But by the 1880s there was again growing concern on the state of the river; raw sewage was detected at Greenwich. Not until 1885 did the Metropolitan Board of Works use chemicals at the outfall to precipitate the sewage, and invest in building special ships to carry the resultant sludge far out to sea.

By this time the board had acquired a wide range of responsibilities beyond sewer-building: the enforcement of building regulations; demolition of condemned houses; maintenance of Thames bridges and tunnels; regulation of slaughterhouses; inspection of cow-sheds; supervision of storage of petroleum and explosives; administration of the fire brigade; prevention of Thames flooding; building of new thoroughfares; and formation of parks. All of this reflected the growing complexities of urban life. The board had all the responsibilities, but without the authority, of a municipal government – its powers had been sanctioned by Parliament, not by local democracy. The board was dissolved in 1889, and its powers were transferred to the real authority of the London County Council.

The board may have spent millions on London's sewers; it was a trifle compared with what was being spent at the same time to transform the capital on the other side of the Channel. But then Paris had a very different form of government.

3.2 Paris: the planned capital

Industrial Paris?

France's Industrial Revolution was a much more gradual and sporadic process than Britain's. The first signs are discernible in the 1780s at Le Creusot, a site little more than a village, situated in a district rich in coal and iron some eighty kilometres south-west of Dijon. Here the new English technique of coke-smelting of iron was introduced by a migrant English entrepreneur, William Wilkinson, who directed operations for five years. Later, in the 1820s, two English entrepreneurs, Aaron Manby and Daniel Wilson, brought the puddling process to Le Creusot, erecting an English-style ironworks. But still in 1850, Le Creusot's population was no bigger than 8,000. By that time two other centres of heavy industry had developed around coalfields. In east-central France, Saint Etienne grew into a modern industrial town with factories, producing firearms and ribbons; its population in 1851 was 56,000. And in the north, close to the Belgian border, and in the extension of the great Belgian coalfields, France's largest industrial zone grew up around Lille and the adjacent towns of Roubaix and Tourcoing, all a few kilometres apart. This was, above all, an agglomeration of steam-powered textile factories, spinning wool, cotton and flax, using the machines the English had invented in past decades. If there was a French city resembling Manchester, then it was Lille, its air so polluted by smoking factory chimneys that there was regular smog. Yet neither the population of Lille (71,000) nor that of Roubaix (46,000) reached 100,000 in the census of 1841;

only three French cities then exceeded that figure: Paris (935,000), Lyon (190,000) and Marseille (147,000). Lyon was the centre of a luxury silk industry, whose growth had been stimulated by one of its citizens, Joseph Jacquard, after his invention in 1801 of a mechanized loom.

France, throughout the nineteenth century, remained predominantly rural and its urban population lived mostly in small towns. But Paris stood out. Its population and growth were unmatched in France, though it was dwarfed by London. From a population of 547,800 in 1801, later censuses recorded 714,000 in 1817; 785,900 in 1831; 935,300 in 1841; 1,053,300 in 1851; 1,174,000 in 1856; 1,696,100 in 1861 (this reflected the extension of the city limits by annexation of surrounding districts); 1,851,800 in 1872; and 2,269,000 in 1881. Immigration from rural provinces explains much of Paris's growth – the attraction of the big city with its higher wages. Paris had nothing of London's status as a great port – it was much too far inland for that (well over 150 kilometres from the Seine estuary) – nor docks to attract masses of casual labour. But the bursts of building in the French capital, which would reach a soaring peak in the 1850s, did act as a magnet – to such a degree that state officials in rural *départements* (administrative divisions of France) sent urgent communiqués reporting depopulation of the countryside and serious shortages of agricultural labourers, and an inability to compete with Paris wages. From the impoverished rural *département* of Creuse, in central France, there was a traditional seasonal exodus of labourers, travelling 320 kilometres to work for six months as stonemasons in Paris, where they could earn up to eight times their wage at home. Their journey took five days until the railways reduced it to twelve hours. They now came in their thousands and with their families, and the trend was for them to settle permanently in Paris. Drastic remedies were contemplated to check the drain on the countryside: restrictions on freedom of movement and even the destruction of railways (Pinkney, 1958, pp.152–63).

When we think of nineteenth-century Paris, we tend to conjure up images of a beautiful city or a city of political revolution, but not of an industrial city. That, according to some recent research, is attributable to the failure of historians to portray the city in its true colours: a manufacturing centre of the first importance (Ratcliffe, 1994).[3] Just as with London, revisionists are beginning to alter the conventional image of the French capital.

Paris in 1801 had a large cotton-spinning factory: Richard-Lenoir's works employed nearly 8,000 and housed 20,000 spindles (Coleman, 1982, p.55). Still in 1821, there were sixty-seven cotton-spinning mills in Paris, but these rapidly declined because of the availability of cheaper labour in the provinces. As in London, the small-scale manufacturing unit – the workshop – predominated in Paris, producing furniture, jewellery and scientific instruments for the city's eminent astronomers and surgeons. Conditions had been particularly favourable for the establishment of small workshops in Paris where, until the seventeenth century, the population, confined by medieval walls, was accommodated in tenements; Paris grew vertically, unlike London's horizontal spread. A single house might contain several floors of accommodation for artisans, a landlord on the first floor, a ground-floor shop, and workshops in the courtyard. Something of this still existed in nineteenth-century Paris. The medieval walls had long since gone and an extensive ring of fortifications (thirty-five kilometres in circumference) had been built in the early 1840s five kilometres from the city centre, but the city's soaring population continued to concentrate in the medieval core, along the Seine (Coleman, 1982, p.42; Sutcliffe, 1970, pp.323–4). The Left Bank, long the area of abbeys and the

---

[3]   Extracts from Barrie Ratcliffe's paper are reproduced in the Reader associated with this volume (Goodman, 1999).

university, had little industry, though the university stimulated printing in the area. The Right Bank, the commercial zone, had metallurgical workshops and clockmaking. But the emphasis was always on the finishing process: Parisians assembled clock components manufactured elsewhere, in Montbéliard (near the Swiss border) or even Switzerland itself, and housed them in elaborate bronze cases (Ratcliffe, 1994, p.268). Restriction of the capital's industry to the small-scale was reinforced by government policy. In 1848 the mayor of Paris, Armand Marrast, declared opposition to any heavy industry within the city; and in 1866 similar sentiments resulted in the imposition of duties on coal entering the city for industrial use. A customs barrier, a wall of about three kilometres radius around the city centre, had been erected in 1784 to collect revenue from duty on goods entering the city. Now it was used to deter heavy industry. To escape taxes on raw materials, and to acquire cheaper land and more space, there had been a movement of industrial entrepreneurs out of the city centre and beyond the customs wall. But when, in 1860, these outlying districts – about five kilometres from the centre and inside the new fortifications – were annexed to the city, taxation was imposed.

For heavy industry in the Paris area in the nineteenth century, one must look further still from the city limits, above all to Saint-Denis, a town situated on the Seine, eight kilometres to the north of the capital and at an important railway junction. By the 1860s, Saint-Denis was entering a period of rapid industrial growth, including three foundries, each with over 500 workers. The greater industrial development on the northern side of Paris has been attributed to communication with the industrialized *départements* of northern France (Bastié, 1964, p.137).

Piecemeal improvement

Political revolution and urban improvement are inseparable in the development of nineteenth-century Paris. It was in this period that the city acquired its present-day aspect, the result of ambitious civil engineering projects executed by regimes of very different political complexion, but having the common aim of embellishing the capital and enhancing its prestige. Amidst the turbulence of drastic political change, some recurring features are discernible: the strategic importance of Paris and its repeated role as the initiator of national insurrection; the rise of dictators with aspirations to assume the status and acquire the power of Roman emperors; the flight to Britain of overthrown heads of state and political refugees; the state training of technical experts for employment in the capital and nation. All of these would change the fabric of Paris.

The French Revolution (conventionally dated 1789–99) overthrew the monarchy, replacing it with phases of varying republicanism and oligarchy. This was not merely an internal upheaval: almost all of Europe would be drawn in. Faced with war on all frontiers, the revolutionary authorities created a national institute to meet the urgent need for engineers to build defences, bridges, roads. The Ecole Centrale des Travaux Publics (Central School of Public Works), later to be called the Ecole Polytechnique, opened in 1794 in Paris, on the Place de la Concorde. Here the rigorous scientific education, especially strong in geometry, provided France with an engineering school unmatched in the world and an abundant reservoir of experts, who in subsequent decades would be available for large-scale public works in Paris.

The revolution ended in 1799 with Napoleon Bonaparte's seizure of power. Under his military dictatorship, the regions of France were now governed by prefects, autocratic officials appointed by him. The municipal council of Paris was replaced by the prefect of the *département* of the Seine. After 1804, when Napoleon crowned himself emperor of the French, Paris soon showed the consequences. He began to plan and erect triumphal arches in the style of the caesars. In the Place Vendôme he built a copy of the Emperor Trajan's column in Rome, similarly glorifying the exploits of his own armies; at the top was a statue of Napoleon in a toga. Other works, motivated by his desire to turn Paris into a prestigious imperial capital and the most beautiful city in the world, resulted in real urban improvements. He removed the slaughterhouses from the city centre, began the important thoroughfare of the Rue de Rivoli, built some other roads, extended the quays along the Seine, put four bridges across the river, and built a canal from the River Ourcq to augment the city's supply of drinking-water. But Napoleon's absorption in the conquest of Europe prevented the realization of a total transformation of the capital. His military campaigning ended in 1815 with defeat at Waterloo and banishment to the remote Atlantic isle of St Helena.

Napoleon was replaced by a restored Bourbon monarchy: Louis XVIII (who reigned from 1814 to 1824), followed by Charles X (1824–30). But Charles's repressive policies provoked the revolution of 1830 and his flight to Edinburgh. Louis-Philippe (1830–48), from the other Orléanist branch of the royal family, was chosen to rule as a more constitutional monarch. The three reigns are notable for the rise of a government-sponsored public health movement in reaction to growing concern over the rapid growth of Paris and the increasingly inadequate provision of sewers and water supply. Here the outstanding contribution came from Alexandre Parent-Duchâtelet, a physician who, from 1821 to his death in 1836, encouraged municipal authorities to undertake sanitary reform. He and fellow hygienists made Paris Europe's leading centre of public health. They founded a journal, *Annales d'hygiène publique*, and undertook research. When cholera struck Paris in 1832, killing 18,000, they made the city their laboratory. They studied the urban environment: the water supply, sewers and the city dump. They called for more and better sewers than the existing types – open or made of rough, small stones difficult to clean. During the 1830s the more sanitary smooth, stone tiling was introduced and the number of sewers in the capital doubled. But they were principally storm sewers, also taking liquid waste. The legacy of the hygienists was to reinforce the collection of solid human waste from cesspools for use as fertilizer (La Berge, 1993, p.184f).

Louis-Philippe imitated the Romans in bringing an obelisk from Egypt and erecting it on the Place de la Concorde. And to defend the capital, he built a new ring of bastioned walls, a circuit of thirty-five kilometres, about five kilometres from the city centre. But the greatest threat came from within – a Bonapartist conspiracy to overthrow him. Napoleon's nephew, Louis Napoleon Bonaparte, led abortive attempts to seize power. Imprisoned, he escaped to London in 1846. His stay left lasting memories of London's parks, perhaps also of sanitary reform and projects to alleviate traffic congestion. He bided his time; he did not have to wait long. In France, economic depression and the refusal to grant electoral reform had brought demonstrations; in February 1848 barricades rose in Paris and Louis-Philippe fled to Surrey. The continuing magic of Napoleon's name led to calls for his nephew to assume power; in December he was elected president of the Second Republic. But Louis Napoleon was no republican. In prison he had written of his longing to be 'a second Augustus', so that he could turn Paris into a great imperial capital of marble, just as Augustus had done in Rome (Pinkney, 1958, p.3). In December 1851, as his presidential term was running out, he first extended his authority by a *coup d'état*, and then, twelve months later, secured a massive vote to assume the imperial title. As Napoleon III he now ruled over the autocratic Second Empire. With peace, centralized power and a strong political will, conditions for a complete rebuilding of the capital were never better.

Master-plan in execution

Pinkney, the author of the standard work on this most formative alteration of Paris (while the more recent study by Jordan, 1995, adds some detail, his book does not supersede that of Pinkney), accurately portrays the magnitude of the project:

> The rebuilding of Paris was an immense and complicated operation, and its history is not a simple narrative of plans, demolitions, and building but a complex story of architecture and engineering, slum clearance and sanitation, emigration and urban growth, legal problems of expropriation and human problems of high rents and evictions, public finance and high politics, dedicated men and profiteers. It involved planning on an unprecedented scale – parts of cities, even entire new cities like Versailles, Karlsruhe or Saint-Petersburg, had been planned and built, but no one before had attempted to re-fashion an entire old city. It posed technical problems for which there were no ready solutions ... The whole operation was constantly complicated by a growing population (the city's inhabitants nearly doubled in numbers in the 1850s and 1860s) that intensified difficulties of housing, provisioning, and sanitation. Costs were enormous ... less orthodox methods of financing ... opened to political opponents of the Empire an avenue of attack upon the whole imperial regime and brought the rebuilding of Paris into national politics. In a democratic regime it would have become a political issue much earlier. The transformation of the city within two short decades probably would not have been accomplished in a state less authoritarian than the Second Empire.
>
> (Pinkney, 1958, pp.4–5)

The emperor had already drawn a general plan for his new capital. Obstacles to its execution were removed. In 1852 his decree abolished existing restrictions on compulsory purchase. The municipal council was made compliant by packing it with the emperor's nominees. And Berger, prefect of the Seine, was dismissed for being too cautious about expenditure on public works. He was replaced by Georges Haussmann, an official with no such compunctions. Haussmann, of Alsatian ancestry, had studied law in Paris before entering the civil service, rising to become prefect of the Gironde *département*. From now on, for eighteen years, he would have the emperor's full support for developing and executing the grand plan, using municipal revenues, state subsidies, loans and other means to amass the huge sums needed.

Pressing needs made the building of a completely new road network in Paris a high priority in the emperor's master-plan. Photographs of the 1850s show the poor quality of the capital's streets then: narrow, dim, airless and with central gutters carrying liquid filth (see Figure 3.36 overleaf). Their medieval narrowness so hindered movement from north to south, from east to west, that Parisians tended to live and work in their neighbourhoods; the city was not an integrated whole. For the government, such lack of communication meant a serious weakening of its powers to maintain public order. Nine times over the years 1825–52, conspirators and rioters had erected barricades in areas whose narrow streets prevented any immediate deployment of troops. The urgency of reform here was an argument Haussmann used to persuade the provinces to subsidise road-building in Paris: security of the capital was a matter of national importance. And he asked the provinces to see these planned roads as really continuations of the national railway network, for they would link railway termini located on the periphery of the capital (Pinkney, 1958, p.183). A law of 1842 required railway lines from France's borders to converge on Paris. Any repetition of the recent insurrections in Paris's crowded east end could be swiftly crushed by the rapid arrival of troops, sent by rail from the frontiers to the capital and then along Haussmann's new thoroughfares to the trouble spots. But the idea that the straightness of Haussmann's broad boulevards was designed solely to provide soldiers with a clear line of fire has now been qualified by historians; in addition to military considerations, which were

Figure 3.36 Rue Bernard de Palissy, photographed around 1850, shows medieval conditions in the heart of Paris (courtesy of Musée Carnavalet; photograph: Lauros–Giraudon)

important, aesthetic motives were also influential, notably Haussmann's insistence on placing a striking edifice at the end of a boulevard to provide a climax to the line of sight (Sutcliffe, 1970, pp.31–2).

Economic motives quickened Napoleon III's resolve to embark on a vast road-building project. Since the revolution of 1848, the building industry had been in depression. His advisers pointed out the advantages of heavy government spending on public works: the seething discontent of revolutionaries would be replaced by stability engendered by fuller employment. That was later borne out by the construction boom of the 1860s, when nearly one in five of the Parisian labour force prospered through employment in the building trade (Pinkney, 1958, pp.6, 37). Sanitary improvement also drove the emperor to provide the capital with new roads. At an elaborate ceremony in March 1858 to mark the opening of a new boulevard in central Paris, his address stressed the importance of creating new streets to 'permit light and air to penetrate the unhealthy quarters', all the more urgent now that the railways were causing an intensified immigration to urban districts already densely populated. Demolition must accompany the building of arterial highways in cities (Van Zanten, 1994, pp.206–7).

The first task was to produce a road map of Paris; no adequate one existed. Here Haussmann called on the capital's experts, reorganizing and enlarging an existing municipal 'technical service' which included men with official titles such as 'geometer-in-chief'. He spoke of them as 'my geometers'. They would determine the exact routes of the new roads, and draw up plans for expropriation of property situated in their path. This was precise geometry in action – alignment – and Haussmann would later record his admiration for experts who had made 'not a single error' (Haussmann, 1890–93b, pp.5–7, 15). The task required a comprehensive topographical survey to be carried out. Haussmann asked his geometers to supervise the triangulation of Paris (triangulation is the established surveying technique of dividing large areas into a network of triangles). Lengths on the ground would have to be measured with exactness, angles determined by sighting instruments, and calculations performed in elementary geometry and trigonometry. This was the type of practical mathematics at which the French excelled. Soon Parisians witnessed the installation of large carpentry structures at numerous sighting stations. The intrusive equipment was around for over a year, bringing, Haussmann said, 'prolonged visual distress to passers-by'; its presence in the city was so prominent that it was 'still remembered' decades afterwards (Haussmann, 1890–93b, pp.13–14). Taller than the highest houses, these structures supported narrow platforms from which the geometers took measurements of angles with precision instruments. Contemporary satirical sketches depict the surveyors on a giraffe's neck, but also convey the number and great height of structures which temporarily altered the face of central Paris (see Figure 3.37).

Figure 3.37 Poking fun at the elevated surveyors (Cham, n.d.)

The product of all this activity was a city map in sheets on a very large scale, 1:5,000, which Haussmann had had engraved for his personal use. Assembled and framed on a wheeled trellis, and positioned behind the armchair of his study,

> it functioned as a huge draught-screen and, at any moment, I could swing round to find a detail, check information, and grasp the topographical relations between the various arrondissements and districts of Paris. With this trustworthy chart before me, I frequently devoted myself to cogitations which bore fruit.
>
> (Haussmann, 1890–93b, p.15)

Subsequently reduced in scale to 1:10,000 to make it less cumbersome, it was even then two and a half metres by one and a half. For the public a still smaller edition, at a scale of 1:20,000, was necessary.

Armed with his faithful chart, Haussmann drove numerous arteries through the fabric of medieval Paris (see Figure 3.38 overleaf). The plan was to establish a central crossroad with links to every outlying district. Twisting narrowness was replaced by straightness and width – the streets were generally twenty metres wide; the widest were forty metres. The demolition was immense (see Figure 3.39), the final results dramatic. On 5 April 1858 a huge veil was lifted to reveal the new Boulevard de Sébastopol: one and a half kilometres long and over thirty metres wide, this was a main part of the new north–south crossing of the city. The *Moniteur universel*, a daily newspaper, excitedly reported the replacement of a labyrinthine slum by elegant houses along a grandiose avenue with an 'endless perspective' more effective than any in Rome. It was a street 'unrivalled in the world, the most gigantic work of slum clearance ever accomplished in the capital of a great empire' (Van Zanten, 1994, p.205). Together, the emperor and his prefect rode up the boulevard which led to a railway station, now the Gare de l'Est.

Figure 3.38 Principal new
streets built in Paris between
1850 and 1870 (Pinkney, 1958;
copyright © 1958 by Princeton
University Press; reproduced by
permission of Princeton
University Press)

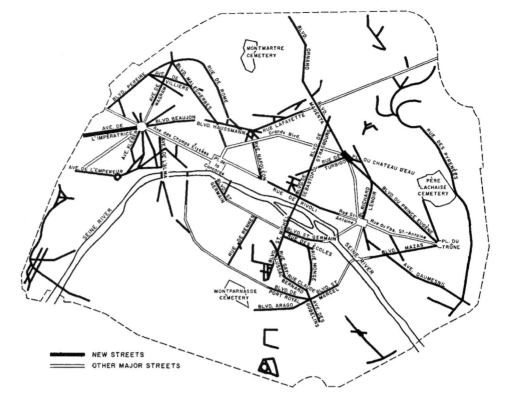

Figure 3.39 The initial effect of the works of the 1850s and 1860s was one of shock and disorientation. Demolition in Rue de la Harpe, in preparation for the Boulevard Saint-Michel, 1857 (photograph: Ladet; Collection des Estampes, Musée Carnavalet; © Photothèque des Musées de la Ville de Paris)

Excavation proceeded on a scale unprecedented in
Paris: 300,000 cubic metres of earth were removed in
eight months to lay the Boulevard Malesherbes.
Inventive engineering was required to improve the
city's east–west communication. The Canal Saint-Martin
stood in the way. A swing-bridge would have
alternately interrupted barges and street traffic. A better
solution was to lower the canal by six metres and carry
the Boulevard Richard-Lenoir over it (see Figure 3.40).
This new road had the strategic value of direct access to
the Faubourg Saint-Antoine, a district with a history of
insurrection; rioters could no longer use the canal as a
defensive barrier (Pinkney, 1958, pp.65–6).

Haussmann's great map may have been reliable but
it was, as he soon discovered, through 'cruel
deception', inadequate for the task ahead. When his
technicians tried to extend the Rue de Rivoli they
encountered a mound surmounted by the Tour Saint-

Figure 3.40 Haussmann's new Boulevard Richard-Lenoir,
covering part of the Canal Saint-Martin (Joanne, 1878)

Jacques – a tower fifty metres high, the remains of a destroyed church. The
terrain of the whole surrounding area had to be levelled. The tower required
delicate and risky operations: the ground beneath it had to be bored through,
the tower temporarily underpinned, and then lowered on to strong supporting
pillars. The lesson was to produce a relief map, showing the changing contours
of Paris: the height of every point in the city had to be determined
(Haussmann, 1890–93b, pp.16–17).

Napoleon III and Haussmann disagreed over road surfaces. The emperor
wanted macadam, but Haussmann objected that such roads were dusty in
droughts and muddy after rain. There were experiments with wooden blocks
and asphalt, but Haussmann rejected them because of their high maintenance
costs and alleged rapid wearing under heavy traffic. He preferred the hard-
wearing porphyry for a smooth surface and absence of traffic noise. But it was
slippery for horses, so the emperor, a horseman, would not have it.
Haussmann's response was to consider coating horses' hooves with a material
to prevent slipping, perhaps hardened rubber; it would be more expensive
than horseshoes but, in the long term, a saving compared with the cost of
resurfacing a road. In the end asphalt, in the 1860s, became the common
material for roads in the capital's central area (Haussmann, 1890–93b, pp.139–44;
Pinkney, 1958, pp.71–2).

Improved lighting brought greater security and ornament to the new streets.
This part of the plan was supervised by Adolphe Alphand, an engineer of the
Ecole Polytechnique. He had been busy working on quays in the port of
Bordeaux when Haussmann summoned him to the capital. Chemists now
worked closely with Alphand, monitoring production of coal gas in the plant
of the concessionary company. Jean-Baptiste Dumas, an eminent chemist,
experimented with burners and gas pressure to ensure efficient combustion
and maximize illumination (Haussmann, 1890–93b, pp.131, 156–7). Alphand
issued regulations for the selection of lighting equipment, in accordance with
the width of the street, and for their positioning. The height of lampposts was
adjusted and improved reflectors introduced to optimize the lighting on the
pavement between two lampposts, so reducing shadowed areas. Elegant cast-
iron candelabras were designed by artists; by 1870 well over 50,000 had been
installed in the city.

Electric arc-lights were used – floodlighting – to allow the hectic road-
building to continue through the night (see Figure 3.41 overleaf), but
Haussmann rejected their general introduction for street lighting because he
believed they damaged the eye.

Figure 3.41   Work continues into the night on the extension of the Rue de Rivoli (engraving by 'Gaildrau', *L'Illustration*, 30 September 1854, p.249; courtesy of Mary Evans Picture Library)

Alphand's unit was also responsible for parks. And here is the strongest sign of English influence in the grand plan. During his exile in England, the emperor may or may not have absorbed something from the debates on relieving London's traffic congestion; but it is practically certain that Hyde Park and the Serpentine sparked his determination to provide Paris with a similar municipal park. In 1852 he transferred the Bois de Boulogne, a state forest, to the city for conversion into an urban amenity – a park for recreation and healthy air. Alphand directed operations, which required surveying, hydraulic engineering and drainage to produce two lakes and cascades. For the approach to the vast park, Haussmann ordered the building, from Place de l'Etoile, of a stately avenue, the broadest in the city (now the Avenue Foch); 140 metres wide and nearly half a kilometre long, it was adorned with 4,000 trees.

Alphand's staff included gardeners as well as engineers. Horticultural techniques were perfected by the massive planting of trees and shrubs in boulevards and parks. Nurseries were established in the Bois de Boulogne to meet the increasing demand for trees in the city. Other trees were transplanted from mature forest stock using a specially designed transport wagon, still to be seen at work in Paris in the 1920s (see Figure 3.42).

So far, the emperor had been as much the architect of the capital's transformation as the prefect. But the provision of an adequate city water supply and the associated installation of an integrated, modern sewer system were Haussmann's initiatives and his crowning achievements.

In his memoirs, Haussmann recalled the inadequacies of the capital's water supply in the early 1850s, when he first took up office. Drinking-water, taken from the Ourcq Canal or River Seine, was then being transported to houses by private operators using barrels on wheels. This water was 'cold in winter,

Figure 3.42 Wagon, of the design invented for the purpose in the Haussmann era, bringing a tree for transplantation in the city of Paris, 1927 (courtesy of Archives du Laboratoire d'essais des matériaux Mairie de Paris Doc. Direction de la Voirie)

warm in summer and always dirty or cloudy'. And it was difficult to find water for washing the streets; reservoirs were few and remote (Haussmann, 1890–93a, p.315). Improvement was hindered by the widespread belief in the virtues of Seine drinking-water, a prejudice that affected Dupuit, chief engineer of the roads and bridges sector of the Ecole Polytechnique, who had been at work on Paris sewers. But Haussmann had different opinions, formed from his service, around 1840, as deputy prefect of a Pyrenean *département*, where he attributed the conspicuous good health of the population to drinking from pure mountain springs. Dupuit wanted steam-engines to pump drinking-water from the Seine; Haussmann rejected this modern technology and instead wanted to imitate the Roman practice of gravity supply by long aqueducts from remote springs. Friction between the two men became intense and Dupuit was removed to another post (Haussmann, 1890–93b, pp.110–11). Haussmann reorganized the city's engineering departments to create a municipal water service. And in March 1855 he appointed as its head Eugène Belgrand, another product of the Ecole Polytechnique, but one temperamentally in tune with Haussmann. The two men had crossed paths in the course of their careers. While prefect of the Yonne *département* (some 110 kilometres south-east of Paris), Haussmann had got to know Belgrand through his installation, in the town of Avallon, of a public fountain providing a continuous supply of clear water. Haussmann questioned Belgrand on the geological formation of the source he had used, and the quality of the springs that filtered across the subterranean seams. The replies formed a bond between the two men:

> I was astonished to find in this large, bald man, whose peasant exterior gave no hint of superior intelligence, a highly accomplished geologist and hydrologist. And since I myself was an adept, or at least a devotee, of these two sciences, devoting my leisure time to them, a mutual sympathy was established between us.
>
> (Haussmann, 1890–93b, p.112)

Belgrand's geological knowledge should not really have surprised; students at the Ecole Polytechnique routinely attended courses on geology and mineralogy to enable them to find seams suitable for making bricks and cement and for soil drainage. On the day of this questioning, Haussmann and Belgrand were travelling to a local cement-works, where the prefect later recalled that he first saw the type of cement and brick that would be used for the reservoirs of Paris. In other conversations over a geological map, they discussed the practicability of sinking deep wells to reach thermal springs. Later, working closely together in Paris, they remembered this in plans, never realized, to bring hot water to the capital's houses (Haussmann, 1890–93b, p.113).

Haussmann instructed Belgrand to undertake systematic research on the quality of spring water in the Seine basin. And he sent A. Mille, another engineer of the Ecole Polytechnique, across the Channel to report on urban water supply and sewer systems in the largest British cities. Mille subsequently expressed amazement that the authorities in London had rejected supply from remote springs (a decision which, in a later technological era, brought relief to a twentieth-century English engineer who reflected that aqueducts from Wales or Cumberland would have been dangerously vulnerable to German bombers in the world wars: Dickinson, 1954, p.109).

Belgrand soon reported that he had found suitable springs with all the requisite qualities: abundance, freedom from pollution, and so situated as to permit flow under gravity all the way to the capital at a height sufficient to serve all floors of the tallest tenements. The source was near Châlons-sur-Marne, some 160 kilometres east of Paris. But the strength of subsequent protests during the emperor's visit to the area persuaded him that this water should not be taken, or provincial support for the regime would be lost. So Haussmann proceeded instead to tap the Marne Valley further upstream, at its tributary with the Dhuis – a much smaller volume of flow. And to supplement this, he bought marshy land on the valley of the River Vanne, about 100 kilometres south-east of the capital (see Figure 3.43), this time defeating the opposition of a small town, Sens, which tried to defend its water rights (Pinkney, 1958, pp.105–26).

In October 1865 spring water from the Dhuis began to flow along an aqueduct 130 kilometres long, mostly consisting of an oval masonry conduit with a cross-section in some parts as large as 1.8 metres. Thirty-two tunnels, more than twenty bridges and many kilometres of siphons, built at a cost of nearly 18.5 million francs, delivered the water to Ménilmontant, a reservoir on the eastern edge of the capital. The other aqueduct from the Vanne Valley, even larger, was not completed until 1874: 173 kilometres in extent, with tunnels over one and a half kilometres long, siphons and bridges, culminating at Arcueil, a southern suburb of Paris, in a bridge of seventy-seven arches rising thirty-eight metres above the valley bottom. The cost of this aqueduct and the city's southern reservoir at Montsouris was over 45 million francs.

Haussmann had doubled the city's water supply, but he failed to achieve his goal of general distribution. He wanted to imitate the English system with its tax levied on property owners for the installation of supply, and where 'in the humblest lodging there was a kitchen tap and other taps for the water-closet and shower bath'. But he judged this impracticable in France because 'with us, it would be difficult to secure acceptance of a tax to make subscription compulsory' (Haussmann, 1890–93b, pp.323–4), evidence that even in an authoritarian regime, some respect remained for individual liberty. Consequently, while Haussmann's term of office saw an increased mains connection from 6,000 houses to 34,000, still in 1870 half the city's houses were without running water and many more lacked it above the ground floor.

The water supply was now distributed in two separate systems: *eau potable* (drinking-water) brought from the distant springs; *eau non potable* (undrinkable water) from the city's river and Ourcq Canal for street-cleaning and fire-hydrants. These two sets of pipes were housed in Belgrand's vast new sewer system, which has continued to serve Paris ever since.[4]

[4]   Extracts from Pinkney's discussion of this sewer system are reproduced in the Reader associated with this volume (Goodman, 1999).

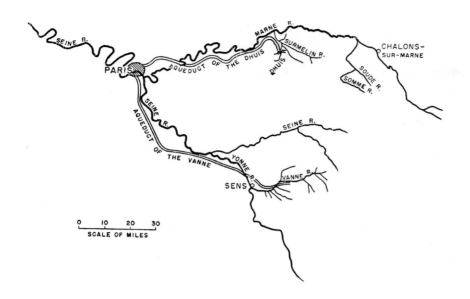

Figure 3.43 Seine river basin and the aqueducts built by Haussmann and Belgrand (Pinkney, 1958; copyright © 1958 by Princeton University Press; reproduced by permission of Princeton University Press)

In his analogy likening a well-run city to the human body, Haussmann revealed the mixture of sanitary and aesthetic motives that drove his great project for providing Paris with a comprehensive sewer system:

> Subterranean galleries are the internal organs of the great city and they function like those of the human body ... Secretions are mysteriously performed and public health maintained without disturbing the running of the city or spoiling its external beauty.

(quoted in Belhoste *et al.*, 1994, p.175)

So, the disgusting and unhealthy open sewers in the streets of Paris must be eliminated. And this liquid waste must no longer spoil the Seine in the vicinity of the capital. Instead, every street must have its covered sewer, concealed underground, the offensive contents of all collecting into larger pipes for discharge beyond the city centre. No human excrement would pass through these sewers, only liquid waste. Why spoil this solid manure, so valuable as fertilizer, by immersing it in liquid sewage? It must be separately collected from the cesspools and deposited on agricultural land around the capital. Yet the sewers must be very large to house the water-mains, gas-pipes, and permit mechanical cleansing by sluice-boats and sluice-cars on rails.

The consequences of Haussmann's specifications were to produce a wholly different design for the sewers of Paris, compared with those of London. Bazalgette in London and Belgrand in Paris were each striving to install radically new systems and at almost exactly the same time. There were some common features. Both employed large collectors, orientated parallel to the banks of the Seine or Thames. In both cities the engineers were forced to pump water out of stretches of subterranean quicksand in order to secure stable terrain for construction. But there the resemblance ends. The designs of the two projects diverged because of differences in the nature of the two rivers and, above all, because Haussmann's preferences posed the engineering problem in different terms. The Thames had a strong tide, the Seine in Paris had none, so that it was safe to locate the sewage outfall just two and a half kilometres from the city centre; in London it had to be much further out – even twenty-two kilometres proved insufficient to avoid the tide bringing sewage back to the city. London's sewers were of much smaller cross-section because there was no intention to introduce cleaning vehicles into them; nor did they contain water-mains and gas-pipes (a feature later envied by some English engineers because it avoided traffic disruption while repairs were carried out to these). Some of Belgrand's elliptical sewers were as large as 4.3 metres in

Figure 3.44 Sewer beneath the Boulevard de Sébastopol: a sluice-car on rails cleanses; the large pipes on either side of the gallery separately carry water for drinking and other uses (Belgrand, 1887; courtesy of Archives Direction de la Voirie – Ville de Paris)

height and had an inner width of 5.5 metres (see Figure 3.44). And in sharp contrast to Bazalgette's sewers, Belgrand's system used no pumps – only a gravity flow. In London, solid as well as liquid sewage was carried away in the sewers; in Paris they were mainly storm sewers, draining excess rainfall and also liquid waste from houses and urinals. Between 1857 and 1870 Belgrand installed over 320 kilometres of sewers in Paris. Ever since their completion, tourists have descended to this underground world, a mirror of the capital with gallery signs repeating the names of corresponding streets immediately above. Visitors to the International Exhibition in Paris in 1867 flocked to the sewers and, still today, curious tourists pay the entrance fee.

For Haussmann, the sewers were a constant reminder of ancient Rome. Agrippa, the minister of Emperor Augustus, had sailed along the sewers he had built for his master. Haussmann saw close parallels with his introduction of sluice-boats: 'I was not the first to make a boat float in a sewer.' And he reflected that, although 'our century is so proud of the level of civilization it has attained', the Paris sewer system had been 'borrowed' from ancient Rome: 'We invented nothing here. We are only imitators. Our only merit is to have dared to imitate, in the face of ignorance, in order to sanitate our Queen City, the imperial Rome of our times' (Haussmann, 1890–93b, pp.350–52).

Haussmann's sewers, boulevards and parks cost a fortune – on one estimate £100 million, four times the sum spent by London's Metropolitan Board of Works in its improvements over a period twice as long as Haussmann's operations (Sutcliffe, 1979, p.80). William Tite, London architect and member of the Metropolitan Board of Works, which had to struggle for funds, explained how Haussmann found the huge sums for street-building:

> The state has come to the assistance of the city in this matter; but it can only be by casting the burthen upon the taxpayers of the country generally – a course which may be tolerated in a highly centralized country, like France, where, in fact, Paris is everything, and the rest of the nation nothing in comparison with it – but which would hardly be tolerated in England, where we pride ourselves on making every place pay for its own improvements.

(quoted in Dyos, 1957, p.265)

Figure 3.45 Paris at the end of the nineteenth century (courtesy of Estate of George R. Collins)

Haussmann's extravagance and financial irregularities contributed to his dismissal in 1870 and to the unpopularity of the regime – that collapsed a few months later, after defeat in the Franco-Prussian war (Napoleon III was taken prisoner at the frontier fortress of Sedan and spent his last years in exile in Kent).

Historians of architecture have blamed Haussmann for his failure to conserve medieval buildings in his bulldozing of central Paris. Urban historians have criticized other results of his demolition: the flight of displaced residents to surviving slums in the centre or to shanty towns on the capital's periphery, to be replaced in the rebuilt centre by those able to pay the soaring rents (Pinkney, 1958, pp.165–6; Sutcliffe, 1970, p.42). But all acknowledge that Haussmann's urban planning was unrivalled in contemporary Europe, and that the new capital which arose was an inspiration to city-planners far and wide: all over Europe and beyond – to Cairo, and the USA. And the look of Paris today (despite the later nineteenth-century additions of the Eiffel Tower and Sacré-Coeur, and Pompidou's recent burst of monumental construction) continues to display the powerful imprint of Napoleon III and his autocratic prefect of the Seine (see Figure 3.45).

From a technological point of view, eighteenth- and nineteenth-century Paris and London had much in common. Both cities were important centres of manufacturing and national centres of transport networks. And as the two capitals experienced soaring population growth, the same pressures for change were felt. Sanitary demands for purer drinking-water and sewage disposal became urgent and were eventually satisfied by large-scale public-works projects begun in the two cities at almost identical times. There were parallel developments also in public transport: Paris supplied London with the idea of an urban omnibus system; but the transference of transport technology across the Channel was in the opposite direction in the case of the railways. Yet there were also striking differences between the two cities. Paris never had anything of London's importance as a world port or the associated massive dock developments. The systems of water supply and sewers were completely different engineering solutions. There were marked differences too in the pattern of urban expansion. London grew outwards, Paris upwards (but also outwards). Above all, Paris was subjected to urban planning on a scale unknown in London. That is attributable to another fundamental difference: the ways in which the two capitals were governed.

References

ALDCROFT, D. and FREEMAN, M. (eds) (1983) *Transport in the Industrial Revolution*, Manchester, Manchester University Press.

AL NAIB, S. (1998, 8th edn) *London Docklands: past, present and future*, London, published by the author.

ARNOT, N. and PHILLIPS KAY, J. (1838) *Fourth Report of the Poor Law Commissioners, Parliamentary Papers*, XXVIII, Appendix A, Supplement 1.

BARKER, T.C. (1989) 'Business as usual? London and the Industrial Revolution', *History Today*, vol.39, pp.45–51.

BARKER, T.C. and ROBBINS, M. (1975) *A History of London Transport: passenger travel and the development of the metropolis*, London, Allen and Unwin for the London Transport Executive, vol.1.

BASTIÉ, J. (1964) *La Croissance de la banlieue parisienne*, Paris, Presses Universitaires de France.

BAZALGETTE, J. (1864–5) 'On the main drainage of London', *Minutes of Proceedings, Institution of Civil Engineers,* vol.24, pp.280–314.

BELGRAND, E. (1872–87) *Les Travaux souterrains de Paris*, Paris, Dunod, vol.5.

BELHOSTE, B., MASSON, F. and PICON, A. (1994) *Le Paris des polytechniciens: des ingénieurs dans la ville 1794–1994*, Paris, Délégation à l'Action artistique de la ville de Paris.

LA BERGE, A. (1993) *Mission and Method: the early nineteenth-century French public health movement*, Cambridge, Cambridge University Press.

BRAUDEL, F. (1974) *Capitalism and Material Life 1400–1800* (trans. M. Kochan), Fontana, Glasgow (first published 1967, in French).

BRIGGS, A. (1968) *Victorian Cities*, Harmondsworth, Penguin Books.

BROODBANK, J. (1921) *History of the Port of London*, London, Daniel O'Connor.

CHAM (pseud. of A.H.C. de Noé) (n.d.) *Croquis contemporains*, Paris, Arnauld de Vresse.

COLEMAN, W. (1982) *Death is a Social Disease: public health and political economy in early industrial France*, Madison, University of Wisconsin.

COONEY, E.W. (1955–6) 'The origins of the Victorian master builders', *Economic History Review*, vol.8, pp.167–76.

DICKINSON, H.W. (1954) *Water Supply of Greater London*, London, Newcomen Society.

DYOS, H. (1957) 'Urban transformation: a note on the objects of street improvement in Regency and Early-Victorian London', *International Review of Social History*, vol.2, pp.259–65.

FLINN, M.W. (ed.) (1965) Chadwick's *Report on the Sanitary Condition of the Labouring Population of Great Britain*, Edinburgh, Edinburgh University Press, pp.1–73.

FOX, C. (ed.) (1992) *London: world city 1800–1840*, New Haven, Yale University Press.

GENERAL BOARD OF HEALTH (1850) *Report by the General Board of Health on the Supply of Water to the Metropolis, Parliamentary Papers*, XXII.

GOODMAN, D. (ed.) (1999) *The European Cities and Technology Reader: industrial to post-industrial city*, London, Routledge, in association with The Open University.

GREEVES, I.S. (1982) *London Docks 1800–1980: a civil engineering history*, London, Thomas Telford.

HAUSSMANN, G.E. (1890–93a) *Mémoires du baron Haussmann*, Paris, Victor-Havard, vol.2.

HAUSSMANN, G.E. (1890–93b) *Mémoires du baron Haussmann*, Paris, Victor-Havard, vol.3.

HOBHOUSE, H. (1971) *Thomas Cubitt: master builder*, London, Macmillan.

HOBHOUSE, H. and SAUNDERS, A. (eds) (1989) *Good and Proper Materials: the fabric of London since the Great Fire*, papers given at a conference organized by the Survey of London at the Society of Antiquaries, 21 October 1988, London, Royal Commission on the Historical Monuments of England in Association with the London Topographical Society.

INSTITUTION OF CIVIL ENGINEERS (1864–5) *Minutes of Proceedings*, vol.24.

JOANNE, A. ([1878], 4th edn) *Paris illustré en 1878*, Paris, Librairie Hachette.

JORDAN, D.P. (1995) *Transforming Paris: the life and labors of Baron Haussmann*, New York, The Free Press.

KELLETT, J.R. (1969) *The Impact of Railways on Victorian Cities*, London, Routledge and Kegan Paul.

LANGTON, J. and MORRIS, R. (1986) *Atlas of Industrializing Britain*, London, Methuen.

LUCKIN, W. (1986) *Pollution and Control: a social history of the Thames in the nineteenth century*, Bristol and Boston, Adam Halgar.

OWEN, D. (1982) *The Government of Victorian London 1855–1889*, Cambridge, Mass., Belknap Press of Harvard University Press.

PINKNEY, D.H. (1958) *Napoleon III and the Rebuilding of Paris*, Princeton, Princeton University Press.

POLLARD, S. (1950–51) 'The decline of shipbuilding on the Thames', *Economic History Review*, vol.3, pp.72–89.

PORTER, R. (1994) *London: a social history*, London, Hamish Hamilton.

RASMUSSEN, S. (1982) *London: the unique city*, Cambridge, Mass., MIT Press (first published 1937).

RATCLIFFE, B. (1994) 'Manufacturing in the metropolis: the dynamism and dynamics of Parisian industry at the mid-nineteenth century', *Journal of European Economic History*, vol.23, pp.263–328.

SCHWARZ, L.D. (1992) *London in the Age of Industrialization: entrepreneurs, labour force and living conditions, 1700–1850,* Cambridge, Cambridge University Press.

SKEMPTON, A.W. (1978–9) 'Engineering in the Port of London, 1789–1808', *Transactions of the Newcomen Society*, vol.50, pp.87–108.

SKEMPTON, A.W. (1981–2) 'Engineering in the Port of London, 1808–1834' *Transactions of the Newcomen Society*, vol.53, pp.73–96.

SMITH, S. (1838) *Fourth Report of the Poor Law Commissioners, Parliamentary Papers*, XXVIII, Appendix A, Supplement 3.

SOMERVELL, D.C. (ed.) (1951) *100 Years in Pictures: a panorama of history in the making*, London, Odhams.

SUTCLIFFE, A. (1970) *The Autumn of Central Paris: the defeat of town planning 1850–1970*, London, Edward Arnold.

SUTCLIFFE, A. (1979) 'Environmental control and planning in European capitals 1850–1914: London, Paris and Berlin' in I. Hammarstrom and T. Hall (eds) *Growth and Transformation of the Modern City*, Stockholm, Swedish Council for Building Research, pp.71–88.

VAN ZANTEN, D. (1994) *Building Paris: architectural institutions and the transformation of the French capital 1830–1870*, Cambridge, Cambridge University Press.

YOUNG, G.M. and HANDCOCK, W.D. (eds) (1956) *English Historical Documents 1833–1874*, London, Eyre and Spottiswoode.

Part 2
EUROPEAN CITIES SINCE 1870

Chapter 4: THE SECOND INDUSTRIAL REVOLUTION AND THE RISE OF MODERN URBAN PLANNING

by Colin Chant

4.1 The changing system of European cities

Why plump for the year 1870 as the great divide in a history of European cities and technology since the Industrial Revolution? Any single year is bound to be an arbitrary choice, but there are sound reasons for regarding the decade of the 1870s as pivotal. After the dramatic surge in economic growth associated with investment in the technologies of coal, iron and steam power, Europe entered a period of economic deceleration during the last quarter of the nineteenth century. But during these very years, European citizens began to experience a new configuration of technological, economic and political forces. A generation of innovations in steelmaking and in electrical and chemical engineering was already in embryo in the 1850s and 1860s; their exploitation under a more corporate form of capitalism underlay the next broad upswing in the European economy, and associated reworkings of the urban fabric. As before, these new technologies were developed and applied within shifting and intersecting urban contexts.

The demographic context

The great population expansion which propelled urbanization in Europe of the nineteenth century carried on for most of the twentieth. The population of Europe rose from 290 million in 1870 to some 750 million in 1950, despite massive migrations, notably to the USA, and the toll of two world wars (Landes, 1969, pp.241, 487). By 1970, the populations of Britain and Italy had more than doubled, and the combined figures for East and West Germany were nearly twice the total for the various German states in 1870. France was less prolific, with population growth of less than 38 per cent in this 100-year period, the proximate cause being its significantly lower birth-rate compared with its neighbours up to the middle of the twentieth century. Thereafter, all the nations of Western and Central Europe experienced a marked slowdown in population growth, as falls in the birth-rate approached those achieved in the death-rate during the first half of the century. Most strikingly, the population statistics for West Germany stood still from the early 1970s; and by the early 1990s, the fertility rates in Spain and Italy were the lowest in the world. The fall in the birth-rate is a complex historical phenomenon, but seems to mark the maturation of any industrialized, urbanized society with improved welfare provision, in which large families come to be seen more as a source of cost than of income. But it also went in step with a dramatic decline in infant mortality: in Italy, a fairly typical example, deaths before the age of one year per 1,000 live births fell from 230 in 1870 to ten in 1985 (Mitchell, 1998, pp.93–128). Declining mortality overall reflected improvements in housing, healthcare and nutritional standards. As in the nineteenth century, the rate of increase in the European population as a whole was exceeded by the rate of increase in the proportion living in towns and cities. By 1950, 53.7 per cent were classified as urban, with significantly more than half in the north and west, and significantly less in the east and south; but by 1970, more than half the population of Eastern and Southern Europe alone were urban dwellers (Department of International Economic and Social Affairs, 1980, p.16). The great reductions achieved in mortality rates meant that in the period since 1870 – probably for the first time in European history – more people have been born in cities than have died in them. Cities might therefore grow not only through in-migration, but from the natural increase of their native inhabitants.

The political context

The urbanization of European society was accompanied by radical changes in international and national politics. In the years since 1870, the state boundaries of Europe have been redrawn several times (see Figure 4.1). The unification of Italy was completed in 1870, and the German Empire was established in 1871. During the intense last chapter of European global imperialism in the period leading up to the First World War, these new nations joined Britain, France, Belgium, Russia, Spain and Portugal in a dash for colonies, especially in Africa and the Far East. The colonies continued to have European styles of building and urban planning imposed on them (the case of India is discussed in Chapter 10). Economic and political rivalry between these colonial powers, in the context of the disintegrating Ottoman Empire in the east, plunged Europe into full-scale war. The trench warfare into which the First World War degenerated partly resulted from a stalemate between the weapons of offence and defence. Although this artillery-dominated style of warfare devastated the countryside along the Eastern and Western fronts, and consumed the lives of millions of young soldiers, with incalculable effects on the populations and economies of cities, the urban physical fabric went relatively unscathed. The worst urban damage was incurred by the cities of north-eastern France and south-western Belgium, including the industrial cities Lille and Liège; many of these cities were faithfully restored to their pre-1914 form. The new threat of aerial bombardment, first unleashed by German airships (Zeppelins), led to the temporary removal of the government of France from Paris to Bordeaux in the first months of the war. London and some other industrial cities of Britain also experienced raids, but not on such a scale that extensive rebuilding was needed.

Figure 4.1 (opposite)   National boundaries of Europe: 1914, 1922, 1949 and 1992. Almost all the changes to national boundaries took place in Central and Eastern Europe. A number of independent states emerged in the Balkans after 1870, as the Ottoman Turks were driven out of the region. Other independent nations were created after the First World War in Central and south-eastern Europe out of the old Austro-Hungarian empire, and Poland regained its independence from Russia. After the Second World War, Germany was divided, and the eastern portion, along with most other nations east of the Iron Curtain (shown by a heavier line, running roughly north–south, in the 1949 map), was dominated by the Soviet Union. Following the ending of Soviet hegemony in Eastern Europe, and then the break-up in 1991 of the Soviet Union itself, Germany was reunified, several new states were created in the Balkans, and in the East independence was achieved in what were formerly subordinate republics of the Soviet Union. In the West, the bulk of Ireland became independent in 1921; thereafter, the name 'United Kingdom' denotes Great Britain and Northern Ireland only

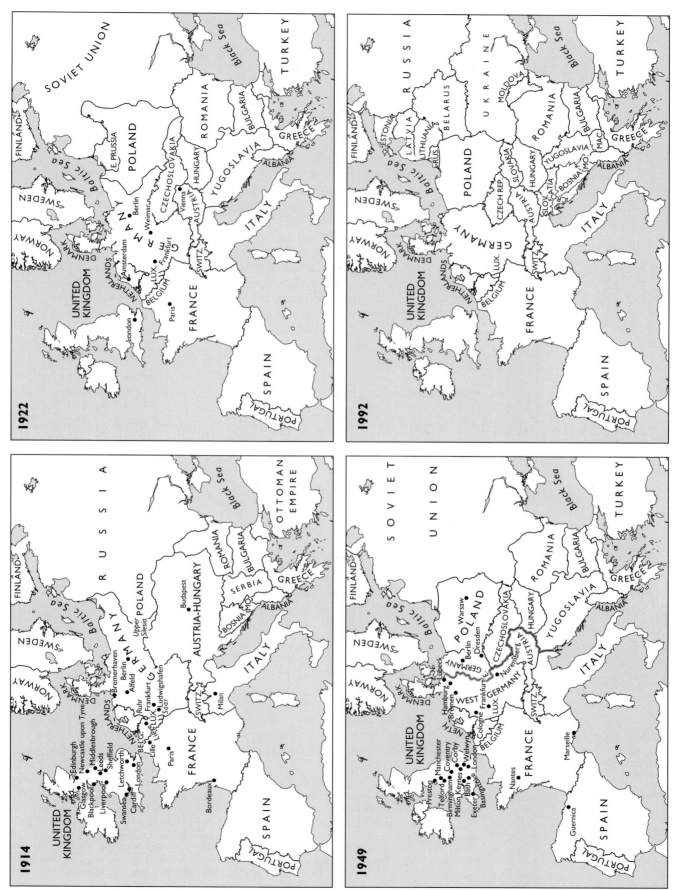

Key LUX.: **Luxembourg**; MAC.: **Macedonia**; MO.: **Montenegro**; RUS.: **Russia**; SLOV.: **Slovenia**; SWITZ.: **Switzerland**

After the First World War, governments resumed the struggle to match urban amenities and infrastructure with urban population growth. Then the victorious Allies' demand for reparations sent the Central European economies spiralling into hyperinflation, and the Wall Street crash in 1929 led to the cessation of US loans to Europe, which also suffered an economic depression. From 1933, a resurgent Germany under the Nazis sought dominion over the entire continent, and, in alliance with Fascist Italy, brought yet more conflict and instability to Europe. The destruction resulting from the Second World War was spread much wider than that of the First, partly through further developments in aerial warfare. Cities, whether as manufacturing centres or as symbols of national pride, or in the end just as concentrations of the enemy population, came to be regarded as legitimate targets; this kind of havoc had already been wrought in 1937, during the Spanish Civil War, when the German *Luftwaffe* destroyed the Basque town of Guernica. Indeed, the evacuation of women and children from London began even before the formal declaration of hostilities marking the start of the Second World War. The 'Blitz' of London in fact began in August 1940, and soon extended to industrial cities such as Manchester and Coventry. There were 'softer' targets, too, notably cities such as Exeter and Bath, bombed during the 'Baedeker' raids of April 1942. Starting in 1941, the RAF retaliated with attacks on German cities, such as Nuremberg, Lübeck, Cologne, Bremen and Hamburg. This was controversial even at the time: the blanket bombing of Berlin in 1944 led to protests in the House of Lords. In April of that year, the Allies dropped more than 81,000 tonnes of bombs on Germany and on occupied Europe. In the final months of the war, before the instantaneous atomic obliteration of the centres of Hiroshima and Nagasaki, the 'conventional' technologies of destruction enabled large cities such as Dresden to be reduced to rubble and ashes within a matter of hours; it has been estimated that 50 per cent of the total built area of all German cities with populations over 100,000 was destroyed (Diefendorf, 1993, p.11) (see Figure 4.2). By no means all the pulverization of cities came from above. In 1944, German 'demolition and annihilation squads' on the ground wielding artillery and dynamite charges set about reducing to rubble what remained of Warsaw; during the course of the war, 80 per cent of the Polish capital's buildings were destroyed (Jankowski, 1990, pp.79–80).

The legacy of this destruction was the challenge of post-war reconstruction, one aspect of which is considered by Nicholas Bullock in Chapter 7. The rebuilding of the continent took place within the context of a new post-war order in Europe, which was now divided into two opposing camps, each dominated by the politics of one of the two superpowers, the USA and the Soviet Union. Underpinning the reconstruction of the West was the Truman doctrine of economic and military support for states seen as threatened by Communism, and in particular the European Recovery Programme (Marshall Aid). The nations of the capitalist West gravitated towards each other, eventually forming the European Economic Community at the beginning of 1958. For forty years and more the urban dwellers of the satellite states of the Soviet Union looked on from their austere cities, empty of traffic jams and neon advertising signs, until the collapse of Soviet hegemony from the late 1980s unleashed the forces of Western enterprise on the East European cityscape (see Chapter 9).

As well as the effects of tumultuous international changes, several national urban systems also bore the stress of internal political upheavals. Generalizations are difficult, as the political history of many a Central or East European nation between 1870 and the end of the twentieth century would readily show; what any technological history of urbanization and the urban built environment needs to consider are the effects of a succession of imperial autocracies, fledgling democracies, Fascist or Socialist totalitarian technocracies,

Figure 4.2 The ruins of Kreuzberg, immediately to the south of the old centre of Berlin, 1945 (photograph: Landesbildstelle, Berlin)

and latterly the mix of parliamentary democracy and private enterprise that some complacent commentators now see as the end-point of historical development. Clearly, these changes affected decision-making about the planning and building of cities. Among the most notable, perhaps, was a reversal of the *laissez-faire* approach to local government which characterized the growth of cities during the Industrial Revolution. This trend toward much greater ambition and intervention by local government reflected the strength of emergent trade unionism, the rising stock of socialism, and extensions of the franchise across Europe. In this period, local authorities became one of the main agents of technological change in the urban built environment.

The economic context

Far-reaching economic changes also had a bearing on the pattern of urbanization. The 1860s and 1870s were a time when many of the restrictions on European companies were lifted, paving the way for a corporate form of capitalism based on the limited liability company, and the increased availability of credit through the proliferation of banking services. The most successful companies tended to become larger and more spatially concentrated as they sought economies of scale, either through the take-over of similar firms (horizontal integration) or through the ownership of all stages of production, from raw materials to distribution (vertical integration). As a result, manufacturing activity became even more concentrated in industrial regions, and the countryside further deindustrialized. Technological change was a necessary condition of these linked economic and urban phenomena. The concentration of population in giant cities became marked from the second half of the nineteenth century, and by 1915 the pioneering Scottish planner Patrick Geddes had coined the term 'conurbation' to cover highly urbanized regions such as Greater London, and the cluster of settlements centring on Birmingham, Cardiff, Glasgow and Manchester (Geddes, 1968, pp.34–42). Such phenomena were rooted in the movement of manufacturing industries from their raw material sources to their largest urban markets, a movement first made possible by steam power in factories, and then by the building of railways.

The disproportionate growth of the main nodes of the urban system was reinforced after 1870 by electric power and a succession of innovations in urban transport. The replacement of industrial steam-engines by electric motors enabled factories to be located even more flexibly. They also reinforced a revolution in the design of factories and in production processes, associated with ideas about 'scientific management', including time-and-motion study, that were put forward at the start of the twentieth century by Frederick Winslow Taylor in the USA. In particular, electric motors powered the conveyor belts that brought components to assembly-line workers responsible for the multiplicity of mass-produced goods that fed twentieth-century consumerism. The introduction of electric railways and tramways, and then motor omnibuses, made the mechanized journey to work from the suburbs increasingly the norm. Electrical communications, principally the telegraph and telephone, also fostered the clustering of metropolitan knowledge-based institutions. The most striking European example of these developments was the capital of the new German Reich, Berlin, the population of which leapt from 0.8 million in 1870 to 4.2 million by 1933 (Mitchell, 1998, p.74). Berlin was a manufacturing as well as a political centre; spurring its growth were the great Siemens and AEG electrical engineering establishments, and the lighting, trams and industrial motors they provided (see Chapter 6).

In the period since the Second World War, the service sector of the economies of the leading West European nations has become dominant, a fundamental economic change ramifying through their urban systems. Many of the industrial cities and conurbations that sprang up in the European coalfields during the Industrial Revolution have declined, because of international competition and technological changes in heavy industry; but the growth of the larger older cities (national capitals, regional centres and ports) has more than compensated. Innovations in transport systems, including wider car ownership, have also helped to reinvigorate many of the provincial towns which were sidelined during the Industrial Revolution, but which, in some respects for that very reason, have become attractive both as centres for new and relocated light manufacturing and service industries, and as workplaces for commuters seeking accessible rural residences.

Patterns of urbanization

In Western Europe in general, the rise of capitalism and the related processes of industrialization and urbanization revived cities as centres of political power, following their relative decline as the map of Europe started, from the Early Modern period, to simplify into nation states with strong centralized governments. But although cities became the engines of industrialization, and their administrations acquired great powers within their boundaries after 1870, they have remained subservient to national or regional governments. These and other high-order generalizations about technology and the European system of cities conceal the diversity of the European experience: some quite distinctive national patterns of urbanization have emerged. The two most powerful and long-established nation states of the early industrial era, Britain and France, continued to be dominated by 'primate cities' (as urban geographers classify cities far bigger than any other in a national urban system; see Chapter 5). There were contrasting developments in Germany and Italy, where urbanization, industrialization and national state-formation developed alongside each other. In the German states, the technological basis of early industrialization was steam power and railways, so that much new manufacturing plant could be installed in established towns and cities. Exceptions were new settlements in mining areas such as the Ruhr, Saar and Upper Silesia, and new ports such as Bremerhaven, Ludwigshafen and Wilhelmshaven. A key political difference was that the German Empire, created in 1871, was a federation, under which cities were directly responsible to their state government, rather than to the imperial government in Berlin. The result of these technological and political considerations was a multi-centred urban system, in which growth was controlled by the first modern zoning legislation and by welfare-minded city engineers.

Throughout Europe, as some regions underwent rapid industrial and urban growth, others declined and deindustrialized. Although by 1870 some other European nations (principally France, Belgium and Germany) were catching up with the indices of British industrialization and urbanization (Germany overtook Britain in steel-production in 1896), in some areas of Southern and Eastern Europe the developments already covered in the first part of this book had scarcely begun. In these areas, some of the peculiarities of their urban development are attributable in part to the wholesale transfer of technologies that in Western Europe may have been adopted piecemeal over the course of a century (see Chapter 9).

4.2 *The Second Industrial Revolution and the urban built environment*

Apart from their implications for the system of European cities, the demographic, political, economic and technological changes just outlined directly affected the form and fabric of individual cities, the central concern of this series. A (sometimes literally) dazzling array of new technological instruments became available to the various public and private agents of city-building during the seminal decades between 1870 and the outbreak of the First World War. First came the electric power station and the urban innovations connected with it – the electric cityscape, urban public transport based on electric traction, and the electrification of industrial plant; second, the new instruments of electrical communication, especially the telephone, which contributed to the upward and outward growth of cities; third, new building materials and construction techniques, notably structural steels and reinforced concrete, which underpinned a range of high-rise and wide-span structures; and finally, the internal-combustion engine which, in the form of the motor bus, began to supplant electric traction in urban transport systems. Along with the proliferation of organic chemicals, these technological innovations, set within the context of sweeping economic changes, were momentous in the history of industrial society; indeed, for commentators of quite diverse historiographical persuasions, this set of changes amounted to a 'Second Industrial Revolution' (Cardwell, 1994; Hobsbawm, 1968; Landes, 1969).

How did the deployment of the technologies of the Second Industrial Revolution affect urban morphology? In retrospect, the prevalent technologies of the periods surveyed in this series seem to have reinforced either centrifugal or centripetal trends in human settlement. The invention of agriculture was centrifugal in effect, in that its adoption led to the dispersal of formerly closely bound, albeit mobile, hunter-gatherer societies; though as we have seen, it also provided the material basis for a concentration of the powerful and the specialist minority in cities. The industrial technology of steam, coal, iron and rail was centripetal, in that it was associated with the concentration of production in factories, and of residence around them. Subsequent technologies based on electricity and the internal-combustion engine, and latterly the silicon chip, have all have been seen as reversing this centripetal trend, and helping to liberate people from the stresses of city life. But these technologies were not as revolutionary in their effects on cities as some expected; their adoption failed in the medium term to bring about the dispersal of population and the demise of cities often predicted. Indeed, the opposite can be argued, at least for the period up to 1950. To put it crudely, the new technological instruments enabled high-density upward growth and/or low-density outward growth; both were manifest in European cities, but not in equal measure or proportion. This is a complex story; one way to unravel it is to consider each technology in turn.

Transport

In Chapter 1, David Goodman described how in Britain the movement of bulk goods by canals and then by railways helped to concentrate people in factories and in housing packed around them. The density of population was often intensified by driving railways into cities, demolishing poor housing in the process and leading to even worse overcrowding in the slums that remained. Apart from their direct effects on the urban fabric, these primary transport innovations stimulated the growth of metropolises such as London to the point where the city centre, the hub of a radial transport system, was threatened with strangulation. Thereafter, secondary forms of transport were increasingly touted as a means of relieving urban congestion, and of offering factory workers a more healthy environment. In practice, they acted to heighten social segregation. As the option of commuting to work became more widespread, the better-off continued to flee their socio-economic inferiors to the fast-receding rural margins of urban development. This dynamic was reflected in the built environment, in the progressive movement of suburban residential areas down the social scale, with some corresponding adaptation of the housing stock.

Before discussion of these effects of urban transport innovations, it should be recognized that the established rail networks of European nations continued to expand and intensify well beyond 1870. The length of the French network increased from 15,554 kilometres in 1870 to a peak of 42,600 in 1933. In Britain, the peak of 32,857 kilometres was reached in 1926; but by 1985, this figure had almost been halved – the most marked run-down of any European system (Mitchell, 1998, pp.673–83). Continuing growth of surface railways up to the 1930s was coupled with important new canal links in the system of waterways; such inter-urban developments were always likely to affect the economic status of individual cities, and therefore their propensity to grow. But railways were seldom directly involved in the physical spread of cities, except in the case of such metropolises as London and Berlin (see Chapters 5 and 6); steam railways continued to be used mainly for inter-urban passenger and freight traffic. There was a technological and economic limit to the steam railway's contribution to suburban growth, as the locomotive's slow acceleration meant that stations could not be much less than two kilometres apart. Where steam railways were used for commuting, the result was nuclei of development along radial lines beyond the high-density area in which people walked to their destinations, or were served by horse-drawn modes of public transport – the private carriage, short-stage coach, hackney carriage and omnibus (see Figure 4.3).

The further extension of this high-density area was indeed achieved through the increased efficiencies available from rail transport, but not in its pioneering steam-driven form. For most rapidly growing cities of the late nineteenth century it was the horse-drawn tram that allowed less-well-paid workers to live at a greater distance from their workplace. The idea of horses drawing a carriage on rails laid into the road surface was introduced in New York as early

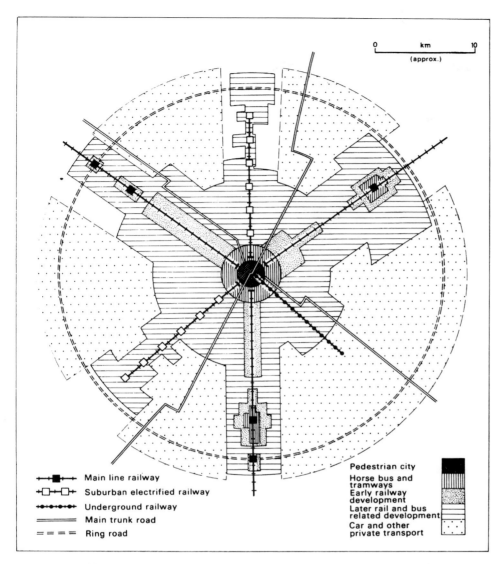

Figure 4.3 The pattern of urban growth associated with urban transport modes (reproduced from Daniels and Warnes, 1980, p.3)

Main line railway
Suburban electrified railway
Underground railway
Main trunk road
Ring road

Pedestrian city
Horse bus and tramways
Early railway development
Later rail and bus related development
Car and other private transport

as 1832, and became widespread in the USA from the 1850s. George F. Train, an extrovert American ex-army officer, introduced the tram to Britain at Birkenhead in 1860, and went on to construct three lines in London in 1861. These were unsuccessful, because of Train's insistence on the step-rail, which interrupted the street surface, and if poorly laid could project above it (see Figure 4.4). According to D. Kinnear Clark, a nineteenth-century British authority on trams, this kind of rail had been tolerated for the early tracks of US cities such as New York and Philadelphia because of their relative lack of other horse-drawn vehicles (Clark, 1992, p.5); indeed, in Philadelphia, the lower part

Figure 4.4 G.F. Train's Birkenhead tramway. *Left* a section of the tramway: the rolled wrought-iron rails, laid to a gauge of four feet eight and a half inches (143.5 centimetres), were spiked to longitudinal sleepers, themselves spiked to transverse sleepers by means of a piece of iron shaped to fit an angle (a 'knee'). *Right* a section of the rail, with the spike-hole shown by dotted lines: each rail was six inches (15.2 centimetres) wide, with the step rising three-quarters of an inch (1.9 centimetres) above the sole. Train modelled his Birkenhead tramway, including the step-rail, on the system introduced in Philadelphia in 1855 (reproduced from Clark, 1992, p.13)

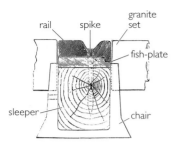

Figure 4.5 A section of the rail and sleeper used in the Liverpool tramway of 1868. The rail was four inches (ten centimetres) wide, and was bedded on timber sleepers laid in cast-iron chairs, which rested on a concrete foundation. The rails were joined by means of four vertical spikes driven into the sleeper through the rails themselves and a wrought-iron fish-plate laid across them to strengthen the joint. (The term 'fish-plate' was possibly derived from the French *ficher*, meaning 'to fix'.) The surface between the rails, and eighteen inches (forty-six centimetres) on either side, consisted of granite sets (adapted with permission from Clark, 1992, p.20)

Figure 4.6 Typical late nineteenth-century London horse tramcar, passing Mornington Crescent on the London Street Tramways Company's Hampstead–Euston route (reproduced from Klapper, 1961, plate 18; © London Transport Museum)

of the rail accommodated what cart and carriage wheels there were (Barker and Robbins, 1963, p.181). But in other US cities and in Britain, the step-rail led to big conflicts with carters, and cab and omnibus drivers, who saw their axles and wheels imperilled. This hostility seems to have set back the adoption of the tram in Britain by several years, as it had in Paris a decade earlier.

The deployment of a narrow-grooved rail flush with the road surface, of a type increasingly preferred in US cities during the 1850s, led to the tram's successful revival in Liverpool in 1868 (see Figure 4.5), and in London in 1870 (see Figure 4.6). The Tramways Act of that year led to the proliferation of horse tramways in Britain; the Act limited a private company's right to operate a tramway to twenty-one years, a clause that would facilitate local authority control of urban transport. The greater efficiency of the tram, due to reduced friction, enabled a pair of horses to draw forty-eight rather than the twenty-five passengers of the omnibus; the tram also offered a smoother ride, and safer mounting and alighting. Its clientele was mainly middle class, though there was a twopenny workmen's return fare on certain lines.

The choice between road and rail was not the only technological issue in urban transport. More fundamental was the shift from animate to inanimate motive power – the replacement of animal traction, with its roots in prehistory, by various engines. A caveat is needed here: for much of the post-1870 period, horse-drawn transport remained viable. Indeed, there were three times as many horse-drawn vehicles in 1900 as in 1830, partly because they made up losses from the railway over the long haul by picking up increased short-haul traffic from railway termini. In many instances, horse-drawn transport complemented rail transport, rather than competed with it; Michael Thompson estimated that the number of horses 'not on farms' rose from about 487,000 in 1811 to a peak of 1,766,000 in 1901 (Thompson, 1976, p.80). Nevertheless, various attempts were made to replace horses, whose dung was a growing nuisance in late Victorian streets, as well as a threat to health. In addition, the exploitation of these beasts, many of whom were literally worked to death on the streets, was offensive to some Victorians; but perhaps there was more concern among the influential that horses, including their feeding and stabling, accounted for more than half the tramway companies' costs.

For road transport, various mechanical alternatives to the horse were tried over the years, including compressed air and clockwork. The first had been steam power; indeed, the immediate predecessor of Richard Trevithick's first steam locomotive was a road vehicle. New steam carriages were designed in the late 1820s; they were tried out in London in the 1830s, but met with objections: they frightened horses, and were too heavy for contemporary roads. In Britain at least, the issue may not have been purely technological; some have seen as decisive the punitive tolls levied by road authorities dominated by coaching interests (Beasley, 1988). The 1865 Locomotive Act restricted steam carriages to four miles (6.4 kilometres) per hour in the country, and two miles (3.2 kilometres) per hour in towns, where a man had to walk sixty yards (twenty-seven metres) ahead, carrying a red flag in daylight and a red lantern at night. Used above all for the carriage of goods, there were actually 8,000 such vehicles on the roads in England and Wales in the 1890s, and some 9,000 by the end of the 1920s, before the motor lorry drove them off the

roads (Barker and Gerhold, 1993, pp.74, 85). Despite the inefficiency of the heavy steam locomotive for urban transport, and objections to the steam carriage, steam trams were eventually introduced in some industrial areas of country, especially in the Midlands, where there was less resistance to their smoke and noise. Other examples were the Wolverton–Stony Stratford line in what is now the Milton Keynes area, and the Swansea and Mumbles tramway (see Figure 4.7).

Figure 4.7 Steam tramway locomotive drawing a Swansea and Mumbles train in 1877. The separate 'dummy' locomotive (with false sides to make it resemble the passenger-carrying trailer cars) was typical of British steam tramways. Combined engines and cars operated in many cities on the European mainland, including Paris (reproduced from Klapper, 1961, plate 10)

The most successful application of steam power to the tramways was through a moving cable, set in a conduit in the road and powered by a stationary steam-engine. For cities that were big enough to warrant the investment, and had the necessary straight streets, cable trams attained speeds of more than twenty kilometres per hour, at half the cost of the horse tram, thereby doubling the distance people could journey to work. By the early 1890s almost every major city in the USA had a cable-tramway system. They were, however, less viable in Britain. The cable tram in Highgate opened in 1884, and another ran from Kennington to Streatham in the 1890s. It was more successful in hilly Edinburgh (Hume, 1983), but was destined to be confined to such specialized niches (see Figure 4.8). With the possible exception of the monorail, the cable tram has been seen as 'the most inflexible form of transport ever devised' (Hilton, 1969, p.125). It could be dangerous, too, as it needed to traverse any curves on the route at maximum speed, and if the grippers on the cars became fouled in the rope-and-steel cable, the cars would run out of control.

The cable and steam trams were soon challenged by an even more cost-efficient competitor, based on the electric motor. This new power source was probably the first innovation based on nineteenth-century theoretical science that had a direct bearing on urban morphology. The physical principle underlying the electric motor was demonstrated by Michael Faraday in 1821, and in 1831 he succeeded in generating a continuous current of electricity (see Figure 4.9 overleaf). But it took several decades of research and development before any motors or generators could be applied for profit.

Figure 4.8 The change at Joppa, from open and covered Edinburgh cable cars, in the foreground (the 'ship's wheel' for working the grip can be seen on the covered car), to an electric tram of the Musselburgh and District Company, on the far right (reproduced from Klapper, 1961, plate 12; © photograph courtesy of Lothian Region Transport plc)

Figure 4.9 Faraday's demonstrations of the principles of the electric motor (1821) and of the electricity generator (1831).
Left Faraday's apparatus for converting electricity into mechanical rotation. On the left a magnet floating in mercury and tied at the bottom of the vessel rotates around the fixed current-carrying wire (the conductor). On the right the conductor is free to move and rotates around the fixed magnet. *Right* production of a continuous electric current by rotating a copper disk between the poles of a powerful permanent magnet (reproduced by permission from Bowers, 1982)

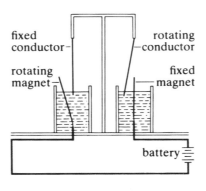
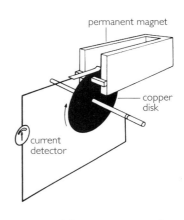

A crucial advantage of an electric tramway over a cable system was the much greater efficiency with which the mechanical energy produced by a steam-engine was conveyed to the tramcar. Instead of turning a cable drum, the energy from the steam-engine was converted into electrical energy by a central generator, and transported by wire to the tramcar's electric motor, which converted it back to mechanical energy (the rotation of the wheels). Electric trams were introduced by Werner von Siemens at the Berlin Industrial Exhibition in 1879, and in 1881 Siemens opened the first commercial electric tramway in the Berlin suburb of Lichterfelde. The first Siemens trams obtained their power from potentially dangerous live rails, though a system of overhead wires was soon introduced; however, it was tramway pioneers in the USA who developed the trolley wheel, a more efficient method of collecting current from an overhead wire.[1] Capable of an average sixteen kilometres per hour, the electric tram was more flexible than the cable car, because a single central generator could power a grid pattern of tramways in a way that was inconceivable using moving cables. Most horse and cable tramway systems in the USA were electrified during the 1890s. Blackpool opened the first British electric street railway in 1885, using an underground conduit, and the first overhead trolley line in Britain was the Leeds–Roundhay Park route, opened in 1891 (see Figure 4.10). But the adoption of electric traction was generally slower in Britain than in other parts of Europe and the USA, and really got under way only in the early twentieth century (McKay, 1976). The reason usually given for this lag is that the 1870 Tramways Act allowed for local authorities to purchase tramways from private companies at cost after twenty-one years. It happened that many of these leases were due to expire during the 1890s, when electrification was a technological option; private concerns declared themselves unwilling to make the large investment needed because of the imminent prospect of municipal take-over. It has, though, been argued that the horse tramways were insufficiently profitable to have raised the capital for electrification, and that blaming the clause was a 'public relations exercise' (Ochojna, 1978, p.138). There was also the question of resistance in many historic European cities to what was seen as an unsightly profusion of overhead wires (see Figure 4.11); this led to the adoption of expensive underground-conduit systems in central London, Paris and Berlin (see Figure 4.12). But once the horse tramways had been electrified, a fare of one penny became economic, and for the first time was affordable by the working-class commuters. By 1913, there were more than 12,000 electric trams, and over 3,300 million passenger-journeys in Britain (Munby, 1978, pp.297, 342); evidently, a considerable latent demand was now being tapped (Sutcliffe, 1988).

---

[1]   Tramway innovations in the USA are discussed in Roberts and Steadman (1999), the third volume in this series.

Figure 4.10 Overhead wires on a single-track branch of the Leeds–Roundhay Park route. The overhead equipment and rolling stock was supplied by the US Thomson–Houston Company. The copper conductor was suspended from transverse steel wires stretched between steel poles. The single-deck car seating twenty-two passengers was standard for US tramways; British lines usually ran double-deckers (reproduced from Clark, 1992, plate IX)

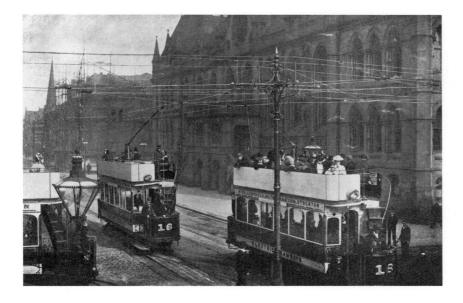

Figure 4.11 Middlesbrough, Thornaby and Stockton Electric Tramways, showing overhead wires suspended from central standards (photograph: National Tramway Museum, Crich; © London Transport Museum)

Figure 4.12 Underground-conduit traction in front of the Brandenberg Gate, Berlin. By 1910, tramway companies had succeeded in overcoming most objections to overhead equipment in the centre of the German capital; this was the only stretch of Berlin's tramways that remained free of wires. In a number of European mainland capitals, including Berlin and Paris, cars had both trolley and conduit contact mechanisms in order to cope with both systems on the same line (McKay, 1976, pp.100, 124; photograph reproduced from ibid., p.103)

The very largest and most congested cities considered a solution more radical than embedding rails in their streets. The alternative was to take the rails above or below street level. Early steam railways in London were often carried on arches through congested districts, but elevated railways were generally opposed in Europe; they were noisy, unsightly and had a potentially devastating effect on property values. There was also the inconvenience of climbing to stations. The Liverpool Overhead Railway, opened in 1893, is notable not only as an exception to Europe-wide resistance to elevated railways, but as the first railway above ground to use electric traction.

Because of its massive capital costs, the underground railway solution was and remains most likely to be adopted in a large, rich city with high-density residential areas separated from its commercial and business district by water, or other natural barriers. London's Metropolitan District Railway of 1863 (described in Chapter 3) had no immediate imitators; the use of deep tunnelling through the London clay, and of electric traction, came much later, with the City and South London Railway, opened in 1890. The proliferation of London's underground electric railways in the early years of the twentieth century was partly stimulated by an innovation due to the US tramway engineer Frank Sprague, who in 1897 demonstrated multiple-unit operation;[2] this allowed a multi-car train, each car with its own motor, but all under the control of a single operator (the 'motorman'). With the very efficient stopping and starting this innovation provided, stations could be located less than a kilometre apart. The first underground railway elsewhere in Europe was opened in Budapest in 1896, and was followed in 1900 by the first line of the great Paris Métro network. In time electric traction came to be seen as the most efficient power source for surface railways, starting in London where they had been long used for commuting.

The decade of the 1870s saw the birth of the power unit that has so dominated transport in the second half of the twentieth century. The internal-combustion engine, whether fuelled by gas, or by petrol or diesel, was not as revolutionary in principle as the electric motor. It was a linear descendant of the steam-engine, with the reciprocal motion of pistons being driven by the combustion of a fuel–air mixture instead of by the expansion of steam. But it had even more far-reaching consequences. Apart from motorized road vehicles, heavier-than-air modes of aviation were unthinkable without the increased power-to-weight ratio first afforded by the internal-combustion engine. The first such engine, the Otto stationary gas-engine of 1876, served as an industrial prime mover, but it was not long before the engine's transport potential was realized. The first commercial motor cars appeared in 1885 in Germany: that of Gottlieb Daimler was powered by a gas-engine; Karl Benz's vehicle, a three-wheeler, used petrol. Benz was the first to produce cars in any quantity (see Figure 4.13). The first British motor car was built in 1895, but it was several decades before it began to affect the shape of cities.

The motor bus was a different matter: it made its first appearance at the very end of the nineteenth century in London. In Britain, it soon began to augment and in many instances supplant electric traction in urban transport systems. By 1913, after some initial mechanical problems had been sorted out, 3,000 motor buses were in operation in London, and most other major cities were using them on routes not covered by electric trams. By 1914, there were 51,000 buses, coaches and taxis (Barker and Gerhold, 1993, p.82) (see Figure 4.14). The development of buses in Britain really got going during the 1920s, with vehicles seating seventy passengers appearing by the decade's end. Although tram-ridership reached its absolute annual peak of 4,140 million in Britain in 1927–8, this figure had probably already been overtaken by the number of bus

---

[2]   For Sprague's own account of this innovation, see Roberts and Steadman (1999), the third volume in this series.

Figure 4.13 The Benz Velo, Karl Benz's first four-wheeled vehicle, produced between 1891 and 1893. It was the first inexpensive passenger car and the first motor vehicle to be produced using standardized parts. The driver is Benz's daughter, Clara (© Mercedes-Benz AG Classic Archives, Stuttgart)

Figure 4.14 Early motor buses, dating from around 1905. In the foreground is a charabanc, the precursor of the motor coach: it was a long vehicle with transverse, forward-facing seats, mainly used for excursions (© The National Motor Museum, Beaulieu)

journeys; certainly by 1931, when statistics for all bus and coach journeys become available, there were more than 5,000 million passenger-journeys, and these were to grow to a peak of 13,544 million in 1955 (Munby, 1978, pp.296, 303–4). Buses affected the relationship between town and country more systematically than did the fixed-rail systems. Bus routes were extended into the countryside after the First World War, enabling villagers to shop in towns, to visit cinemas, and to attend larger schools: in other words, to be drawn into urban ways of life. Buses had distinctive implications for urban form, mainly because of their much greater flexibility of routing; after the First World War, they took over from trams as the vehicle of suburbanization in the country as a whole.

During the 1930s, city transport managers embarked upon the wholesale replacement of trams with motor buses; this is a very controversial issue among transport enthusiasts, many of whom still lament the passing of the tram, and would argue that the scrapping of the investment they represented was an act of official vandalism. The trolleybus rather nicely captures the ambivalence of the issue; it was a design compromise, extending the life of the electric tram's overhead-supply installations, and using rubber wheels to avoid further investment in track (see Figure 4.15). Among the first trolleybus

Figure 4.15 Trolleybus at Hampton Court, 1953. The first trolleybus service in London, between Twickenham Junction and Teddington, began in 1931, and by 1939 there were more trolleybuses than trams running in London. This model was somewhat wider than normal at eight feet (2.4 metres), and was primarily intended for the South African market (reproduced from Barker and Robbins, 1974, plate 95; © London Transport Museum)

services were those begun at Bradford and Leeds in 1911, and at Middlesbrough in 1919. More passenger-journeys were made by trolleybuses than by trams in Britain during the 1950s and 1960s, though both modes were by then in decline (Munby, 1978, pp.303–4); not surprisingly, trolleybuses proved a temporary expedient in the transition to a motorized society.

In Britain, the transition was gradual. The well-heeled early patrons of the motor car had been rewarded in 1896 by the Locomotives on Highways Act, which repealed the red-flag legislation of 1865, and raised the speed limit to twelve miles (nineteen kilometres) per hour. But it was not until the 1930s that the motor car began to strain the fabric of European cities: there were more motor cycles than cars in Britain between 1920 and 1924, as there were in Germany up to 1938. Motor-cycle ownership in Britain rose from 288,000 in 1920 to 724,000 in 1930, though by 1939 it had fallen to 418,000 (Barker, 1993, p.159). Motor cycles were used by the better-off among the working classes, and could serve as a cheap form of private passenger transport: there were pillion riders, and sidecars. However, they were used mainly for pleasure, and were often licensed only for the summer months. Journeys to work were usually made on foot, by public transport, or by bicycle (employers during the inter-war period being unlikely to provide parking space for motor cars). The bicycle, in its reliable chain-driven form designed by J.K. Starley in 1885, and with Dunlop's pneumatic tyres from 1888, had initially sparked a middle-class leisure craze (see Figure 4.16); but by the inter-war period, mass-production

Figure 4.16 Cyclists in Hyde Park, London, 1895 (photograph: Mary Evans Picture Library)

techniques and rising incomes had brought the machine within the reach of the working classes, and it became an important means of urban transport.

In Britain, all modes of public transport (as well as private alternatives to the motor car) began to decline during the 1950s, as car ownership rapidly took off. The motor car is situated at a junction in the history of industrial societies: the transition from a society typified by a sense of community based on occupation, to one in which individuality as expressed by consumer goods is the main source of identity. But although it is a supremely private mode of transport, the motor car is heavily dependent on the public provision of a road network. A big early problem was that the car's pneumatic tyres drew dust and stones from roads; a British solution was Tarmac, a brand name for a kind of

Figure 4.17 Gravelly Hill interchange ('Spaghetti Junction'), linking the centre of Birmingham with the M6 motorway, under construction. Opened in 1972, this link was one aspect of 'a frenetic period of road-building, fully reflective of the auto-oriented culture of the city' (Cherry, 1994, p.201; photograph: Aerofilms)

tarmacadam, or tar-coated aggregate, invented in 1902 by a Nottinghamshire county surveyor (Barker and Gerhold, 1993, p.80). The first European road specifically for cars was a four-lane speedway of nine and a half kilometres in Berlin's Grunewald (Green Forest) opened in 1919. The Nazis followed this up with their country-wide network of concrete *autobahnen*, constructed between 1935 and 1941. But it was not until the second half of the twentieth century that investment in roads came to match in its direct urban effects the investment in fixed-rail systems in the nineteenth. The first British motorway was the Preston bypass of 1958; the first section of the M1 motorway was opened in 1959. Over the following decades, the urban fabric has been reworked to accommodate high-speed junctions, flyovers, slip roads, bypasses and inner ring-roads (see Figure 4.17). Urban streets designed in 1870 for pedestrians and horses continued to be transformed, as petrol stations, car parks, traffic lights, roundabouts, zebra crossings and street signs abounded. One response of the British government to escalating urban problems caused by widening car ownership was the commissioning of the Buchanan Report, *Traffic in Towns*, which appeared in 1963. Partly as a result of the Buchanan Report, the inequality of the contest between motorist and pedestrian began to be recognized in urban planning, with underpasses, and latterly walkways over streets, and with pedestrianized city centres and enclosed shopping arcades (see Figure 4.18).

Figure 4.18 Basingstoke, north Hampshire, one of several towns expanded after the Town Development Act of 1952, which provided for the accommodation of 'overspill' population from the clearance of former slum areas in large cities – in this case London. The multi-level town-centre extension, opened in 1969, was built on a large platform spanning the Loddon Valley, and included areas reserved for pedestrians, such as this shopping centre. Space for delivery vehicles was provided below the pedestrian level, and motorist shoppers and employees gained access from multi-storey car parks (Burke, 1971, pp.171–3; photograph: Robert Brown)

Among other easily visible effects of these transport innovations on the fabric of cities were changes in the design of houses and neighbourhoods. At the beginning of the nineteenth century, London was noted for its noble estates, with big houses set around a central square, the less well-off living in adjoining streets, and stables and the poor concealed in the back streets: this was a configuration built around a set of social and technological relations, including the exclusive ownership by the rich of horse-drawn transport. The development of public transport allowed people to live in more remote suburbs, as it dispensed with the need to provide for horses: this was the origin of the detached and semi-detached suburban villa, without stables or mews. There were further changes in middle-class house design in the first half of the twentieth century. A general reduction in size occurred as domestic appliances took the place of increasingly scarce domestic servants; and as car ownership became more widespread, there was a demand for a garage to be incorporated into the house. In the post-war period, the need to accommodate the car was integral to the move away from the terraced house that opened directly on to the street, to the planned housing estate, in which houses were set back from roads to make space for cars at the expense of social interaction. This change has inspired conflicting reactions. For some, the motor car not only reinforced the breakdown of a sense of community fostered by pedestrian streets and urban public transport, but also played its part in the breakdown of communication within communities. For others – notably the US planner Melvyn Webber,[3] whose ideas influenced the design of the new city of Milton Keynes – one kind of community was replaced by a more dispersed kind held together by modern transport and communications technologies (see Chapter 8).

The wider availability of public and then private transport underlay a revolution in retailing, which left its mark on the form and fabric of European cities. Department stores and chain stores began in the 1860s, but real growth occurred after 1870; department stores such as Whiteley's and Harrods, and multiples such as Sainsbury's and Thomas Lipton, sought economies of scale, which depended partly upon the ability of buyers to spend on their own transport in order to make savings in purchases. The large department store required a dispersed customer base made possible by urban transport; indeed Whiteley's, the best-known London store of the second half of the nineteenth century, ran its own omnibus service for a while (Barker and Robbins, 1963, p.202). In the late twentieth century, supermarket retailing presupposes private motor transport, both for supply and for sale: the movement of goods by lorry has contributed to the relocation of warehouses and stores to the urban periphery; and, at the consumer end, the additional transport cost borne by the buyer has to some extent been balanced by the retailer's provision of ample free parking space on relatively inexpensive land. A seemingly patternless edge-of-town distribution of superstores and retail warehouses, as well as industrial parks and offices, has been one result of the motorized society (Haywood, 1996); there has been a corresponding decline in traditional city-centre retailing.

These quite specific changes in the urban built environment raise a more general question about the link between transport innovations and urban morphology, from specific features such as street layout, block form and plot size, to general models of urban form. Two of the most influential models are those devised by the US sociologists Gideon Sjoberg and E.W. Burgess. The Sjoberg model describes the compact pre-industrial walking city, with the rich and powerful at the centre, and poorer residents on the periphery; the Burgess model depicts the modern two-part city, with a central business district,

---

[3]   There is an extract from Webber's writing in Roberts (1999), the Reader associated with the third volume in this series.

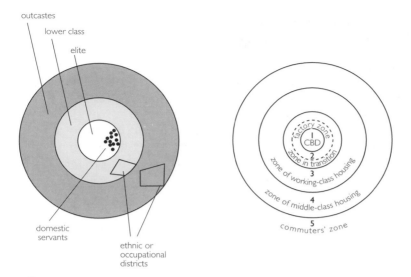

Figure 4.19 The Sjoberg (left) and Burgess (right) models of urban form. In the Sjoberg model, the political and religious elite occupy the city centre, around which live a lower class of merchants and artisans, sometimes concentrated in occupational districts. On or near the extremities are the outcastes: slaves, and minority ethnic and religious groups (adapted with permission from Short, 1984, p.9). The Burgess model focuses on the central business district (CBD), where retailing, financial and administrative buildings have largely replaced elite residences. Next is a 'zone in transition', with an inner belt of factories and businesses, and an outer belt of deteriorating housing, often occupied by immigrants. Beyond this are mainly residential rings, the character of which is determined by what their inhabitants can afford for both accommodation and transport (adapted with permission from Johnson, 1972, p.171)

surrounded by concentric rings of suburban residential and industrial development, this time with the richest living furthest away from the centre (see Figure 4.19).[4] Not surprisingly, Burgess has been criticized for over-simplifying the form of modern cities; British historians in particular have raised doubts about the applicability of his model outside North America, and about the specific contribution of transport innovations to the development of the two-part city (Cannadine, 1977). The relations between transport and urban form are indeed complex and open to debate. Dense urban settlement, for example, may initially enhance the economic viability of urban transport, even though such transport may in another context be seen as the key to achieving a more socially desirable lower-density urban environment. But for many scholars (especially those from the USA) taking a macro-historical perspective, the development of the modern city is readily explained by successive transport innovations:

> The modern metropolis in both its good and bad features is peculiarly a product of transportation technology, or rather of the use we have chosen to make of transportation technology.
>
> (Rae, 1968, p.299)

> The replacement of rail passenger transport by the automobile had an impact on the urban pattern greater than any which had preceded it … The development of the automobile gave Americans a lateral mobility they had never had before; it initiated a decline of central business districts, relatively in all cities, and absolutely in most.
>
> (Hilton, 1969, p.129)

Each transport mode can be associated with a characteristic pattern of urban growth: the steam railway with its separate nodes of residential growth outside the compact pedestrian city; the various forms of tramway with ribbon development along fixed radial routes; the more flexible motor bus with the filling in of the lines of radial development, and more generally with suburban sprawl (see Figure 4.3, p.129). With hindsight, it seems as though the transition from a fixed-rail, centrally powered urban transport system to one with roving, independently powered units was part of an inevitable move to a motorized society, in which cities become ever more spread-out and less dense.

---

[4]   For discussion of the Sjoberg model, see Chant and Goodman (1999), the first volume of this series; for an extract from Sjoberg's writing, see Chant (1999), the Reader associated with the first volume. For an extract from Burgess's writing, see Roberts (1999), the Reader associated with the third volume.

Consideration of the varied experiences and policies of the nations of Western Europe gives the lie to this simple scenario (Wagenaar, 1992). In some parts of Europe, notably in France and Germany, something akin to the Sjoberg model persisted into the industrial era. There was undoubtedly rapid urbanization in these regions as their populations grew and an increasing proportion chose to live in cities; but the outer suburbs of these cities were more notable for working-class apartment blocks than middle-class houses in low-density developments. What needs to be considered is whether this greater density of settlement on the urban fringe was simply a matter of social preference. Other explanations are that it was due to higher construction costs than in Britain, the high cost of rural land and/or legal obstacles to obtaining it, and the persistent effects of fortification rings. One explanation balances technological change with the contingencies of continental European industrialization: the simultaneous adoption of steam-powered machinery and railways resulted in the rapid emergence of factory belts which, not unlike the fortification rings, formed a barrier to any middle-class exodus. The richest preferred to keep a town house in the centre, with a country residence some distance away (Sutcliffe, 1981, pp.3–4).

In due course, developing transport links led to the removal of much bulk manufacturing from increasingly expensive inner-city locations in all parts of Western Europe, including London and Berlin (see Chapters 5 and 6). This centrifugal movement also led to the location of industry in the Parisian 'red belt' outside the fortifications, and the extramural industrial zones of Milan. Evidently, these varied morphological phenomena cannot be explained by transport innovations alone. And even if transport were the main *technological* cause, the fact that nations had different transport policies needs to be considered; indeed, such differences in approach may shed some light on contrasting urban morphologies. Across Europe there were different mixes of urban transport. Some urban authorities were more supportive of public transport in general than others (Barke, 1992; Morgan, 1986), and there were different fare policies with different implications for urban form; there was greater suburbanization in Amsterdam, where a universal fare was applied, than in Vienna, where fares went up with distance (Capuzzo, 1998). Some, as in Berlin in the 1920s, preferred fixed-rail installations (a mix of trams and underground railways) and the compact, planned peripheral settlements they served; the flexible motorized solutions favoured in many British cities from the 1930s led to a more dispersed urban sprawl. Glenn Yago has compared the USA and Germany, and in particular Chicago and Frankfurt-am-Main, to show how differing economic and political institutions at various levels of society resulted in different transport developments. The decline of public transport took place in both countries, but in different ways because of different forms of corporate power and differences in the timing and form of state intervention; a key part of his explanation is the tradition of public investment in Germany (Yago, 1984).

A specific issue that has exercised British historians is the relation between urban transport and suburban growth. It would be impossible to deny that a close causal link exists between suburbanization and the lengthening journey to work made possible by transport innovations. According to a recent study by Colin Pooley and Jean Turnbull, at the turn of the last century 50 per cent of the British population walked to work, an average distance of five kilometres; between 1920 and 1950, 20 per cent walked, another 20 per cent cycled, and 20 per cent went by tram or bus; by the 1990s, 60 per cent were travelling by car, and only 7 per cent going on foot. Since the 1930s, the average commuting time has remained around thirty minutes, but the distance covered has risen from fewer than eight to more than sixteen kilometres – and to more than thirty kilometres in the London area; also, it is now only in London that public transport brings the majority to work (*Guardian*, 8 January 1998, pp.10–11). A British historian of the railways has taken an ostensibly technological-determinist position on this relationship:

> by the [1870s] the railways had brought a whole new suburban world into being – or rather, two worlds: that of the closely built-up suburb, comprising miles of two-storeyed houses, mainly in terraces … and the outer suburbs, where the houses spread themselves out with sizeable gardens.
>
> (Simmons, 1973, pp.299–300)

Despite Simmons, most British historians, especially those in the H.J. Dyos tradition of urban history, have taken a different line: that transport responded to outward pressures, it did not create them (Dyos, 1961; Simpson, 1972[5]). The largest English towns were suburbanized and socially segregated half a century before the arrival of cheap mass transit; the omnibus and suburban train permitted rather than created the suburb (Cannadine, 1977; Thompson, 1982). For Theo Barker, the doyen of British transport historians, transport in the first half of the nineteenth century was an 'important adjunct' to but not an 'essential element' in London life. In the railway phase, it became the 'prerequisite of continued outward growth'; nevertheless, Barker recognized that universal access to public transport depended on the wider economic context – rising incomes, and shorter working hours, allowing more time to commute. In addition, the near wholesale replacement of residential by commercial functions in the centres of big cities propelled some of the movement of the population into the suburbs (Barker and Robbins, 1963, pp.xxvi–xxx).

In a study of the middle-class journey to work in Newcastle upon Tyne, Michael Barke insisted that the dissociation of workplace and residence was well under way before the first horse trams were introduced in 1878, and before the expansion of suburban railways during the 1880s. It was a development that in no way depended upon innovations in urban transport: he discounted the commuter use of the Newcastle and North Shields Railway, opened in 1839; and although horse omnibuses were introduced in the 1850s, they were moribund by the 1870s. Nevertheless he admitted that an impetus was given to suburbanization by the decision of the Newcastle City Council in 1899 to take over the trams and electrify them, and also by the electrification by the North Eastern Railway Company of the coastal line to Whitley Bay in 1904. Barke's conclusion was that transport did not initiate changes but enlarged them and heightened their significance. He also insisted that housing development preceded the provision of transport in any given suburban extension (Barke, 1991). This was not invariably the case, as some London Underground stations were built before housing (see Figure 4.20 overleaf). A question remains: is it really fatal to the thesis of technological determinism that speculators were prepared to gamble on the provision of transport links?

However this question is decided, a wider consideration of the issue suggests that there may be a peculiarly English preference for suburban living that cannot be reduced to the effects of transport innovations. The social and spatial relations of the Georgian public square and terraced house reflected not only the prevailing form of horse-drawn transport, but also the hierarchies of a society dominated by the landed aristocracy. The Victorian middle class in their turn emphasized privacy for the individual and the family, and so looked favourably on low-density suburban sprawl as a solution to the problems of tearaway urban growth. This attitude found expression in their dislike of formal Regency architecture and the 'brick cancer', and of tall buildings and 'immoral' Parisian blocks of flats, with their common entrances and stairways.

---

[5]   An edited version of Simpson's article can be found in Goodman (1999), the Reader associated with this volume.

Figure 4.20 *Top* Burnt Oak Station in 1924. The fort-like building – actually the London Underground station – was the first at Burnt Oak, over fourteen kilometres from central London. The station is the penultimate on the Northern Line to Edgware, and was deliberately planned by the Greater London Council to attract builders. *Bottom* Burnt Oak in 1987. The footpath across the planned railway line in the old picture is now a major road to the A5 trunk road (Gardiner, 1989, pp.20–21; photographs: Aerofilms)

The suburbs were seen as strengthening family life, and the urban culture of pubs, music halls, clubs, theatres and restaurants as weakening it. Lurid pictures of decadent inner-city life in the nineteenth century were peddled by crusading journalists, a sensationalist press and reactionary intellectuals. This anti-urban ideology persisted in various invocations of *rus in urbe* ('countryside in the city') in the promotion of suburban estates in the twentieth century, despite the objections of socialists such as H.G. Wells, who condemned the suburb as an instrument for the subjection of women (Olsen, 1976) (see Figure 4.21).

Figure 4.21 London Underground poster of 1908 promoting a semi-rural suburban ideal (reproduced from Weightman and Humphries, 1983, p.124; © London Transport Museum)

A final objection to any account of suburbanization that sees it flowing inexorably from transport innovations is that transport technologies themselves may also be moulded by pre-existing social divisions. In Britain, class and gender differences were literally built in to new modes of transport (first-, second- and third-class carriages, workers' trains, 'women only' compartments, 'modesty rails' on tram staircases), as they were to building types (saloon and public bars in public houses). Railway companies sometimes operated cheap fares to concentrate working-class commuters in certain areas, and protect more lucrative traffic; clearly, the social effects of technologies are often decided by those who control them.

Building construction

Ostentatious verticality has always been a feature of cities, from the ziggurat onwards, and the desire to erect the tallest structure yet has continued to animate architects and their city government sponsors in the post-1870 period. The obvious example is Paris's Eiffel Tower, built in 1889 of wrought iron. It marks a turning-point in the ways the social relations of technology were expressed in the urban built environment: it was the first such city virility symbol to express not religious or political power, but engineering expertise. A more routine verticality marks the twentieth-century city, in which functional buildings on confined urban plots have been raised well beyond the usual five- or six-storey limits of the mid-nineteenth-century tenement, factory, hotel or department store. Apart from building regulations, there were two main considerations limiting the height of such structures. First, there was a practical and economic limit: people were only prepared to mount so many stairs to get to upper floors; hence, the further a storey was from the ground, the lower was the rent. The second was primarily technological, though again with economic implications: the entire weight of brick or stone buildings was transmitted through the walls to the foundations; the higher the building, the thicker the walls had to be on the lower floors, and the less economic the use of this prime space.

A solution to the first problem was that vertical transport innovation, the lift or elevator. The first safe version, incorporating an automatic brake in the event of the rope's failure, was invented in the USA by Elisha G. Otis in 1852. From the 1870s, elevator blocks in traditional materials reached up to nine and more storeys. The solution to the second problem was the development of the metal-framed building. This was a structural innovation of mixed US and European parentage. Cast-iron columns were introduced from the early nineteenth century in the attempt to fireproof British textile mills (see Chapter 1), and cast iron and wrought iron were increasingly deployed in European and North American buildings. They were complementary, in that cast iron, strong in compression, could be used for vertical columns, and wrought iron, strong in tension, was suitable for horizontal beams. In fact, buildings incorporating these materials never realized the full potential of the metal frame to bear the full load of a building. There is no fundamental technological reason for this; it happened that at a time when financiers and builders in the USA were looking to maximize the returns from valuable inner-city plots, a new construction material came on to the market. This was bulk steel, which combined most of the advantages of wrought and cast iron, and was cheaper than wrought iron and traditional steels. Bulk steel covered a variety of materials, the carbon content of which ranged from a point intermediate between that of wrought iron and cast iron, to a material with as little carbon as wrought iron (mild steel).

The structural applications of wrought iron had been limited by the capacity of the puddling furnace. Puddling was a skilled, labour-intensive operation made necessary by the physical fact that pure iron melts at a

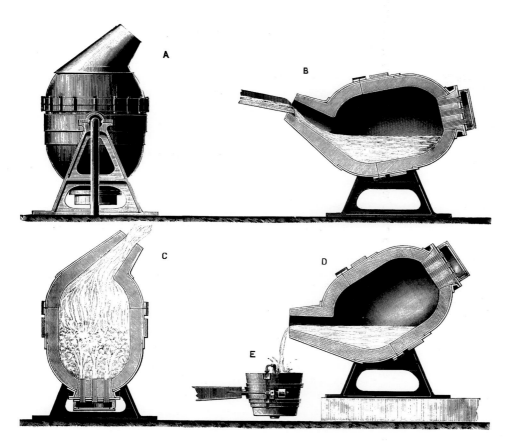

Figure 4.22 Bessemer's first tilting converter. The refining process was dramatic and violent, lasting less than thirty minutes: B is the position for running in the charge of molten cast iron, C for blowing cold air through the charge and D for pouring out the refined metal. The metal was then carried in the ladle, E, to ingot moulds (adapted from Bessemer, 1905, plate xv, figure 43)

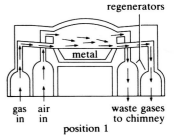

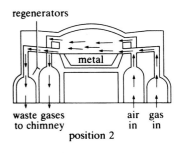

considerably higher temperature than the carbon-rich form (cast iron) which issues from the mixture of iron ore and coke in the blast furnace. Cast iron had to be refined, or decarburized, to make wrought iron, a material with more tensile strength; but in the temperatures obtainable in the puddling furnace, the metal gradually solidified as it was worked. The secret of producing bulk steel was the achievement of much higher temperatures in the refining process, thereby keeping the metal molten throughout. In the converter invented by Henry Bessemer in Britain in 1856, such temperatures resulted from the simple expedient of blowing cold air through a charge of molten cast iron; 'exothermic' reactions – chemical reactions that give off heat – between the oxygen in the blast and elements in the cast iron actually raised the temperature of the charge (see Figure 4.22). The more controlled and versatile open-hearth furnace was invented in 1864 by the Germans William and Charles Siemens (brothers of Werner, the electrical pioneer); in this case, extra heat was gained by recycling waste gases from the primary combustion of the furnace fuel (see Figure 4.23). Tonnes of refined metal could now be poured into ingot moulds, dispensing with the need to weld together lumps of puddled iron; the product was purer, more consistent in quality, and with much greater all-round strength. Further down the production line, a mill for rolling structural-steel girders from steel ingots was invented by the Englishman Henry Grey in 1877 (see Figure 4.24 overleaf). A frame made from these girders was strong enough to bear the entire weight of the roof, floors and internal partitions. All that was required on the outside were 'curtain' walls – walls with no load-bearing function, serving solely to keep out the weather and to let in light. The patenting in Germany in 1901 of a continuous process for drawing sheet glass from a Siemens-type regenerative furnace opened the way to the manufacture of glass components strong and large enough for curtain walls. The steel-and-glass

Figure 4.23 Principle of the open-hearth furnace. This was a much more controllable set of reactions than those in the Bessemer converter, and longer in duration (typically some fourteen hours). In position 1 the flow is from the left-hand firebrick regenerators, which heat the coal-gas fuel and air, through the furnace to the right-hand regenerators, which extract heat from the waste gases. In position 2, the flow has been reversed; the right-hand regenerators give up their stored heat, and those on the left-hand are being reheated (reproduced from Chant, 1989, p.141; © The Open University)

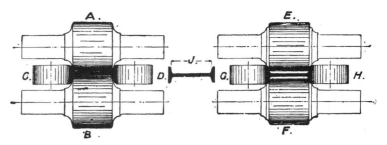

Figure 4.24 Grey's mill for rolling girders. Traditional rolling mills for the production of iron or steel bars had horizontal rolls with a succession of grooves that after a number of passes gave the desired cross-section. Grey's mill was an adaptation of the universal mill, in which an additional pair of rolls in the vertical plane supplemented a flat horizontal pair, an arrangement allowing steel plates to be rolled. Grey's mill consisted of two stands, one in front of the other. The thickness of the girder's web was determined by the horizontal rolls A and B in the first stand, and the thickness of the flanges by the faces of the vertical rolls C and D, and the ends of rolls A and B. The width of the flanges was governed by the horizontal rolls E and F in the second stand; the vertical rolls G and H prevented the flanges from being turned over by rolls E and F (reproduced from Harbord and Hall, 1918, p.635)

towers that have dominated cityscapes in the second half of the twentieth century owe much to the float-glass process, invented by the British firm Pilkington in 1952, in which a perfectly plane bath of molten tin shapes the underside of the glass sheet. Since the 1960s, developments in mountings, and in reflective materials reducing the build-up of solar heat, have led to façades made entirely of glass.

An alternative to bulk steel in combining the strengths of cast and wrought iron was a composite material: reinforced concrete. Its combination of concrete and iron was not in itself new; iron tie-bars were sometimes used to strengthen ancient Roman vaults, and iron or steel columns and beams were encased in concrete for fireproofing purposes in some late nineteenth-century industrial buildings (see Chapter 1). But in reinforced concrete, the marriage of the materials was consummated; tensile strength was added to the compressive strength of concrete by reinforcing bars, rods or wires, initially of wrought iron, and then of the new bulk steel. The first patents for the material date from the mid-nineteenth century, and a reinforced-concrete house was built by the plasterer W.B. Wilkinson in Newcastle upon Tyne as early as 1865. European engineers, above all in France, Germany and Denmark, took the lead in the material's development. These were regions in which phosphorus, originating from the available iron ores, impaired metal produced by the new refining methods, though a solution to the problem, in the form of special linings in the Bessemer converter and the open-hearth furnace, became available from the late 1870s. Reinforced concrete was first used for foundations, and then as an alternative to the metal frame. But from the 1890s, French engineers, starting with François Hennebique and followed by Auguste Perret and Eugène Freyssinet, began to demonstrate the remarkable plasticity of the material. Hennebique vigorously promoted his own system of reinforced-concrete construction, licensing it in Britain and elsewhere in Europe (Cusack, 1987). Fittingly, one of the first reinforced-concrete buildings in Europe was itself a new building type: the Garage Ponthieu in Paris, designed in 1905 by Perret (see Figure 4.25); and Freyssinet's great work was a huge, vaulted airship hangar at Orly on the southern outskirts of Paris (Hitchcock, 1958, p.314). Reinforced-concrete components could either be pre-cast in factories, or prepared on site, using wooden formwork. The material could take on shapes impossible to achieve with rolled steel. In time, new structural units such as slabs, shells and thin ribs liberated architects from the cellular regularities of the steel frame; their potentials were displayed to striking effect in structures such as bridges and sports stadiums (Addis, 1997).

Figure 4.25 Garage Ponthieu, designed by Auguste Perret, Paris, 1905–6. Perret regarded this garage, with its clearly stated concrete frame and glazing infill, as the first attempt in the world at a 'concrete aesthetic' (Abram, 1987, p.86). The building had a central nave with two multi-storey galleries, to which cars were carried by a lift. The geometric rose window indicates the height of the nave. Office space was provided on the fourth floor, as shown by the frieze-like series of vertical windows beneath the cornice (photograph: F. Stoedtner/ Bildarchiv Foto Marburg)

These new materials, and the other new technologies of city-building, were embraced with fervour by Modernist architects. Modernism was European in origin, though after it was taken up by US architects in the 1930s, it was known for a while as 'the International Style'. For the first time since the concrete architecture of the Romans, and the Gothic structures of the medieval master masons, radically new approaches to European public buildings were embraced, but this time in the name of technological innovation. Rejecting the highly ornamented styles of the late nineteenth century, early Modernists advocated functional designs for mass-produced objects and industrial buildings; these were pioneered by the group known as the Deutscher

Werkbund, founded in 1906. Its most influential member was Peter Behrens, who designed some buildings for the German electrical giant AEG that are now seen as landmarks in the history of industrial architecture (see Figure 4.26).

Another member of the Werkbund, Walter Gropius, became the first director of the Bauhaus, a school of design, building and crafts founded at Weimar in 1919, and generally regarded as the heart of Modernist design and architecture. After an initial approach that had much in common with the anti-industrial Arts and Crafts movement in Britain, the Bauhaus turned to machine-age designs in its quest to better the world; its mature emphasis on plain surfaces and geometrical shapes has inspired much of twentieth-century public and corporate architecture. A classic early example of Gropius's work is the Fagus factory (see Figure 4.27). The Bauhaus championed in particular the application of mass production and standardized parts to working-class housing, and it influenced the design of many post-war European housing projects consisting of uniform 'slab blocks' of apartments surrounded by open spaces. (Slab blocks are multi-storey structures greater in length than in height, like a tower block on its side.) This 'super-block' (*Zeilenbau*) approach was also identified with the socialist architect Ernst May, who oversaw the construction of 15,000 such blocks on the fringes of Frankfurt-am-Main between 1925 and 1933 (Mullin, 1977, p.10) (see Figure 4.28). The Bauhaus's third director, Ludwig Mies van der Rohe, took the quest for simple lines and exact geometric proportions to its limit, and after the disbanding of the Bauhaus by the Nazis, he moved on to become the most influential post-war designer of North American corporate steel-and-glass skyscrapers.

Figure 4.26 AEG turbine factory in Moabit, an industrial suburb of Berlin; building designed by Peter Behrens, 1909–10. The use of an unadorned poured-concrete façade, exposed steelwork and expansive fenestration was remarkable for its time (photograph: F. Stoedtner/ Bildarchiv Foto Marburg)

Figure 4.27 Fagus factory, Alfeld-an-der-Leine, 1911–14, designed by Walter Gropius and Adolf Meyer. This factory, which made shoe lasts, has been called 'one of the most important buildings of the twentieth century' (Hitchcock, 1958, p.365). It went much further than Behrens' turbine factory in the direction of a fully glazed curtain wall; among its novel features was the omission of the building's narrow brick piers at its corners (photograph: W. Gropius/ Bildarchiv Foto Marburg)

Figure 4.28 Siedlung Westhausen, Frankfurt-am-Main, in the 1920s. Westhausen was one of the first *siedlungen* (settlements) planned for land on the fringes of Frankfurt, and the first to be laid out entirely on the basis of May's *Zeilenbau* ('super-block') concept. It represented May's main goals: to apply industrial methods to produce cheap, functional, standardized housing (based on the use of prefabricated reinforced-concrete panels); and to set such housing in a healthy, spacious, 'green' environment, protected from through traffic. Care was taken to orient the blocks to obtain the maximum benefit from sunlight (photograph: Frankfurt Stadtarchiv)

Probably the best-known exponent of Modernism was the Swiss architect Charles-Edouard Jeanneret, known by his pen-name, Le Corbusier. It is noteworthy that the building many reckon to be his greatest, the *Unité d'habitation*, built in Marseille between 1946 and 1952, was a functional self-contained reinforced-concrete apartment block (see Figure 4.29). Le Corbusier was a fierce advocate of the machine aesthetic, and among the principal inspirations for his designs were engineering products such as ocean liners, aeroplanes, motor cars and North American grain elevators. Many of the features he insisted on for his buildings were those made possible by modern materials and technologies, and in particular reinforced concrete (as a young man he worked for both Perret and Behrens): the use of stilts (*pilotis* in French), to allow free access beneath buildings in a motorized age; ribbon windows running the length of the building; flat roofs, now that reinforced concrete and central heating had rendered the gabled roof obsolete; and open floor plans, now that load-bearing walls were unnecessary. The persistence of these features in contemporary public and corporate building is evident to all.

For all the technophilia of the Modernists, can it really be asserted that these new technologies drove change in the urban fabric? There are two obvious objections to this notion. One is that to begin with there was considerable resistance in some parts of Europe to Modernist designs, and in particular to steel and concrete. In Britain, for much of the first half of the twentieth century, frames of either material were concealed by traditional brick or masonry façades, and neo-classical styles prevailed for buildings such as banks and government offices. Totalitarian governments in Germany and Italy in the 1930s also celebrated their power with variants of neo-classicism. Many cities enforced strict fire regulations limiting the height of buildings, so that there were no US-style skyscraper-dominated skylines in Europe until late in the century: 'most Americans accepted tall buildings but wanted them fireproof ... Britons forbade tall buildings for the sake of fire safety' (Wermiel, 1993, p.23). Another objection is that non-traditional materials and designs proved particularly unacceptable for certain kinds of building, especially housing, in certain European regions. Le Corbusier famously regarded houses as 'machines for living', and thought his design principles perfectly suited to residential structures; but in many parts of northern Europe, the consumer has remained stubbornly attached to the detached or semi-detached family house in traditional brick, timber and tile, in assorted rustic-vernacular styles.

This disassociation of producers' and consumers' ideals of housing recalls an earlier bifurcation in the morphological history of European cities: the low-rise nature of housing in English and many north European cities contrasted with the high-rise nature of, for example, Berlin, Edinburgh and cities of southern Europe, well before the advent of Modernism. Such a contrast is not simply due to the general availability of new building materials and construction methods. Why were high-rise dwellings always favoured in certain regions of Europe, including Scotland (Gordon, 1990)? Why were they resisted, along with other non-traditional housing construction methods and materials, in English cities until the 1950s, and why were they then embraced, only to be rejected again (Glendinning and Muthesius, 1994; Sutcliffe, 1974)? These issues will be explored further in some of the following chapters.

But the relations between these technological innovations and buildings are not all about increased verticality. Among the outstanding building types of the twentieth century are picture palaces, supermarkets, petrol stations, prefabricated houses and concrete- or steel-framed, electric-powered, scientifically managed factories, such as those that sprang up on the Great West Road in London during the inter-war years (see Figure 4.30). By no means are all of these new, or completely reworked, building types high-rise. This is where transport innovations have to be returned to the causal mix; the

Figure 4.29 *Unité d'habitation*, Marseille, 1946–52. This tall slab block, carried on a double row of central supports, was designed by Le Corbusier to accommodate a self-contained community. A storey halfway up was intended for shops, and other communal facilities were provided on the roof. The finish of the poured-concrete surfaces, and of the prefabricated members of the sunbreaks in front of the apartment windows, was deliberately rough. Further *unités* were built at Nantes (1953–5) and Berlin (1956–8) for the Interbau Exhibition (photograph: Giraudon)

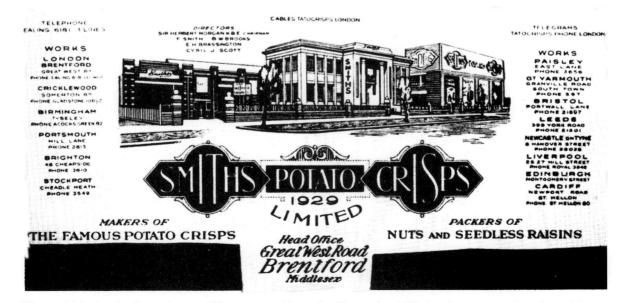

Figure 4.30 Smiths Crisps factory. When it moved from Cricklewood in 1927, Smiths was one of the first of the firms that colonized the Great West Road. It was followed by firms such as Macleans, Firestone Tyres, Curry's Cycles and Radios, and Gillette UK Ltd. The Great West Road became a 'gallery of modern architecture', marked at night by floodlit façades and neon lighting (Weightman and Humphries, 1984, p.59; illustration reproduced from ibid.; © Gunnersbury Park Museum Collection)

move from rail to rubber wheels in the movement of goods has been linked to a shift from multi-storey commercial and industrial buildings in big cities near rail termini, to more rambling single-storey structures, with wide-span loading bays, on cheaper land away from city centres but conveniently situated for the motorway network.

Even so, the availability of new building materials plus transport innovations remain insufficient to explain developments in urban fabric and form. The framework of broad economic and social changes also has to be considered – above all, the rise of the business corporation, with funds to invest in prestigious city-centre buildings. The increasing importance of the service sector also found expression in the inner suburbs, as banks, theatres, cinemas and department stores tended to replace small-scale manufacturing. The final decades of the nineteenth century saw the development of the department store and chain store, reflecting innovations in retailing, such as packaged, brand-name goods and advertising. One of the most obvious differentiating features of cityscapes before and after 1870 is the emblazoning of the latter with various advertising devices, including billboards big and brash enough to be noticed by the speeding motorist. The urban fabric reflected the increasing scale of production and widening markets of building materials and components, factors which tended to diminish the local character of cityscapes; increasingly, national rather than local agents shaped the urban built environment, most obviously in the case of commercial and retailing premises. This raises an important general question that has brought forth opposing responses from historians: did European (and North American) cities from the mid-nineteenth century start to become more alike, as they applied the same array of urban technologies to the same set of urban needs (Cannadine, 1977; Tarr and Dupuy, 1988)?

Electric lighting and electrical communications

As well as its upward and outward growth, one of the most striking features of the twentieth-century European city has been the continuing extension of its activities into the night. Street lighting was far from new, even in pre-industrial cities, and the beginnings of a metropolitan night life date to the gaslights of the mid-nineteenth century (Schivelbusch, 1988; see Chapter 1). But the ubiquity, variety and intensity of the electric light almost literally put any of its predecessors into the shade. The first form of lighting derived from the generator's electrical energy was the arc-light, obtained by making an electric current jump the gap between two carbon electrodes. Urban applications of the arc-light were pioneered by C.F. Brush in the USA during the 1870s, and from 1878 soon transferred to the streets of European capitals (see Figure 4.31). Remarkably, in that same year, the power of these lights was demonstrated in Sheffield at the first floodlit football match. The harsh, flickering arc-light gave way to the incandescent light bulb, invented independently by Thomas Edison in the USA and Joseph Swan in England. Electric lighting in general was slower to catch on in Britain than in the USA, partly because gas was more competitive in price; but by the end of 1903, all but two British cities with a population over 100,000 had at least one central electricity-generating station, and 6–7 per cent of the urban population possessed electric lighting (Byatt, 1979, p.25).

Electric lighting had notable effects on the built environment and the urban landscape: in time, the improving efficiency of interior electric lighting, and the absence of the toxic fumes generated by gaslights, meant that work spaces could be deeper; this trend was accelerated by the development of the fluorescent tube in the 1930s, which greatly reduced the heat generated by a given amount of light (Bijker, 1995). The much reduced fire risk associated with electric lighting also allowed shop-window displays to be illuminated

Figure 4.31 Arc-lights on the Potsdamer Platz, Berlin, c.1880; a painting by C. Saltzmann (photograph: courtesy of Siemens Forum Museum Archive, Munich)

from the inside. Another nineteenth-century scientific discovery, that of the inert gases making up a small proportion of the earth's atmosphere, lent colour as well as brightness to the cityscape. The red neon lamp was invented by the French physicist Georges Claude in about 1910, and was widely used in advertising from the 1920s. Since the 1960s, signs using neon, other inert gases – helium, argon, krypton and xenon – and elements such as mercury, have generally given way to backlit signs of brightly coloured plastic. Initially, the power of the electric light to attract the attention of the consumer was confined to the commercial centres of large cities – places such as Piccadilly Circus in London. But as car ownership has grown, and retailing has accordingly disseminated, these devices have become a standard feature of suburban main streets and shopping precincts, as well as city centres.

The telegraph and the telephone, and subsequently radio and television, represented further exploitations of the most pregnant scientific discovery of the nineteenth century, the interconvertibility of forms of energy. Provided that information, including sound and vision, could be converted into some electrical or electromagnetic equivalent or analogue, and then converted back again, it could be conveyed over great distances virtually instantaneously. The first practical application of electricity was the telegraph, an instrument invented as early as 1837 by Cooke and Wheatstone. The telegraph played its part in the development of the system of cities, both by conveying business information instantaneously, albeit in coded form, and enhancing the efficiency of the railway network. By conveying sound, the telephone acted as a far more effective version of the telegraph. First patented by the Scottish-born Alexander Graham Bell in the USA in 1876, the device played its part in the upward and outward growth of North American and European cities. There was a telephone exchange working in London as early as 1879, and although telephones were expensive compared with their cost in the USA, there were 3.3 million in use in the UK by 1939; in the rest of Europe, only in Germany were there more working telephones during the inter-war period (Mitchell, 1998, pp.765–72).

The other great electromagnetic innovations, the phonograph, radio and television, have made their mark on modern cities because they have become the business of large companies with imposing offices, recording and transmitting installations and retailing outlets in metropolitan centres. Otherwise, their specifically urban effects are less clear. Their products are associated with the demise of many forms of urban street culture, and with the packaging of entertainment for domestic consumption; if there are urban implications in this, it can only be that low-density, suburbanized settlements are encouraged. They have certainly been more technological grist for the 'end-of-the-city' mill, now happily grinding away on the personal computer. There is an exception to this centrifugal scenario: latterly closed-circuit video cameras have joined the repertory of street furniture, in the attempt to make city centres, especially at night, less threatening.

That offspring of photography, the cinema, has been quite another matter. Cinematography is really a mix of chemistry and electricity as a medium of communication: chemistry to fix images and, later, sound on rolls of film; and electricity to project the film and play back the sound. It was another innovation of mixed North American and European parentage: Thomas Edison in the USA patented his kinetoscope in 1889, a system enabling individuals to view images fixed on George Eastman's new celluloid film; the Lumière brothers in France developed the first cinema projector in 1894, and the following year demonstrated the equipment in Paris. Before the arrival of the domestic video recorder, the consumption of feature films was for obvious commercial reasons primarily an urban activity, displacing the nineteenth-century music hall from prime central locations. Cinema-going became the principal leisure activity in the USA and Europe in the years before television; in 1920, there were 5,000 cinemas in Britain, 3,200 in Germany, 2,700 in France and more than 1,000 in Italy. In Britain during the boom years between 1925 and 1945, the audience figures worked out at a daily attendance of some six million (Moore, 1989, p.226). The cinemas themselves became perhaps the most prominent new urban building type of the first half of the twentieth-century. Apart from the cinematographic wizardry they housed, lurking under the ornate façades and interiors with which they seduced their patrons was the steel or concrete frame, and guaranteeing the patrons beneath the balcony an unblocked view of the silver screen was a rolled-steel girder or a reinforced-concrete beam (Sharp, 1969; see Figure 4.32).

Figure 4.32 The Kensington, in Kensington High Street, London, 1926, designed by Julian Leathart and W.F. Granger, reckoned to be the largest cinema in England at the time, with a seating capacity of 2,300. *Top* the auditorium, with its sweeping balcony and neo-classical decorative details, including a coffered ceiling. Lighting came from visible sources; within a few years, concealed and coloured lights became standard. *Bottom* the steelwork structure (reproduced from Sharp, 1969, pp.95, 164; photographs: Julian Leathart)

4.3   The rise of the modern planned city

Any discussion of the relationship between technology and urbanization in this European period must give due weight to the rise of *modern urban planning* as a profession and as an established element of public policy (Sutcliffe, 1980). Planning has been defined as:

> the deliberate ordering by public authority of the physical arrangements of towns or parts of towns in order to promote their efficient and equitable functioning as economic and social units, and to create an aesthetically pleasing environment.
>
> (Sutcliffe, 1981, p.viii)

According to this definition, planning champions certain social goals of equity and beauty, and should therefore be regarded as part of the shaping context of technology. However, the activity of the planner in seeking to achieve the efficient ordering of the physical environment is also compatible with this series' broad definition of technology as 'all methods and means devised by humans in pursuit of their practical ends'. Planning might therefore qualify as a branch of technical knowledge. On balance, it may be best to see it, not unlike architecture, as an area of theory and practice seeking to associate technologically feasible means with politically, economically and aesthetically desirable ends.

One way of exploring the relationship between technological innovation and planning is to examine briefly the technological preconceptions in the ideas of some of the leading theorists of urban planning. Theorists had varying attitudes to the city as it had developed after the Industrial Revolution. From the first impulse in the late nineteenth century, the early decades of modern

Figures 4.33a and **b**
The role of transport in
Howard's Garden City concept

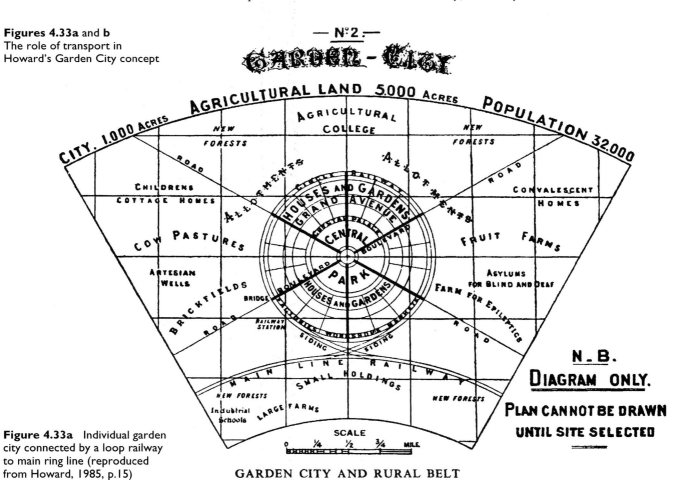

Figure 4.33a Individual garden
city connected by a loop railway
to main ring line (reproduced
from Howard, 1985, p.15)

GARDEN CITY AND RURAL BELT

urban planning were marked by a desire to beautify the city, expressed above all in the writings of Camillo Sitte, the Austrian co-founder of *Der Städtebau* ('City-building'), the first journal of town-planning. Sitte was in favour of asymmetrical and curvilinear streets, and irregular clusters of buildings around squares (Sitte, 1979). An alternative to city beautification, though with a design outlook probably influenced by Sitte, was put forward by Ebenezer Howard in 1898 in his tract *To-morrow: a peaceful path to real reform*, retitled *Garden Cities of To-morrow* in its many subsequent editions. Howard looked for a means of escape from the nineteenth-century industrial city and its profiteering landlords through the establishment of new 'garden cities' founded on the municipal ownership of land (Howard, 1985). Howard's Garden City Association, founded in 1899, was the model for similar movements on the Continent: the Deutsche Gartenstadtgesellschaft (German Garden City Society) was founded in 1902. Howard's aim was to combine the best of urban and rural life by planting a city of 1,000 acres (400 hectares) in 5,000 acres (2,000 hectares) of land, thereby making it virtually self-sufficient. The population of each settlement was to be limited to 32,000. Howard's ideas were hardly driven by technological innovation, though they did depend to some extent on innovations in urban transport: his ideal city included a fringe of factories and warehouses connected by a loop railway, and inter-city links were by electric railways (see Extract 4.1 and Figures 4.33a and b).

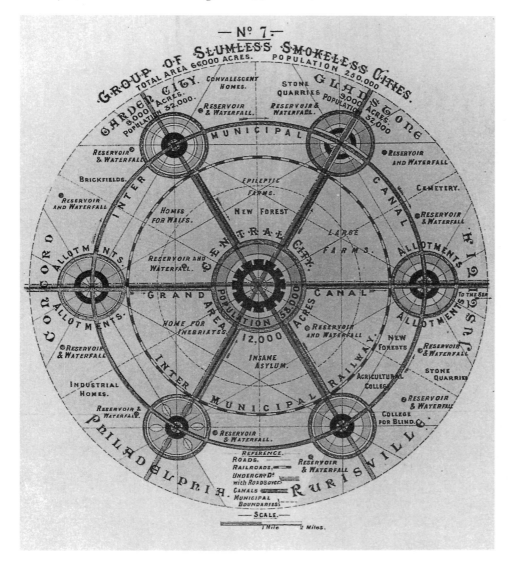

Figure 4.33b 'Group of slumless smokeless cities', a diagram that appeared in the first edition only of Howard's book. It shows an agglomeration comprising an existing city surrounded by new garden cities, connected to each other by roads, rail and canal, in radials, rings and loops (reproduced from Hall, 1992, p.38)

Despite Howard's proposed use of transport innovations, these pioneering planners were clearly not looking for primarily technological solutions to the problems of the modern city; rather, they were looking for solutions to the urban problems *created by* the adoption of modern technologies. They were followed, however, by an approach to planning reflecting attitudes to public administration that merited that 1930s' neologism 'technocracy': in its most hubristic original meaning, this was the belief that society should be governed by scientists and engineers (as Saint-Simon and his followers had advocated in France as far back as the 1820s and 1830s); more insidiously, it is the notion that all dysfunctions in the social system can be corrected by the application of expert, technical knowledge. This conviction was certainly manifest in the fields of architecture and urban planning in the middle decades of the twentieth century. Its best-known prophet was Le Corbusier, whose book *The Radiant City*, first published in 1933, celebrated large cities, and the building and transport technologies that made them possible, and explicitly rejected garden cities as a 'pre-machine-age utopia' (see Extract 4.2). Le Corbusier aspired to sweep away all old buildings and streets, and replace them with skyscrapers and terrace blocks, and wide avenues intended to be 'traffic machines'. The sixty-storey skyscrapers he envisaged for offices and apartments would actually increase the population densities of large cities such as Paris; but by setting buildings in radial lines, he would at the same time greatly increase the amount of open space in cities. He also intended to solve the problem of congestion by giving over the centre to transport facilities, including an airport (Le Corbusier, 1964; see Figure 4.34). Needless to say, Le Corbusier's vision was hardly geared to political realities; but there are post-war echoes of his technicist manifesto, as well as of Bauhaus planning ideals, in the spaced residential tower and slab blocks of the fringes of many a European city (see Figure 4.35).

The actual planning of cities by central authorities goes back practically as far as cities themselves; but in Europe after 1870, there was a notable increase in the scale of such public initiatives, as its large cities grappled with a combination of unprecedented population growth and rapid industrialization, and associated social upheavals and environmental problems. These initiatives were a departure from the early mode of industrial and urban growth in Britain during the first half of the nineteenth century, a mode characterized by the primacy of private interests and ownership. They would have been impossible without the empowering of planners to limit the rights of individuals and companies to own land and to conduct business. Planning in the industrial nations of Europe became for the first time, especially since the Second World

Figure 4.34 Le Corbusier's vision of the Radiant City (reproduced from Le Corbusier, 1964, p.204; © FLC/ADAGP, Paris, and DACS, London, 1998)

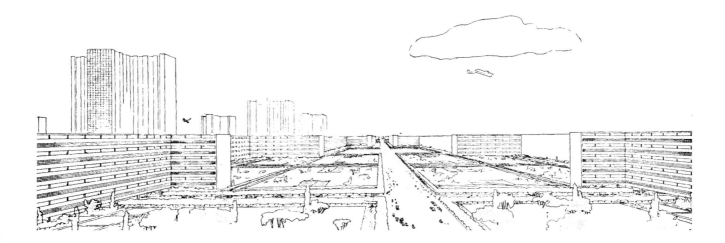

War, a universal rather than a contingent feature of urban development. As a result, city-building in the twentieth century can be characterized by the development of systematic and regulated relationships between planners and other agents, notably architects, engineers, builders and property developers, and their corporate sponsors (Whitehand, 1992). The universality of planning procedures has also been accompanied by their standardization across much of Europe, and indeed the wider world. This 'assembly-line' approach to planned urban features such as land-use zones, road types and neighbourhood layouts needs to be borne in mind when the question arises about the relationship between the diffusion of modern technologies and the homogeneity of modern cities (Relph, 1987, p.140).

The pioneering nation in town-planning theory was Germany, which had to cope with concomitant industrialization and rapid urban growth from the mid-nineteenth century (Ladd, 1990; Sutcliffe, 1981). The municipal authorities responded to this challenge at first with the traditional German pre-industrial approach of town-extension planning, which focused on the laying out of new streets. Many cities followed the example of the extension plan for Berlin, drawn up in 1862 by James Hobrecht; in the Hobrecht plan, wide uniform streets and multi-storey blocks of immense dimensions predominated. Water supply, sanitation and housing remained inadequate in German cities, and these problems were exacerbated by the economic crisis that hit Europe in 1873. Various state governments responded with laws allowing for the compulsory purchase of land for new streets; the costs were spread to nearby landowners, who formerly enjoyed a big unearned increase in property value from streets laid at the expense of the municipal authorities.

Figure 4.35 Housing in the western suburb of Slotervaart in Amsterdam, c.1960 (reproduced from Benevolo, 1981, p.904)

Urbanization in Germany really took off between 1890 and 1910, by which point 60 per cent of the population were living in towns and cities. This period was marked by the appearance in 1890 of Joseph Stübben's book *Der Städtebau* (not to be confused with Sitte's journal), in which he advocated for every rapidly growing city a comprehensive development plan, including street widths and routes, transport, open spaces and public buildings, and water-supply and drainage systems. Another pioneer of German planning, Franz Adickes, the *Oberbürgermeister* of Frankfurt-am-Main, successfully challenged the traditional rights of landowners by devising a set of differential building regulations in 1891. These divided the city into an inner and an outer building zone, each subdivided into residential, mixed and industrial districts; the regulations for the outer zone were intended to create a much lower and less dense built environment. In the early 1900s, town-planning emerged as a profession, and most states enacted planning laws. These initiatives met with great resistance from builders and landowners; but the relative success of social reformers in Germany can be put down to the domination of urban administration by salaried officials, many of whom were the product of Germany's system of technical education, and a tradition of the active participation of states such as Prussia in early German industrialization.

In Britain, where the growth of industrial cities during the first half of the nineteenth century had taken place in a climate of minimalist local government, the first public initiatives took the form of the municipal control of services, starting with water, and later extending to transport and power supply. There was also tighter regulation of private building from the late nineteenth century; local authorities enacted by-laws setting minimum standards for window size and doorway heights, and then of street widths, backyard size and building heights. The result of these by-laws, however, was a new kind of urban blight: row on row of terraced houses, without open spaces, schools or shops. Urban planning in Britain was in part a reaction against such developments, and essentially set out to let some air into these congested urban districts (see Figure 4.36)

The need to improve housing provision was met in several ways. One was the continuing tradition of model industrial communities: Port Sunlight, near Liverpool, and Bournville, near Birmingham, both date from the 1890s. At the turn of the century, there was a move to decentralize the city, by the establishment of spacious garden suburbs and cities, as advocated by Ebenezer Howard. The first attempt to realize Howard's vision was the garden city at Letchworth to the north of London, begun in 1903, and designed by Raymond Unwin and Barry Parker; a second garden city was started at Welwyn in 1919. The most influential aspect of the garden city concept was the avoidance of a formal grid street pattern that incorporated crossroads in the layout of residential areas. In Letchworth, mixed-class, vernacular-style houses were set in a variety of patterns, such as crescents and culs-de-sac (see Figure 4.37). The garden city movement bequeathed a number of durable design principles for the layout of suburban residential developments, but the redemptive vision of a new kind of urban life was never fully realized. A more practicable solution to the housing problem was the construction of planned working-class housing within cities, initially by private philanthropic bodies, but increasingly by local authorities, which built the great majority of new houses in Britain during the inter-war years.

The urban fabric is usually slow to respond to political, economic and technological innovations, but the aerial bombardment of so many cities during the Second World War provided an opportunity for radical reconstruction. This important episode in the history of European urban morphology encapsulates this series' main concerns. A mix of policies, values and technological preconceptions was involved in the planned reconstruction

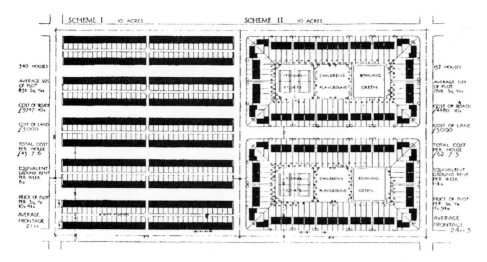

Figure 4.36 Diagram attempting to prove the inefficiency of by-law housing, from Raymond Unwin's tract 'Nothing Gained by Overcrowding', 1912. The illustration glosses over the fact that more than twice the ten-acre (four-hectare) plot at more than twice the overall cost would be required for Unwin's preferred Scheme II if it were to have the same number of houses as Scheme I (Miller, 1992, pp.128–9; diagram reproduced from Unwin, 1912, p.220)

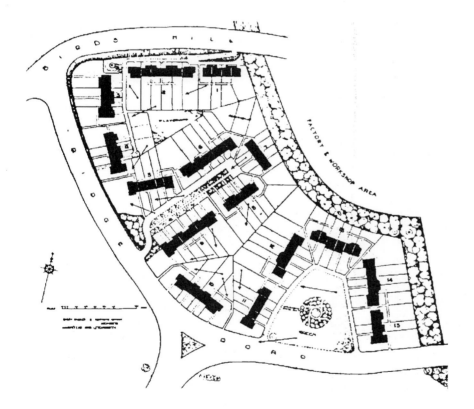

Figure 4.37 Layout of Bird's Hill Estate, Letchworth. Here, Unwin created 'a design kit, capable of varied assembly to fit specific site conditions'; it included a cul-de-sac, a playground and a planted buffer zone demarcating the estate from the adjoining factory area (Miller, 1992, p.71; diagram reproduced from Unwin, 1909, p.348)

of war-damaged European cities – not least the tension between a desire to restore historic centres as accurately as possible, and the opportunity to adapt a weakened physical fabric to new technological imperatives (Diefendorf, 1989,[6] 1990, 1993; Esher, 1981). Examples of the former approach can be seen in Warsaw, Dresden and Cologne; and of the latter in Coventry (Hasegawa, 1992; Mason and Tiratsoo, 1990) (see Figure 4.38 overleaf), where the pedestrianized shopping precinct was pioneered, and in London and Paris, where great tracts of bomb-damaged and slum housing were cleared, and replaced with Bauhaus-style mixes of tower blocks, slab blocks and row housing (see Chapter 5). The size of this

---

[6] An edited version of this article by Diefendorf can be found in Goodman (1999), the Reader associated with this volume.

Figure 4.38 Coventry in 1946 (top) and 1987 (bottom). Much of Coventry's medieval and Elizabethan quarter was swept away by an air raid on 14–15 November 1940, though the adjacent spires of Holy Trinity and of St Michael's Cathedral escaped, as can be seen from the upper right-hand quadrant of the 1946 photograph. The new Modernist cathedral (visible between the two spires near the top of the 1987 photograph, just to the right of centre) was completed in 1962. By 1987, the city centre was dominated by the shopping precinct (centre left), car parks and high-rise office and apartment blocks, enclosed within an elevated ring-road, a section of which runs along the top of the lower picture (Gardiner, 1989, pp.92–3; photographs: Aerofilms)

challenge, as well as the experience of central planning in the wartime years, and the return of a socialist government in 1945, help explain why planning has become such an integral feature of all aspects of urban development in post-war Britain. It was Britain that provided a lead for Europe with the Town and Country Planning Act of 1947, which gave local authorities the responsibility to prepare a development plan, and increased their powers to control development within its terms. These powers included the right to expropriate private property under certain circumstances, including slum conditions and war damage.

The patterns of urban growth in Britain and the USA diverged after the Second World War. As the USA continued to give full rein to the motor car, and its cities became even more decentralized, Britain was frozen into the two-part Burgess model of the city by the political decision to limit the expansion of cities by means of 'green belts', like that imposed on London in 1938 (see Chapter 5) (Fishman, 1987). The solution espoused by post-war planners to the problem of the excess population of the big cities was to decant it to new planned settlements. In 1946 the Reith Committee reported on these proposed new towns, and the New Towns Act of that year allowed for them to be built outside the London Green Belt. As a result of the Act, fourteen new towns were constructed, including the steel town of Corby (in Northamptonshire) in 1950, and Hatfield New Town (in Hertfordshire) in 1952. In 1961, the New Towns Commission was established under Andrew Duncan, and a further fourteen new towns designated; two of these, Telford and Milton Keynes, took the planning of new towns to another level, by integrating existing villages and towns into a regional complex. Similar state-sponsored new-town programmes, but with Modernist rather than traditional styles of building and planning, were adopted in other parts of Europe, including France, Holland and Sweden. Like all planned settlements throughout the entire urban era, these 'greenfield' developments enabled their progenitors to express, in a way untrammelled by a pre-existing built environment and the vested interests it represents, the potentials of its city-building technologies to realize an urban ideal; in the case of the grid layouts of Telford and Milton Keynes, accommodating motor cars had become the prime planning desideratum (see Chapter 8).

These settlements made an important contribution to the urban system; by the 1970s, more than two million Britons lived in designated new towns, or in similarly planned extensions to some fifty old towns. But by this time, changes of equal magnitude were taking place in the established British urban landscape with the vogue for urban renewal. A notable example is the redeveloped centre of Birmingham; one of its best-known features is the unloved Bull Ring, a nine-hectare multi-level shopping and commercial complex with large car parks and pedestrian thoroughfares, officially opened in 1964 (Cherry, 1994, pp.213–14; see Figure 4.39 overleaf). The main motivation for these assaults on existing city centres and street patterns was again to accommodate increasing car ownership.

These technocratic 'clean sweep' approaches, however, appear to have had their day, at any rate in Britain. A reaction against Modernism and the motor car as the mainstays of urban planning got under way in the 1970s. An eclectic and often tongue-in-cheek mix of older decorative styles under Charles Jencks's portmanteau heading 'post-modernism' began to influence the appearance of public and corporate buildings (even though these styles were underpinned by Modernist steel and reinforced-concrete structures, and many of the most garish and humorous examples depend on state-of-the-art structural engineering). To be overtly 'modern' was no longer glamorous; post-modernism goes in step with a new sensitivity to the architectural heritage, prefigured in the 1966 Historic Preservation Act, which ironically now

promises protection for some early 'Modernist classics'. Renewed reverence for
the past is also manifest in the renovation and 'gentrification' of run-down
areas of working-class housing; the resistance of the urban fabric has once
again stiffened against the siren call of technological progress. Considerations
like these serve as a reminder that the social history of the urban built
environment is infinitely more complex than can be conveyed by the
contextual treatments of the principal technologies of city-building – the
subject of this particular series.

Figure 4.39 Central Birmingham, 1973. The Inner Ring Road, built between 1957 and 1971, was now complete. Smallbrook
Ringway, one of the first sections to be completed, runs left–right across the centre of the photograph. At the time of writing, the
concrete jungle of the Bull Ring and some elevated parts of the Inner Ring Road are due for demolition, and a new city centre
development project is under way (photograph: Aerofilms)

4.4 *Conclusion*

The innovations of the Second Industrial Revolution undoubtedly placed powerful new instruments in the hands of the movers and shakers of the urban built environment; but their effects were clearly modified by local and national politics. In addition, the weight of existing fixed-capital investment in the built environment acted as a brake on innovations in urban form, and nowhere was this more evident than in the case of Britain, the first industrial nation, as urban interests reacted apparently sluggishly to the potentials of innovations in power, transport and construction. A general contrast in approaches to the processes of city-building is sometimes posited between British individualism and localism, and continental state involvement. But even in the British case, after the first decades of generally unregulated urban building had exacted their tribute of overcrowding, congestion, mortality and pollution, there followed increasingly regulated and planned development overall, especially from the last third of the nineteenth century, as city authorities increasingly aspired to bring the urban infrastructure (transport networks, power and communications lines, water, sewage and gas mains) under municipal control and ownership. The question arises: was a more interventionist state the natural result of the availability of technological solutions to the problems of urban growth (Sutcliffe, 1981, p.5; 1982)?

That brings us back to the question of technological determinism. There is plenty of evidence to suggest that technologies such as transport, even if they were a vitally necessary condition of urban growth, were themselves socially shaped, or even constructed. Social constructionism is a salutary corrective to the fatalistic implications of technological determinism, or of oversimplified historical overviews. There is a danger, though, of throwing the baby out with the bath-water, and losing the contribution of the technologies themselves to the social changes with which they were involved. An example is the freedom of movement offered by motor transport in comparison to fixed-rail systems.

Perhaps historians of technology need to re-examine assertions about the potency of technology. *Proactive* innovation might be distinguished from *reactive* changes and contextual conditions. Some kind of new typology of causation is surely necessary to clarify thinking about these issues. What might make a given technology a fundamental source of social change in one sense is its releasing from nature a new source of energy, often whole orders of magnitude greater than that hitherto available. This property of an innovation cannot be reduced to any social components, though the ways in which such forces are channelled and exploited surely are open to such an analysis. Perhaps there also needs to be more emphasis on the differential effects of particular innovations. At some risk of over-simplifying the relations of technology and society, it could be argued that certain technologies, most obviously sources of power and their applications, are because of their intrinsic physical properties *geared* differently to society: to put it another way, the adoption of one technology, partly for reasons *inherent* to that technology, may be more likely to change a society radically than the adoption of another. Thus encouraging citizens to buy their own speedy and powerful motor car opens up the prospect of more far-reaching changes to urban form and fabric than providing a succession of electric trams plying a number of fixed routes, even though all city authorities have subsequently been obliged to enact measures limiting the car's intrinsic speed and freedom of movement.

Each technology, although socially shaped in its micro-histories, has its own natural powers and limitations. There must be an ultimate limit on the capacity of the horse, even though there is a great deal of scope for exploring its limits:

notably, the doubling of its drawing power when attached to a tram rather than to an omnibus. It is difficult, however, to prejudge the limits of any technology. For most of the nineteenth century, the steam-engine was dogged by the apparently fundamental problem of its low power-to-weight ratio, given the need for it to carry bulky coal-fuel. It was both feasible and economic to deploy it on rails, in a locomotive with a long train of carriages, or on water, in a large vessel, but neither practical nor economic, it would seem, in an individual steam carriage on roads designed (at best) for light, horse-drawn vehicles. But by the 1880s, higher boiler pressures, the substitution of oil for coal, and the Serpollet flash boiler (for acceleration) permitted its application to much lighter road vehicles. Nevertheless, it was still eventually outperformed by the petrol-engine. The electric motor has a good power-to-weight ratio, but the nature of the technology is such that it was economic only for intensive urban traffic; current electricity is essentially a medium for the transmission of energy, and there is a need for a permanent connection to a generator. Stored electricity in portable batteries is the obvious alternative, but nature has so far resisted efforts to make this option competitive with the internal-combustion engine.

Even if we accept that there are natural physical limits to the capacity of a power source, this still leaves plenty of scope for the play of social forces in decisions about its deployment. Take the internal-combustion engine. It has a history of refinements approaching ever more closely the maximum that could be extracted from it as a power unit. However, there is no technological consideration that decides whether the unit should power a four-seater motor car or a sixty-eight-seater bus. In order to resolve this contradiction between social constructionism and technological determinism, a distinction is needed between the social shaping of innovation, and its natural propensity to be shaped.

Extracts

4.1   Howard, E. (1985) *Garden Cities of To-morrow*, Builth Wells, Attic Books, pp.101–3, 105–8

The problem with which we have now to deal, shortly stated, is this: How to make our Garden City experiment the stepping stone to a higher and better form of industrial life generally throughout the country. Granted the success of the initial experiment, and there must inevitably arise a widespread demand for an extension of methods so healthy and so advantageous; and it will be well, therefore, to consider some of the chief problems which will have to be faced in the progress of such extension.

It will, I think, be well, in approaching this question, to consider the analogy presented by the early progress of railway enterprise … Railways were first made without any statutory powers. They were constructed on a very small scale, and, being of very short lengths, the consent of only one or at the most a few landowners was necessary; and what private agreement and arrangement could thus easily compass was scarcely a fit subject for an appeal to the Legislature of the country. But when the 'Rocket' was built, and the supremacy of the locomotive was fully established, it then became necessary, if railway enterprise was to go forward, to obtain legislative powers. For it would have been impossible, or at least very difficult, to make equitable arrangements with all the landowners whose estates might lie between points many miles distant; because one obstinate landlord might take advantage of his position to demand an altogether exorbitant price for his land, and thus practically stifle such an enterprise. It was necessary, therefore, to obtain power to secure the land compulsorily at its market value, or at a price not too extravagantly removed from such value; and, this being done, railway enterprise went forward at so rapid a rate that in one year no less than £132,600,000 was authorized by Parliament to be raised for the purpose of railway construction. *

Now, if Parliamentary powers were necessary for the extension of railway enterprise, such powers will certainly be also needed when the inherent practicability of building new, well-planned towns, and of the population moving into them from the old slum cities … is once fairly recognized by the people. To build such towns, large areas of land must be obtained. Here and there a suitable site may be secured by arrangement with one or more landowners, but if the movement is to be carried on in anything like a scientific fashion, stretches of land far larger than that occupied by our first experiment must be obtained. For, just as the first short railway, which was the germ of railway enterprise, would convey to few minds the conception of a network of railways extending over the whole country, so, perhaps, the idea of a well-planned town such as I have described will not have prepared the reader for the later development which must inevitably follow – the planning and building of town clusters – each town in the cluster being of different design from the others, and yet the whole forming part of one large and well-thought-out plan [see Figure 4.33] …

Garden City has, we will suppose, grown until it has reached a population of 32,000. How shall it grow? How shall it provide for the needs of others who will be attracted by its numerous advantages? Shall it build on the zone of agricultural land which is around it, and thus forever destroy its right to be called a 'Garden City'? Surely not. This disastrous result would indeed take place if the land around the town were, as is the land around our present cities, owned by private individuals anxious to make a profit out of it. For then, as the town filled up, the agricultural land would become 'ripe' for building purposes, and the beauty and healthfulness of the town would be quickly destroyed. But the land around Garden City is, fortunately, not in the hands of private individuals: it is in the hands of the people: and is to be administered, not in the supposed interests of the few, but in the real interests of the whole community. Now, there are few objects which the people so jealously guard as their parks and open spaces; and we may, I think, feel confident that the people of Garden City will not for a moment permit the beauty of their city to be destroyed by the process of growth. But it may be urged – if this be true, will not the inhabitants of Garden City in this way be selfishly preventing the growth of their city, and thus preclude many from enjoying its advantages? Certainly not. There is a bright, but overlooked, alternative. The town *will* grow; but it will grow in accordance with a principle which will result in this – that such growth shall not lessen or destroy, but ever add to its social opportunities, to its beauty, to its convenience …

Garden City is built up. Its population has reached 32,000. How will it grow? It will grow by establishing – under Parliamentary powers probably – another city some little distance beyond its own zone of 'country', so that the new town may have a zone of country of its own. I have said 'by establishing another city', and, for administrative purposes there would be *two* cities; but the inhabitants of the one could reach the other in a very few minutes; for rapid transit would be specially provided for, and thus the people of the two towns would in reality represent one community.

And this principle of growth – this principle of always preserving a belt of country round our cities would be ever kept in mind till, in course of time, we should have a cluster of cities, not of course arranged in the precise geometrical form of my diagram [see Figure 4.33], but so grouped around a Central City that each inhabitant of the whole group, though in one sense living in a town of small size, would be in reality living in, and would enjoy all the advantages of, a great and most beautiful city; and yet all the fresh delights of the country – field, hedgerow, and woodland – not prim parks or gardens merely – would be within a very few

* (Clifford, 1885, p.88)

minutes' walk or ride. And *because the people in their collective capacity own the land* on which this beautiful group of cities is built, the public buildings, the churches, the schools and universities, the libraries, picture galleries, theatres, would be on a scale of magnificence which no city in the world whose land is in pawn to private individuals can afford.

I have said that rapid railway transit would be realized by those who dwell in this beautiful city or group of cities [see Figure 4.33] ... There is, first, an inter-municipal railway, connecting all the towns of the outer ring – twenty miles [thirty-two kilometres] in circumference – so that to get from any town to its most distant neighbour requires one to cover a distance of only ten miles [sixteen kilometres], which could be accomplished in, say, twelve minutes. These trains would not stop between the towns – means of communication for this purpose being afforded by electric tramways which traverse the high roads, of which, it will be seen, there are a number – each town being connected with every other town in the group by a direct route.

There is also a system of railways by which each town is placed in direct communication with Central City. The distance from any town to the heart of Central City is only three and a quarter miles [5.2 kilometres], and this could be readily covered in five minutes.

Those who have had experience of the difficulty of getting from one suburb of London to another will see in a moment what an enormous advantage those who dwell in such a group of cities as here shown would enjoy, because they would have a railway *system* and not a railway *chaos* to serve their ends. The difficulty felt in London is of course due to want of forethought and pre-arrangement. On this point, I may quote with advantage a passage from the Presidential address of Sir Benjamin Baker to the Institute of Civil Engineers, 12th November 1895: 'We Londoners often complain of the want of system in the arrangement of the railways and their terminal stations in and around the Metropolis, which necessitates our performing long journeys in cabs to get from one railway system to another. That this difficulty exists, arises, I feel sure, chiefly from the want of forethought of no less able a statesman than Sir Robert Peel, for, in 1836, a motion was proposed in the House of Commons that all the Railway Bills seeking powers for terminals in London should be referred to a Special Committee, so that a complete scheme might be evolved out of the numerous projects before Parliament, and that property might not be unnecessarily sacrificed for rival schemes. Sir Robert Peel opposed the motion on the part of the Government, on the grounds that "no railway project could come into operation till the majority of Parliament had declared that its principles and arrangements appeared to them satisfactory, and its investments profitable. It was a recognized principle in these cases that the probable profits of an undertaking should be shown to be sufficient to maintain it in a state of permanent utility before a Bill could be obtained, and landlords were perfectly justified in expecting and demanding such a warranty from Parliament." In this instance, incalculable injury was unintentionally inflicted upon Londoners by not having a grand central station in the Metropolis, and events have shown how false was the assumption that the passing of an Act implied any warranty as to the financial prospects of a railway.'

But are the people of England to suffer for ever for the want of foresight of those who little dreamed of the future development of railways? Surely not. It was in the nature of things little likely that the first network of railways ever constructed should conform to true principles; but now, seeing the enormous progress which has been made in the means of rapid communication, it is high time that we availed ourselves more fully of those means, and built our cities upon some such plan as that I have crudely shown. We should then be, for all purposes of quick communication, nearer to each other than we are in our crowded cities, while, at the same time, we should be surrounding ourselves with the most healthy and the most advantageous conditions.

4.2 Le Corbusier (1964) *The Radiant City* (trans. P. Knight, E. Levieux and D. Coltman), London, Faber and Faber, pp.23–4, 38–9, 94, 131–3

City planning

1. Throughout the ages city planning has always made use of the most effective means which technique can provide.

2. Today, iron and reinforced concrete give us convenient means of carrying out the sort of city planning which responds to the profound social and economic revolution caused by the machine age.

   The social and economic revolution caused by the machine age confronts us with utterly new problems, affecting every part of every country.

3. There is no *central agency* to guide and corollate the analyses which must be made and the initiatives which must be undertaken. There is general confusion, chaos prevails, danger is everywhere. It is imperative that in every country or region there be a plan to create *a permanent agency*, directed by a competent and responsible figure, able to give the country *its new statute*.

   For reforms are extended simultaneously to all cities, to all rural areas, across the seas.

4. The new programs are of wider scope than anything that has gone before. And it happens that the nation's land, through successive sales and inheritances, is more than ever parceled out, along the most arbitrary lines. Countless sites, endlessly split every which way, *hinder any job of city planning.*

   So first there must be a *law decreeing the reconstitution of land, in city and country.*

5. This operation is directly linked to that of enhancing land values through works of public interest. There should be a law by which the basic price of a given property may be established and determining how, following works of public interest, the enhanced value may be shared between the owner of the property involved and the agency which took the initiatives in those works and bore their cost *(Law of the recuperation of enhanced values).*

6. Since new means now exist (steel or concrete), state or municipal ordinances setting limits on construction must be completely revised so as to allow maximum development of the resources provided by new techniques.

   The first limit involved is that on the *height of buildings.*

7. But simultaneously the question involves the areas devoted to traffic and the areas planted (hygiene) between these buildings. So we may establish *the ratios* between areas used for traffic, for planting, and for housing, and another scale dividing cities into zones of population density per *hectare* (approximately 2.5 acres). This allows a flexible confirmation of the principle by which *the center of the city should be very densely populated, with a very great area devoted to traffic and plantings.* This will determine the types of tall structures which are to be built in the center of the city.

8. In some places the law already allows the city to develop freely by buying back a *protective green belt* circling the city and separating it from its suburbs.

9. In view of the urgent need to industrialize building, city planning will proceed to regroup blocks of buildings and traffic arteries along the orthogonal principle.

10. In view of the country's general economy, technical resources, traffic and public health requirements, city planning will make the roof garden compulsory and also, where needed, the street raised on pilotis.

11　In every instance, city planning should respect sports (hygiene, recuperation of nervous energy, etc.) by leaving room for them to be practiced very near the house.

12　Faced by hesitant authorities, always inclined to take short-range opportunistic measures, city planners will be guided in their layouts by the so-called 'surgical' principle (layouts cutting across existing streets and blocks or boundaries), instead of the 'medical' principle (aiming merely to widen existing streets or roads) ...

***

So we come to take the implemental decision: the use of tall buildings. How tall will they be? 30, 40, or 50 meters; some people even propose a height of 150 meters, and more.

The question of elevators comes up as soon as a house is two stories high; the elevator is in fact the keystone of all modern urbanization, be it in proletarian sections or in wealthy sections (and moreover, such distinctions should disappear: the city, having become a human city, will be a classless city). It is a crime to make anyone walk up more than three flights of stairs. Now, by raising the question of the compulsory elevator, the elevator-cum-means of public transportation, we are also asking for a complete reorganization of the way building lots are divided, of the number of streets and of the way they are laid out. Until now, the custom has been to combine an elevator and a stairway to serve a maximum of two, three, or four apartments per floor. In that case, the elevator is run by the people who use it ... If the elevator is to be truly efficient, the running of it should be entrusted to elevator operators, *both night and day*. At the rate of two or four apartments per floor, this is a witless, ruinous system. Suppose we adopt a healthy conception: professional elevator operators, day and night; fast elevators and, with this circulating material, this 'vertical transportation' in existence, let's have a considerable number of floors; but instead of the two apartments which on every floor traditionally give on to a slow and perilous vertical system, we will suppose a system of intense vertical traffic, swift and sure, serving a great many apartment doors on each floor: 20, or 40, or 100. We say: 20, or 40, or 100 apartments on each floor, because we have arrived at the notion of 'interior streets.' So we have, within the apartment building, the introduction of corridors which become veritable *interior streets*,' streets in the air, 12 or 24 or 50 meters above the ground. On the basis of a detailed study I propose, for example, to group 2,400 persons around a single vertical shaft with four elevators; each of these persons will have to walk 100 meters at most from the elevator to the door of his apartment. From then on instead of forty elevators accompanied by forty stairways, and perhaps by forty concierges, there would be only one stairwell with four elevators. The result is crucial, and obvious: we will *no longer have forty doors opening from the building* onto the street but rather, *a single door*; instead of having to make those forty houses open onto the street itself and consequently *having to build the houses directly on the street*, we will have only one door, and *from the street itself an approach road with 'auto-ports'* (parking lots) will lead off to the single door. Thereafter, the problem of automobile traffic is on its way to solution. This means a complete reversal of the economy of streets, of the situation of houses and streets in respect to each other, of the hitherto interdependent function of house and street. The house no longer needs to rise above the street; the street is no longer at the foot of the house! ...

The problem is to create the Radiant City ... The city of light that will dispel the miasmas of anxiety now darkening our lives, that will succeed the twilight of despair we live in at present, exists on paper. We are only waiting for a 'yes' from a government with the will and the determination to see it through! ...

The general characteristics of the plan are as follows: the city (a large city, a capital) is much less spread out than the present one; the distances within it are therefore shorter, which means more rest and more energy available for work every day. There are no suburbs or dormitory towns; this means an immediate solution to the transportation crisis that has been forced upon us by the paradox of the city + garden cities.

The garden city is a pre-machine-age utopia.

The population density of the new city will be from three to six times greater than the idealistic, ruinous and inoperative figures recommended by urban authorities still imbued with romantic ideology. This new intensification of population density thus becomes the financial justification for our enterprise: *it increases the value of the ground.*

The pedestrian never meets a vehicle inside the city. The mechanical transportation network is an entirely new organ, a separate entity. The ground level (the *earth*) belongs entirely to the pedestrian.

The 'street' as we know it now has disappeared. All the various sporting activities take place directly outside people's homes, in the midst of parks – trees, lawns, lakes. The city is entirely green; *it is a Green City.* Not one inhabitant occupies a room without sunlight; everyone looks out on trees and sky.

The keystone of the theory behind this city is the *liberty of the individual.* Its aim is to create respect for that liberty, to bring it to an authentic fruition, to destroy our present slavery … The cost of living will come down. The new city will break the shackles of poverty in which the old city has been keeping us chained.

Its growth is assured. It is the Radiant City. A gift to all of us from modern technology …

Here is the solution … to the problem of the business center of a great city: superdensity: 3,200 occupants to the hectare (allowing 10 square meters of office floorspace per worker).

Skyscrapers built in quincunx or checkerboard pattern, one every 400 meters. The distance between these buildings will therefore be much the same as the average distance between our Paris métro stations. It is worthwhile here to try to visualize what huge, stupendous spaces the introduction of these skyscrapers will create. Their horizontal projection, by which is meant the plan of any given story, will represent no more than 5 (five) per cent of the ground area allotted to each building.

Despite these unexpected empty spaces, it should be noted also that the internal distances of the city center will be *four times less great* than those at present existing in even the most overpopulated of Parisian districts (800 inhabitants to the hectare).

These skyscrapers are all built in the shape of a cross in order to avoid central courtyards: *there are no courtyards anywhere.*

This form provides the maximum possible area of façade, therefore the maximum area of windows, therefore the maximum quantity of light. The offices are never more than 7 meters in depth, measured from the totally glazed surface of the façades: therefore there are no dark offices.

The cruciform skyscraper also provides a maximum of stability in relation to the thrust of high winds.

It is constructed of steel and glass. The vertical girders will spring up from the foundations to a height of 220 meters. The first floor will occur not less than 5 to 7 meters above ground level. Between the piles thus left to form a veritable forest in certain areas of the city's surface area, it will be possible to move about quite freely. Apart from five entrance halls for pedestrians, the space underneath the skyscraper is left vacant. Here again, as in the residential neighborhoods, the pedestrian has *the entire ground surface at his disposal.* He never meets a motor vehicle: all motorized traffic is provided for elsewhere.

All motor vehicles are up on the highway network, which again, as in the residential neighborhoods, is based on a unit measuring 400 by 400 meters. Each skyscraper is built in the center of one of these 400 by 400 meter squares. The highways run 5 metres up in the air … From each of the four highway sections making up the square, a branch road leads off to the road network serving that particular building (thus avoiding any two-way traffic or the need for vehicles ever to meet). These four branch roads lead in to four separate auto-ports. Opening into each auto-port is a loading and unloading bay, of which there are likewise four per building. These four auto-ports provide parking for a thousand cars at the level of the bay itself, a thousand more on the ground beneath, and another thousand still in the auto-port basement. Total: 3,000 cars per skyscraper – far more than will be needed!

The roundabout linking the four parking areas will permit access to any of the four bays in each building from any of the road sections making up the 400 meter by 400 meter square.

At ground level, in the parks, a network of diagonal and orthogonal 'landscaped' paths for pedestrians. Beneath the highways, the 400 by 400 meter square is repeated at ground level, though here enclosed by iron fences. Between these fences, underneath the highway, are the traffic lanes for heavy trucks … This is also where the streetcars run.

In the Radiant City, the streetcar (either in its present form or in that of small trains) has been restored to its former eminence (economy and efficiency). The streetcar network does not coincide with the 400 meter by 400 meter highway network; it consists simply of a series of parallel tracks at 400 meter intervals, and therefore includes no intersections. Every 400 meters, the streetcars stop opposite two skyscrapers. There are breaks in the iron fences at these points occupied by sheltered platforms. Simple but functional.

There are wide underground passages, 20 or 30 meters in length, running underneath the streetcar lines and the heavy traffic lanes. (For pedestrians.)

Underground: the subway network of the Radiant City will then take the passengers on to particular buildings, the basements of which will all include a subway station. The line itself will follow one of the branches of the cross; on either side of it will be the platforms, and beyond the platforms will be located the communal services provided for the personnel working in the building: restaurants, shops, etc.

We do not yet know whether, before long, we shall have air-taxis from the Radiant City airport landing on the tops of the business center skyscrapers. It is possible. There will be runway platforms available 25 meters wide by 150 to 200 meters in length. The problems involved have already been largely solved by naval aircraft carriers.

References

ABRAM, J. (1987) 'An unusual organisation of production: the building firm of Perret Brothers, 1897–1954', *Construction History*, vol.3, pp.75–93.

ADDIS, B. (1997) 'Concrete and steel in twentieth-century construction: from experimentation to mainstream usage' in M. Stratton (ed.) *Structure and Style: conserving twentieth century buildings*, London, E. and F.N. Spon, pp.103–42.

BARKE, M. (1991) 'The middle-class journey to work in Newcastle upon Tyne, 1850–1913', *Journal of Transport History*, vol.12, no.2, pp.107–43.

BARKE, M. (1992) 'The development of public transport in Newcastle upon Tyne and Tyneside', *Journal of Regional and Local Studies*, vol.12, pp.29–52.

BARKER, T.C. (1993) 'Slow progress: forty years of motoring research', *Journal of Transport History*, vol.14, pp.142–65.

BARKER, T. and GERHOLD, D. (1993) *The Rise and Rise of Road Transport*, Basingstoke, Macmillan.

BARKER, T.C. and ROBBINS, M. (1963) *A History of London Transport: passenger travel and the development of the metropolis*, London, Allen and Unwin for the London Transport Executive, vol.1.

BARKER, T.C. and ROBBINS, M. (1974) *A History of London Transport: passenger travel and the development of the metropolis*, London, Allen and Unwin for the London Transport Executive, vol.2.

BEASLEY, D. (1988) *The Suppression of the Automobile: skulduggery at the crossroads*, New York and London, Greenwood.

BENEVOLO, L. (1981) *The History of the City* (trans. G. Culverwell), London, Scolar Press.

BESSEMER, SIR H. (1905) *An Autobiography*, London, Offices of Engineering.

BIJKER, W.E. (1995) 'The majesty of daylight: the social construction of fluorescent lighting' in W.E. Bijker, *Of Bicycles, Bakelites, and Bulbs: toward a theory of sociotechnical change*, Cambridge, Mass., MIT Press, pp.199–290.

BOWERS, B. (1982) *A History of Electric Light and Power*, Stevenage, Peter Peregrinus.

BURKE, G. (1971) *Towns in the Making*, London, Edward Arnold.

BYATT, I.C.R (1979) *The British Electrical Industry: the economic returns to a new industry*, Oxford, Clarendon Press.

CANNADINE, D. (1977) 'Victorian cities: how different?', *Social History*, vol.4, pp.457–87.

CAPUZZO, I.P. (1998) 'The defeat of planning: the transport system and urban pattern in Vienna', *Planning Perspectives*, vol.13, pp.23–51.

CARDWELL, D. (1994) *The Fontana History of Technology*, London, Fontana Press.

CHANT, C. (1989) 'Materials: steel and concrete' in C. Chant (ed.) *Science, Technology and Everyday Life, 1870–1950*, London, Routledge.

CHANT, C. (ed.) (1999) *The Pre-industrial Cities and Technology Reader*, London, Routledge, in association with The Open University.

CHANT, C. and GOODMAN, D. (eds) (1999) *Pre-industrial Cities and Technology*, London, Routledge, in association with The Open University.

CHERRY, G.E. (1994) *Birmingham: a study in geography, history and planning*, Chichester, John Wiley.

CLARK, D.K. (1992) *Tramways: their construction and working*, Chetwode, Adam Gordon (first published 1894).

CLIFFORD, F. (1885) *A History of Private Bill Legislation*, London, Butterworth.

CUSACK, P. (1987) 'Agents of change: Hennebique, Mouchel and ferro-concrete in Britain, 1897–1908', *Construction History*, vol.3, pp.61–74.

DANIELS, P.W. and WARNES, A.M. (1980) *Movement in Cities: spatial perspectives on urban transport and travel*, London, Methuen.

DEPARTMENT OF INTERNATIONAL ECONOMIC AND SOCIAL AFFAIRS (1980) *Patterns of Urban and Rural Population Growth*, New York, United Nations (Population Studies, No. 68).

DIEFENDORF, J.M. (1989) 'Artery: urban reconstruction and traffic planning in postwar Germany', *Journal of Urban History*, vol.15, no.2, pp.131–58.

DIEFENDORF, J.M. (ed.) (1990) *Rebuilding Europe's Bombed Cities*, Basingstoke, Macmillan.

DIEFENDORF, J.M. (1993) *In the Wake of War: the reconstruction of German cities after World War II*, Oxford, Oxford University Press.

DYOS, H.J. (1961) *The Victorian Suburb: a study of the growth of Camberwell*, Leicester, Leicester University Press.

ESHER, L. (1981) *A Broken Wave: the rebuilding of England, 1940–1980*, London, Allen Lane.

FISHMAN, R. (1987) *Bourgeois Utopias: the rise and fall of suburbia*, London, Basic Books.

GARDINER, L. and A. (1989) *The Changing Face of Britain: from the air*, London, Michael Joseph.

GEDDES, P. (1968) *Cities in Evolution: an introduction to the town planning movement and to the study of civics*, London, Ernest Benn (first published 1915).

GLENDINNING, M. and MUTHESIUS, S. (1994) *Tower Block: modern public housing in England, Scotland, Wales and Northern Ireland*, New Haven, Yale University Press.

GOODMAN, D. (ed.) (1999) *The European Cities and Technology Reader: industrial to post-industrial city*, London, Routledge, in association with The Open University.

GORDON, G. (1990) 'The morphological development of Scottish towns from Georgian to modern times' in T.R. Slater (ed.) *The Built Form of Western Cities: essays for M.R.G. Conzen on the occasion of his eightieth birthday*, Leicester, Leicester University Press, pp.210–32.

HALL, P. (1992, 3rd edn) *Urban and Regional Planning*, London, Routledge.

HARBORD, F.W. and HALL, J.W. (1918, 6th edn) *The Metallurgy of Steel*, London, Charles Griffin, vol.2.

HASEGAWA, J. (1992) *Replanning the Blitzed City Centre: a comparative study of Bristol, Coventry and Southampton 1941–1950*, Buckingham, Open University Press.

HAYWOOD, R. (1996) 'More flexible office location controls and public transport considerations: a case study of the city of Manchester', *Town Planning Review*, vol.67, no.1, pp.65–86.

HILTON, G.W. (1969) 'Transport technology and the urban pattern', *Journal of Contemporary History*, vol.4, no.3, pp.123–35.

HITCHCOCK, H.-R. (1958) *Architecture: nineteenth and twentieth centuries*, London, Penguin Books.

HOBSBAWM, E.J. (1968) *Industry and Empire*, London, Weidenfeld and Nicolson.

HOWARD, E. (1985) *Garden Cities of To-morrow*, Builth Wells, Attic Books (first published 1898).

HUME, J.R. (1983) 'Transport and towns in Victorian Scotland' in G. Gordon and B. Dicks (eds) *Scottish Urban History*, Aberdeen, Aberdeen University Press.

JANKOWSKI, S. (1990) 'Warsaw: destruction, secret town planning, 1939–44, and postwar reconstruction' in J.M. Diefendorf (ed.) *Rebuilding Europe's Bombed Cities*, Basingstoke, Macmillan, pp.77–93.

JOHNSON, J.H. (1972) *Urban Geography: an introductory analysis*, Oxford, Pergamon.

KLAPPER, C. (1961) *The Golden Age of Tramways*, London, Routledge and Kegan Paul.

LADD, B. (1990) *Urban Planning and Civic Order in Germany, 1860–1914*, Cambridge, Mass., Harvard University Press.

LANDES, D.S. (1969) *The Unbound Prometheus: technological change and industrial development in western Europe from 1750 to the present*, Cambridge, Cambridge University Press.

LE CORBUSIER (1964) *The Radiant City* (trans. P. Knight, E. Levieux and D. Coltman), London, Faber and Faber (first published 1933).

MASON, T. and TIRATSOO, N. (1990) 'People, politics and planning: the reconstruction of Coventry's city centre, 1940–53', in J.M. Diefendorf (ed.) *Rebuilding Europe's Bombed Cities*, Basingstoke, Macmillan, pp.94–113

MCKAY, J.P. (1976) *Tramways and Trolleys: the rise of urban mass transport in Europe*, Princeton, Princeton University Press.

MILLER, M. (1992) *Raymond Unwin: garden cities and town planning*, Leicester, Leicester University Press.

MITCHELL, B.R. (1998, 4th edn) *International Historical Statistics: Europe 1750–1993*, London and Basingstoke/New York, Macmillan/Stockton Press.

MOORE, J.R. (1989) 'Communications' in C. Chant (ed.) *Science, Technology and Everyday Life, 1870–1950*, London, Routledge, pp.200–49.

MORGAN, R.H. (1986) 'The development of an urban transport system: the case of Cardiff', *Welsh History Review*, vol.13, pp.178–93.

MULLIN, J.R. (1977) 'City planning in Frankfurt, Germany, 1925–1932', *Journal of Urban History*, vol.4, pp.3–28.

MUNBY, D.L. (1978) *Inland Transport Statistics: Great Britain, 1900–1970*, vol.1: *Railways, public road passenger transport, London's transport* (ed. and completed by A.H. Watson), Oxford, Clarendon Press.

OCHOJNA, A.D. (1978) 'The influence of local and national politics on the development of urban passenger transport in Britain 1850–1900', *Journal of Transport History*, new series, vol. 4, no.3, pp.125–46.

OLSEN, D.J. (1976) *The Growth of Victorian London*, London, B.T. Batsford.

RAE, J.B. (1968) 'Transportation technology and the problem of the city', *Traffic Quarterly*, vol.22, pp.299–314.

RELPH, E. (1987) *The Modern Urban Landscape*, London, Croom Helm.

ROBERTS, G.K. (ed.) (1999) *The American Cities and Technology Reader: wilderness to wired city*, London, Routledge, in association with The Open University.

ROBERTS, G.K and STEADMAN, J.P. (1999) *American Cities and Technology: wilderness to wired city*, London, Routledge, in association with The Open University.

SCHIVELBUSCH, W. (1988) *Disenchanted Night: the industrialization of light in the nineteenth century* (trans. A. Davies), Oxford, Berg.

SHARP, D. (1969) *The Picture Palace and Other Buildings for the Movies*, London, Hugh Evelyn.

SHORT, J.R. (1984) *An Introduction to Urban Geography*, London, Routledge and Kegan Paul.

SIMMONS, J. (1973) 'The power of the railway' in H.J. Dyos and Michael Wolff (eds) *The Victorian City: images and realities*, London, Routledge and Kegan Paul, vol.1, pp.277–310.

SIMPSON, M. (1972) 'Urban transport and the development of Glasgow's West End, 1830–1914', *Journal of Transport History*, 2nd series, vol.1, no.3, pp.147–60.

SITTE, C. (1979) *The Art of Building Cities: city building according to artistic fundamentals* (trans. C.T. Stewart), Westport, Conn., Hyperion Press (translation first published 1945).

SUTCLIFFE, A. (ed.) (1974) *Multi-storey Living: the British working class experience*, London, Croom Helm.

SUTCLIFFE, A. (ed.) (1980) *The Rise of Modern Urban Planning*, London, Mansell.

SUTCLIFFE, A. (1981) *Towards the Planned City: Germany, Britain, the United States and France, 1780–1914*, Oxford, Basil Blackwell.

SUTCLIFFE, A. (1982) 'The growth of public intervention in the urban environment during the nineteenth century: a structural approach' in J.H. Johnson and C.G. Pooley (eds) *The Structure of Nineteenth-century Cities*, London, Croom Helm, pp.107–24.

SUTCLIFFE, A. (1988) 'Street transport in the second half of the nineteenth century: mechanization delayed?' in J.A. Tarr and G. Dupuy (eds) *Technology and the Rise of the Networked City in Europe and America*, Philadelphia, Temple University Press, pp.22–39.

SUTTON, R. (1996) *Motor Mania*, London, Collins and Brown.

TARR, J.A. and DUPUY, G. (eds) (1988) *Technology and the Rise of the Networked City in Europe and America*, Philadelphia, Temple University Press.

THOMPSON, F.M.L. (1976) 'Nineteenth-century horse sense', *Economic History Review*, 2nd series, vol.29, pp.60–81.

THOMPSON, F.M.L. (ed.) (1982) *The Rise of Suburbia*, Leicester, Leicester University Press.

UNWIN, R. (1909) *Town Planning in Practice: an introduction to the art of designing cities and suburbs*, London, T. Fisher Unwin.

UNWIN, R. (1912) 'Nothing gained by overcrowding', *Garden Cities and Town Planning*, vol.11, no.10, p.220.

WAGENAAR, M. (1992) 'Conquest of the center or flight to the suburbs? Divergent metropolitan strategies in Europe, 1850–1914', *Journal of Urban History*, vol.19, pp.60–83.

WEIGHTMAN, G. and HUMPHRIES, S. (1983) *The Making of Modern London 1815–1914*, London, Sidgwick and Jackson.

WEIGHTMAN, G. and HUMPHRIES, S. (1984) *The Making of Modern London 1914–1939*, London, Sidgwick and Jackson.

WERMIEL, S. (1993) 'The development of fireproof construction in Great Britain and the United States in the nineteenth century', *Construction History*, vol.9, pp.3–26.

WHITEHAND, J.W.R. (1992) 'The makers of British towns: architects, builders and property owners c.1850–1939', *Journal of the History of Geography*, vol.18, pp.417–38.

YAGO, G. (1984) *The Decline of Transit: urban transportation in German and US cities, 1900–1970*, Cambridge, Cambridge University Press.

Chapter 5: LONDON AND PARIS

by Colin Chant

5.1 Introduction

For most of the twentieth century, London and Paris continued to be the two largest cities in Europe (Mitchell, 1998, pp.74–6). They warrant a chapter to themselves in this post-1870 section, partly to ensure some continuity of treatment in the European sections of this series; more importantly, they exemplify the trend for the biggest cities to grow at a faster rate than smaller settlements, a trend facilitated by the development of policies and technologies for managing gargantuan urban systems. This phenomenon has inspired mixed feelings. For a pessimist such as Lewis Mumford, for whom the medieval city came nearest to the ideal urban settlement, the dominant 'metropolis' (literally 'mother city') of a nation state had become a voracious 'megalopolis' ('great city'), sucking people in and devouring the surrounding countryside. Pursuing his organic metaphor to its conclusion, he predicted that the giant city would enter a terminal phase as a 'necropolis' ('city of the dead') (Mumford, 1948). For an optimist such as the planner Sir Peter Hall, these great urban centres have become, in one of those generic urban terms coined by the planner Patrick Geddes, 'world cities'; they have transcended their role as the primate city in a national system, and in their higher functions now play a specialized role in a global system of commodity and financial flows (Hall, 1977). This is not a recent phenomenon: even by the end of the nineteenth century, London and Paris had grown big, not only because they were large manufacturing and commercial centres and hubs of a national transport network, and because each dominated its nation's political and cultural life; in addition, they were both capitals of overseas empires, and international commercial and financial centres (though by the late 1990s Paris had fallen behind London in this respect; Noin and White, 1997, pp.13–14, 251–3). Both were ports; Paris, however, was a river port only, while London was the biggest maritime port in the world, its great docks and warehouses filled to bursting with foreign and colonial products. After the Second World War, great international airports (Heathrow, Orly and Roissy-Charles de Gaulle, opened in 1946, 1961 and 1974 respectively) were new technological insignia of their world-city status (Brooks, 1957).

Both pessimistic and optimistic views on the development of the largest cities are compatible with the belief that the major European capitals have grown more similar to than different from each other, in their choice of technologies to manage increasingly complex urban systems, and in the development of their built environments. This raises again the question of the causal power of technologies; the idea that certain urban characteristics flowed inevitably from new technologies goes back to the early decades of the Second Industrial Revolution, in both Britain and France. H.G. Wells in

The author of this chapter is grateful for detailed advice and corrections from Professor Anthony Sutcliffe, and for the comments of colleagues at The Open University. The overall approach adopted, however, is the author's own.

1903 forecast the coalescing of London's traditionally independent local communities into a more 'scientific' Greater London administrative entity:

> Every tramway, every new two-penny tube, every improvement in your omnibus services, in your telephone services, in your organization of credit increases the proportion of your local delocalized class, and sucks the ebbing blood from your old communities into the veins of the new.
>
> (quoted in Garside, 1984, p.242)

By contrast, the French demographer Paul Meuriot, writing in 1897, saw the Second Industrial Revolution as bringing about the end of the city:

> Cheap and speedy transport, unknown to our predecessors, is already permitting the partial removal of an overcrowded population from our cities. However, another innovation, the transmission of energy by electricity, may yet produce effects of quite a different order. The most important of them would be to liberate the industrial worker and abolish those concentrations of human population which today's equipment requires. The implications of a revolutionary technical change of this order are boundless. We can sum it up in this way: just as the extreme development of urban centres sprang from one scientific revolution, so may a second revolution cure the evils which are inseparable from that extreme development.
>
> (quoted in Sutcliffe, 1984, p.6)

It is evidently even harder for visionaries to foretell the urban implications of powerful new technologies than it is for historians to agree on what they were in retrospect. What remains to be seen is how much the differing political and economic institutions of the two cities influenced the deployment of the new technologies of urbanization. The main foci of this comparative study will be the reworking of the urban fabric of both cities in the decades leading up to the First World War, and the ways both met the challenge of urban renewal after the Second World War.

5.2 *London*

The population of London increased from 3.9 million in 1870 to 6.6 million in 1900, when it was still by far the largest city in the world; by 1938, the capital had 8.7 million inhabitants (Mitchell, 1998, p.75). The birth-rate had declined dramatically by the 1930s, though the expectation of London's planners, based on the inter-war years, that the population had stabilized was temporarily confounded by the post-war baby boom. The growth after 1900 took place in the suburbs (see Figure 5.1 overleaf); the population of the County of London area peaked in 1901 at 4.5 million, and started to decline significantly during the 1930s; by 1971 it stood at 2.7 million (Barker and Robbins, 1974, p.2). By this time, the entire population of London had peaked, and had begun to decline; the post-war preoccupation with London's 'congestion' would soon be replaced by concerns about inner-city regeneration (Hebbert, 1995, p.91). But it was not only, or necessarily, rising numbers that propelled the outward and upward growth of the physical fabric of London and other British cities. In the first half of the twentieth century, other kinds of demographic change redoubled the pressure of population growth on the urban fabric: greater mobility and earlier departure from the parental home raised the ratio of households to the population, and more bedrooms were demanded in dwellings, even though family size was reducing. Near the end of the twentieth century, politicians and planners are now debating the need for more dwellings, as the number of single-person households increases, through higher rates of divorce and greater life-expectancy.

Figure 5.1   The growth of
London, 1880–1961.
Top   In 1880, outside the built-up
area, in which many still walked
to work, there were a number
of discrete suburbs served by the
well-developed steam surface
railway, including the distinctively
elongated working-class railway
suburb of Tottenham in the
north. This pattern is generalized
in Figure 4.3 in Chapter 4.
Bottom   By 1961, most of the
former green space surrounding
the outer suburbs had been filled
in, and the suburban railway
network was correspondingly
denser. The outward, low-rise
momentum of Greater London's
growth, however, allowed a
proportion of green space to
be preserved that was much
higher than in Paris (reproduced
from Barker and Robbins, 1963,
plate xxix; 1974, p.5)

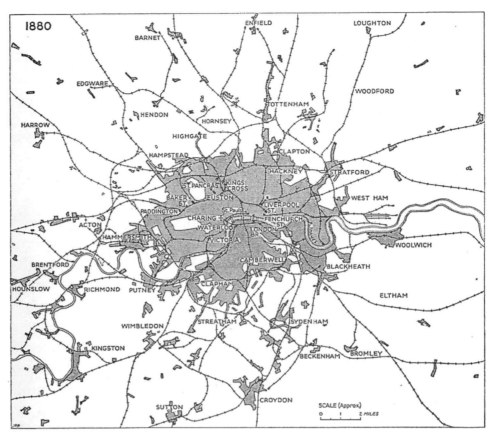

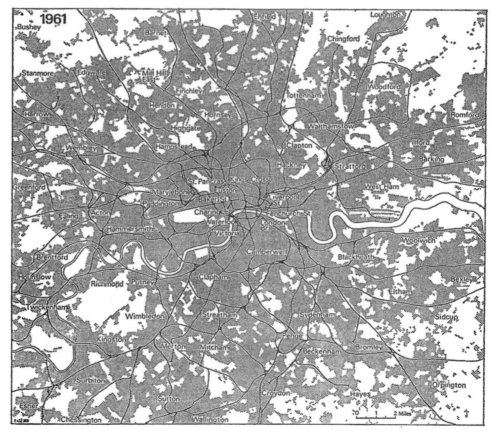

Another impetus to London's suburban growth was changing land-use in the centre of the city; in the second half of the nineteenth century, there was already a marked population decline in the inner city, which was increasingly devoted to retailing, entertainment and administrative premises, with a residue of very poor housing. At first, the dormitory nature of the new suburbs only exacerbated central congestion and the strain on public transport. During the inter-war period, there was some relief of this pressure through the increasingly suburban location of commerce and industry.

In the period since 1870, London continued to grow as an administrative, distributive, financial and cultural centre. The increasing prominence of services, and the disappearance of much of its traditional manufacturing industry, left the city relatively unscathed by the economic depression of the 1930s. But London's deindustrialization, though pronounced since the 1970s, has not been constant: during the inter-war years, new mass-production consumer-goods industries (vacuum cleaners, canned foods, radios, tyres) were attracted by London's vast market, and the proportion of British manufacturing jobs in London rose from 13.4 to 17.6 per cent between 1923 and 1939 (Barker and Robbins, 1974, p.3). The motorization of transport encouraged companies to build their factories along the new arterial roads of the outer suburbs, especially the A40 (Western Avenue) to the Midlands, and the upper Lea Valley, at Enfield and Edmonton (Garside, 1984, p.246) (see Figure 5.2).

Figure 5.2 The Park Royal industrial estate of West London, to the south of Wembley Stadium, part of what was by the late 1930s one of the biggest concentrations of industry in Britain. The main products of this area were consumer goods, including household appliances, motor cars and car accessories, and pharmaceuticals. Originally the site of a First World War munitions factory, Park Royal was favoured not only by its situation near the junction at Hanger Lane of the new Western Avenue and the North Circular Road, but also by its access to existing main railways (the Great Western and the London and North Western). The massive Guinness brewery, built between 1933 and 1936 and the company's first outside Dublin, is in the left foreground (Armstrong, 1996; photograph: Aerofilms; reproduced from Weightman and Humphries, 1984, p.55)

After 1870, the public grew increasingly intolerant of the social and environmental problems of London and other British cities, forcing central government to overcome its reluctance to infringe local autonomy. A measure reflecting this political sea-change was the 1875 Public Health Act, intended to raise standards in all boroughs to those of the most public-spirited. The key administrative change for London followed the Conservatives' Local Government Act of 1888, which established a new tier of government at county level. In 1889, the London County Council (LCC) replaced the ineffective Metropolitan Board of Works (Saint, 1989). The new council was keen to do more than restrain the development of the city through building regulations, and embarked on a more interventionist programme of municipal enterprise. A feature of the LCC's first decades was a struggle over this policy between the Progressives, whose 'gas and water socialism' dominated until 1907, and the more moderate Municipal Reformers.

A microcosm of the old tension between central and local government persisted in the capital. In 1899, another Conservative national administration finally replaced the old parish vestries and boards with twenty-eight new metropolitan borough councils within the area of the LCC (see Figure 5.3). After the First World War, the LCC's growing ambitions to control the city's sprawling outward growth were limited by the independence of numerous suburban authorities outside its boundary. In fact, there were more than 100 planning authorities in the Greater London area. During the inter-war years, planners increasingly perceived the need for planning on a regional scale. The 1930s were a turning-point, partly because London and other British cities were beginning to feel the strain of increasing motor traffic; a report by Sir Charles Bressey and Sir Edwin Lutyens, published in 1937, proposed a system of concentric ring-roads linked by radial highways cutting into the heart of London. There were also widespread and well-grounded fears that the concentration of one-fifth of its population in the capital city left Britain unduly at risk from aerial bombardment (Garside, 1984, p.253). In tune with these concerns, the Advisory Joint Town Planning Committee for Greater London, which included Raymond Unwin, planner of Letchworth Garden City, outlined a road network and recommended an annular green belt; beyond the belt, garden cities and satellite communities would facilitate a planned decentralization of the region. In 1938, a Labour-controlled LCC established the Green Belt at about 20–25 kilometres from the centre. The Royal Commission on the Distribution of the Industrial Population (the Barlow Commission) reported in 1940, and repeated the call for the dispersal of population and industry from London to smaller planned centres separated by belts of open land.

Various plans for London were developed during the Second World War, especially by underemployed architects. They included the plan of the Modern Architectural Research (MARS) Group, and plans by the Royal Academy and the Royal Institute of British Architects (Gold, 1995, 1997; Marmaras and Sutcliffe, 1994). The MARS plan of 1942, inspired by Le Corbusier's Radiant City, envisaged a London composed of sixteen north–south linear settlements separated by open spaces, and arranged in pairs on either side of a main east–west axis along the line of the Thames, where heavy industry, metropolitan services and the main transport arteries were to be located. The other plans were more traditional architecturally, but shared a primary concern with transport. They were eventually overshadowed by the *County of London Plan 1943*, written by J.H. Forshaw and Patrick Abercrombie, and by Abercrombie's *Greater London Plan 1944*. These plans were 'two sides of a seamless web of cloth' and were in essence a 'combination of roads and open space' (Hall, 1995, p.230). Abercrombie, like all planners of the 1930s, saw the car as a liberator,

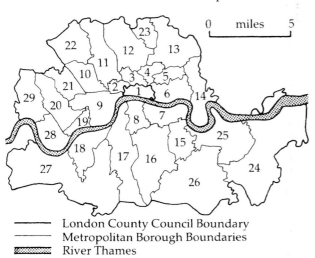

London County Council Boundary
Metropolitan Borough Boundaries
River Thames

Figure 5.3 The London County Council and the twenty-eight new metropolitan boroughs outside the City of London

Key I City of London;
2 Holborn; 3 Finsbury;
4 Shoreditch; 5 Bethnal Green;
6 Stepney; 7 Bermondsey;
8 Southwark;
9 City of Westminster;
10 St Marylebone;
11 St Pancras; 12 Islington;
13 Hackney; 14 Poplar;
15 Deptford; 16 Camberwell;
17 Lambeth; 18 Battersea;
19 Chelsea; 20 Kensington;
21 Paddington; 22 Hampstead;
23 Stoke Newington;
24 Woolwich; 25 Greenwich;
26 Lewisham;
27 Wandsworth; 28 Fulham;
29 Hammersmith (reproduced from Sutcliffe,1984, p.235;
© Patricia L. Garside)

and thought that London should be restructured to accommodate it. He took over the Bressey–Lutyens regional highway plan, and in addition, influenced by the US planning concept of the neighbourhood,[1] proposed a hierarchy of arterial, sub-arterial and distributor roads, to run in the open spaces between established communities, thereby freeing them of all through traffic (see Figure 5.4).

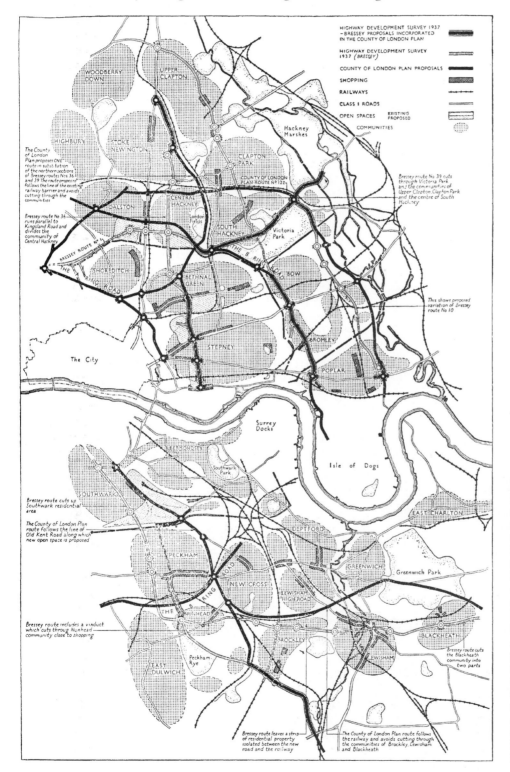

Figure 5.4 Road-planning in relation to communities, The drawing, from the *County of London Plan 1943*, shows how certain routes proposed by Bressey had been dropped or revised where they would have cut through existing communities and public parks, such as Victoria Park and Hackney in the north, and Southwark in the south. According to the authors, 'this principle has been followed in order to encourage the development of community and help to protect the lives of the inhabitants' (Forshaw and Abercrombie, 1943, p.53; drawing reproduced from ibid., p.54; © London Metropolitan Archives)

[1] This concept is discussed in Roberts and Steadman (1999), the third volume in this series.

These plans exemplified to different degrees the mix of technological rationality and efficiency, and yearning for the countryside, that had animated the Garden City movement. They struck a popular chord. But although the 1947 Town and Country Planning Act reduced the London region's 145 planning authorities to twelve counties and county boroughs, planners still had to negotiate London's vested political and economic interests. They were also faced with a change in 1951 from a Labour national government committed to national planning and the public provision of housing, to a Conservative administration that quickly removed controls on private building.

The wartime plans undoubtedly affected the development of the Greater London region. London itself was contained by the Green Belt, much extended after the Second World War, and excess population was relocated to satellite towns beyond it (see Figure 5.5). But the result was a much-diluted version of Howard's Garden City ideal and of subsequent planners' hopes for a more rational transport system, little of which was actually implemented. Part of the problem was the lack of a regional planning authority throughout the period of post-war reconstruction. A new era of planning beckoned in April 1965, when the Greater London Council (GLC) was established over an area reaching out to the Green Belt, and thirty-two large new boroughs came into being. But the political reality was that the Labour majority traditionally enjoyed by the LCC was now challenged by the Conservatives of the outer suburbs, and frequent changes in the political hue of the council prevented the continuity that large-scale regional planning requires. But there were other considerations: the GLC's advocacy of a series of three ring motorways in its Greater London Development Plan of 1969 foundered on the arguments of the increasingly vocal anti-motorway lobby, and on the difficult economic circumstances of the 1970s. The abolition of the GLC in 1985 was symptomatic of a new political tenor, much opposed to regulation and planning. It was within these shifting political and planning contexts that new transport and building technologies were applied.

Figure 5.5 London's Green Belt and new towns. Each of the Home Counties invoked their new post-war planning powers to prevent the continuation of the outward sprawl of London that had marked the inter-war period. As a result, the Green Belt doubled and even quadrupled in some areas after the Second World War (reproduced from Humphries and Taylor, 1986, p.101)

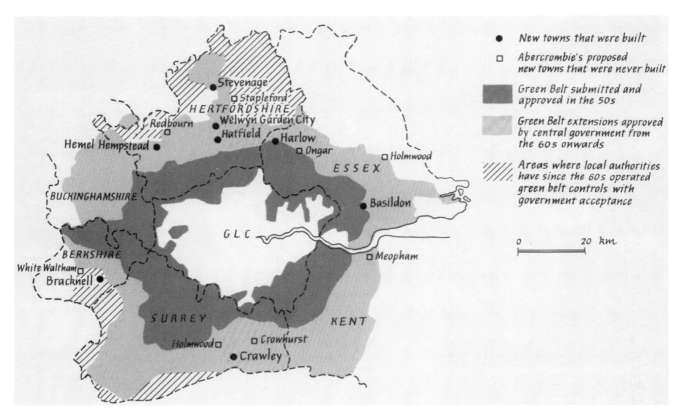

Transport

According to the eminent planner Sir Peter Hall, 'Modern London is largely a creation of its transport system, and that system is the result of several key events and non-events' (Hall, 1989, pp.119–20). However, consideration of London in the middle of the nineteenth century cautions against the view that transport innovations were *necessary* for the growth of any large urban settlement to metropolitan dimensions.

As can be seen from Table 5.1, London was still a walking city in the 1850s, and there were many barriers to traffic in the form of toll bridges, and tollbars on turnpike roads. By 1880, all of London's fifteen Thames bridges were free, and the last tollbar had been removed in 1871, but throughout the city, particularly in the noble housing estates of the West End, barriers on private streets still excluded traffic at night and controlled it by day. These street barriers, between 200 and 300 in number, took the form of swing- or drop-bars, iron posts, wood or iron gates, wooden fences and stone or brick walls. They had been the subject of controversy since the 1860s. In the case of the Bedford Estate, they were manned by intransigent uniformed former prison officers; one fracas at the Gordon Street gate in 1874 led to the death of a cab-driver. The matter came to a head in the 1890s, soon after the founding of the LCC. A private bill went to Parliament in 1890 calling for the removal of four gates in the area of the Vestry of St Pancras, which had long fought the wealthy residents of the protected estates over this issue. Another Act in 1893 led to the removal of fifty-nine other obstructions, and in 1894 the London Building Act outlawed any new unauthorized obstructions (Atkins, 1993a, b). The politics of London's built environment had clearly changed. The property values and convenience of the wealthy carriage-owning tenants of the estates of the landed aristocracy had formerly been dominant; they now yielded to the interest of the metropolitan commercial and financial elite, and their employees, in the more efficient functioning of the congested metropolitan transport system.

During the late 1870s and 1880s, the Metropolitan Board of Works made attempts – modest enough by Haussmannite standards – to improve London's traffic flow by widening certain thoroughfares and building new roads, notably Shaftesbury Avenue (opened in 1886) and Charing Cross Road (1887). New bridges were built at Putney in 1886, and Hammersmith in 1887. Tower Bridge, opened in 1894, was a modern version of the drawbridge: its steel-framed towers housed steam-powered machinery for raising the two bascules that made up the lower level, thereby allowing the passage of tall ships. The last great inner-city highway project (until the Hyde Park Boulevard in 1962) was completed in 1905: it consisted of the semicircular Aldwych, and Kingsway running from it to the north (see Figure 5.6 overleaf). This new avenue incorporated an underground railway extension, and a tramway tunnel to the Victoria Embankment, part of the LCC's plan to link up the tramways in the north and south of the city; this remained the only underground tramway route in Britain (Schubert and Sutcliffe, 1996, p.132). Thereafter, London's inadequate response to the challenge of motor transport took the form of local road-improvement schemes, the replacement of several bridges, and traffic-management measures, such as one-way streets: 'In the main, London was determined to push most of the new wine of its traffic through old bottle-necks' (Barker and Robbins, 1974, p.9).

There were economic as well as physical barriers to mobility. Before the First World War, commuting to work by rail was beyond the pockets of the majority of British employees. But London saw a big increase in suburban lines during the second half of the nineteenth century, especially in the third quarter, a phenomenon that reflected London's exceptional size and its unusually high proportion of white-collar workers. At this time, the railways

Table 5.1 Movement into the City of London, 1854 (derived from Barker and Robbins, 1963, pp.57–8)

Mode of transport	Number of journeys
Foot	200,000
Omnibus	20,000
Steamboat	15,000
Rail	6,000

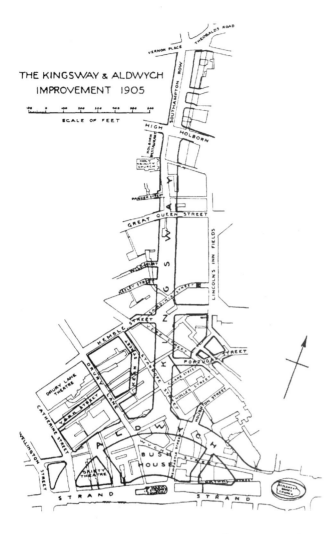

THE KINGSWAY & ALDWYCH
IMPROVEMENT 1905

SCALE OF FEET

Figure 5.6 Plan of Kingsway
and Aldwych improvement,
1905. Following Haussmann's
lead in Paris, the LCC had eleven
hectares cleared to allow this
new link between Holborn and
the Strand, including some of
London's worst slum areas, old
thoroughfares such as Wych
Street and a number of old
theatres. Kingsway, named in
honour of Edward VII, was
fronted with imposing buildings,
mostly with neo-classical
façades, some with steel frames
(Schubert and Sutcliffe, 1996,
pp.135–8). The main Aldwych
block, Bush House, was only
completed in 1935
(reproduced from Garside,
1984, p.237; © London
Metropolitan Archives)

were instrumental in the development of the outer
suburbs. In the west, lower-middle-class residential
estates sprang up in places such as Fulham, Acton
and Ealing on the Metropolitan District line, and
Hammersmith on the Great Western Railway. The
railway companies were less enthusiastic carriers of
working-class passengers. The 1883 Cheap Trains Act
made this service an obligation, but was resented by
most railway companies, as they feared that working-
class commuters would drive out their more lucrative
middle-class clientele. Among the exceptions was the
Great Eastern Railway, the 'Poor Man's Line', which,
along with trams, provided cheap commuting for the
respectable working class of the East End. The
lopsided response to the Act may well have
reinforced the tendency for working-class railway
suburbs such as Tottenham and Walthamstow to
develop in the north and east.

Omnibuses and horse trams came into their own in
the last quarter of the nineteenth century, facilitating
the movement of Londoners between the centre and
the inner suburbs. Horse tramways quickly ramified
in the suburbs after the passing of the Tramways Act
and the opening of London's first three lines; they
were generally welcomed both by poorer workers
and by middle-class residents of the wide suburban
streets on which they operated. It was a peculiarity of
central London, however, that the omnibus, as well
as the hackney carriage and hansom cab, continued
to flourish there right to the end of the nineteenth
century. There were several reasons: imported maize
reduced the cost of horse-feed in the 1880s; fares
came down because of greater competition among
omnibus companies, as well as from other transport modes; and there were
enough people with the income and working hours to patronize the higher-
priced modes of horse transport. But the main reason was the resistance of
wealthy, carriage-owning frontagers; in contrast to the residents of suburban
streets, they strongly objected to tramways being laid in their central streets. As
a result, there were still more omnibus than tram journeys in 1896: 300 million
and 280 million, respectively (Barker and Robbins, 1963, p.263).

The railways remained the means of the further flight of the middle class
from the influx into the inner suburbs of the tram-riding working-classes. The
trams initially had a limited effect in relieving inner-city congestion in the
nineteenth century; it was greater after the new LCC municipalized the
tramways (starting in the 1890s) and then electrified them (at the start of the
twentieth; see Figure 5.7). In fact, the first electric street tramway in the County
of London was operated by a private company, London United Tramways; it
ran between Shepherd's Bush and Acton, Hammersmith and Kew Bridge from
1901. The LCC came to electrification a little later, in 1903, mainly because of its
rejection of overhead wires in the central area, and the consequent expensive
decision to lay underground electric conduits in the streets (see Figure 5.8,
p.188). However controversial at the time, municipalization served to remedy
the uneven provision of transport under private operation. In 1906, after what
became known as the 'Battle of the Bridges', the LCC overcame parliamentary
objections and secured the right to run trams along the Embankment and
Westminster Bridge, a considerable step forward in integrating London's

Figure 5.7 The London electric tramway system. Although their lines mostly had underground conduits, the LCC lacked the powers to overcome the resolute opposition of the central metropolitan boroughs to tramways on their main thoroughfares. This left a void at the heart of the system, and obstructed the connection of its northern and southern parts. Outside the County of London, the electric tramways from the north-east to the south were owned and operated by municipalities – the biggest such tramway networks were those of the West Ham and Croydon local authorities. Suburban lines from the south-west to the north were run by private companies – London United Tramways and Metropolitan Electric Tramways. The dates in 1901 given for certain routes in the Shepherd's Bush, East Ham and Croydon areas reveal that electrification based on overhead equipment was under way some two years before the LCC introduced its underground conduits south of the river on the line from Westminster and Blackfriars Bridges to Totterdown Street, Tooting, by way of the Elephant and Castle. A surface contact system, with live studs in the middle of the road was installed in 1908 by the LCC on the Mile End Road from Aldgate to Bow Bridge; after several accidents, it was quickly scrapped (Barker and Robbins, 1974, pp.28–34, 91–5; map reproduced from ibid., 1974, pp.100–1)

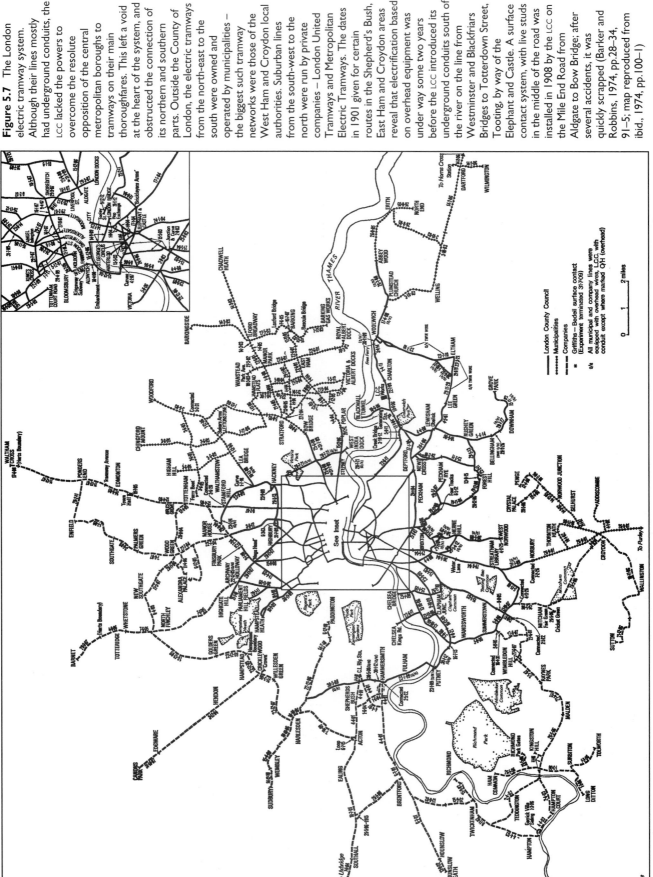

transport system (Barker and Robbins, 1974, p.95) (see Figure 5.9). By the 1900s, electric trams were seen as the principal means by which London was swallowing up the surrounding countryside:

> the railways are innocent of mischief compared with the electric trams which have conquered, or are in process of conquering, the great main roads of Middlesex. Their utility, of course, is beyond question, for the service is frequent, rapid, comfortable, cheap and sure. They have done more than many laborious Acts of Parliament towards the solution of the housing problem of London. Houses spring up along their routes, not in small groups but in large colonies … What for centuries has been a rustic hamlet is suddenly transformed into a suburb of London … The insistent clang of the electric tramcar's bell, and the constant sizzle and rattle of the wires, sound the knell of rusticity wherever they are heard.
>
> (Firth, 1906, p.1070)

Figure 5.8 The LCC's conduit system. Relaying a crossover at the complex intersection of Whitechapel High Street, Commercial Road, Commercial Street and Leman Street in the East End. The depth to which the road had to be excavated to accommodate the conduit yokes (one of which is seen lower right) made track-laying a tedious and expensive task (reproduced from Higginson, 1993, p.41; © London Transport Museum)

Figure 5.9 Trams and buses crossing Westminster Bridge in the late 1920s. County Hall, the vast new headquarters of the LCC, is on the right. The covered trams offered better all-weather transport than the open-topped buses. There was no clear path for southbound traffic: the bus and horse-drawn van in the foreground have taken a kerbside path, whereas the van in the middle distance has chosen a route on the offside of the tram tracks. Such confusion was held against trams, and helped bring about their demise (reproduced from Higginson, 1993, p.10; © London Transport Museum)

As Table 5.2 shows, trains were the principal means of travel in London during the third quarter of the nineteenth century, roads in the last quarter. The 1896 figures are equivalent to 165 journeys per head of population per year.

London was growing big enough to warrant investment in the extension of its pioneering underground lines. The Metropolitan and District Railways temporarily overcame their differences, and an underground Inner Circle line was completed in 1884, linking up the main railway termini. Like the first underground railway (see Chapter 3), this was constructed using cut-and-cover methods, and steam-operated, with all the noxious consequences for passengers, which persisted despite the fitting of condensers to the steam locomotives (Barker and Robbins, 1963, pp.235–7). A cleaner source of power was urgently needed, and a method of construction that avoided disruption of the urban fabric. In 1886 one solution was tried: deep tunnelling began through the clay under the Thames for a cable-operated tube railway (see Figure 5.10). But within two years it was decided to use the recently developed electric motor (unusually, as it would turn out, of British make) instead of cable traction. In 1890 the world's first electrically powered underground railway, the City and South London, began to run between King William Street in the City and Stockwell, south of the Thames. Technological and design shortcomings restricted its financial returns, and there was no further tube line until 1900. This time, the Central London Railway did prove a financial success; it ran to Shepherd's Bush, beneath the great east–west road artery of Cheapside, Holborn, Oxford Street and Bayswater Road, thereby linking the West End to the City. Thereafter, rapid development took place, and the main London Underground network was completed in the Edwardian period, much of it through US investment and enterprise, spearheaded by Charles Tyson Yerkes, the builder of Chicago's elevated railroad.[2] The main-line railways also sought to recover some of the passengers lost to the electric trams by means of the electrification of their suburban London lines, beginning with the Southern Railway's London Bridge to Victoria line in 1909 (Haywood, 1997, p.49). Although there were some surface suburban extensions to lines on both sides of the river in the inter-war period, and a spectacular new underground station

Table 5.2 Passenger-journeys in London, 1875 and 1896 (derived from Barker and Robbins, 1963, p.271)

Mode of transport	1875 (millions)	1896 (millions)
Train	c.160	c.400
Omnibus/ tram	c.115	c.600

Figure 5.10 Excavating a tube railway: at work on the 'shield' (Great Northern and City Railway). Tunnelling shields of this period generally had a cast-iron cutting edge that was forced into the clay by screw-jacks; behind the cutting edge was a face plate with a watertight door, through which the clay was removed by manual labour (Sims, 1903, p.151)

[2]   An account of Yerkes' activities in Chicago can be found in Roberts and Steadman (1999), the third volume in this series.

Figure 5.11 The extraordinary labyrinth of tunnels for the new Piccadilly Circus Station, which opened in 1928, and attracted world attention. The station was also noted for its Art Deco booking-hall, subdued lighting and shopping arcade. Along with several other stations on the Piccadilly line, it was designed by the architect Charles Holden (reproduced from Weightman and Humphries, 1984, p.93; © London Transport Museum)

at Piccadilly Circus in 1928 (see Figure 5.11), it was not until the Victoria and the Jubilee (opened in 1968 and 1979 respectively) that any new tube lines were added. The establishment of an underground system encouraged the 'westward march of business' (Garside, 1984, p.236); its surface extensions were also a new conduit of suburban growth, into areas now to be known as 'Metroland', a term coined in 1915 by the Metropolitan Railway in pursuit of its housing development interests (Haywood, 1997, p.53).

The motor bus made its début at the very end of the nineteenth century. Steam-driven lorries and vans were being produced by the late 1890s, and entrepreneurs were investigating the possibility of a steam bus in London. The London Steam Omnibus Company was registered in 1898, but in the end, internal-combustion engines were chosen, and the name changed to the Motor Traction Company. From October 1899, two buses with Daimler engines ran from Kennington to Victoria, but the enterprise lasted little more than a year. Not long after, in 1905, regular bus services started in the capital, quickly supplanting the omnibuses and carriages of the City and West End (see Figure 5.12). The move from horse-drawn to motor transport prompted the establishment of the Royal Commission on London Traffic in 1903. That was followed in 1907 by the creation of the Board of Trade's London Traffic Branch, which was given responsibility for road improvements. But the Commission's recommendation of a London Traffic Board was resisted both by local authorities and by the private operators involved in railways, trams and buses:

Figure 5.12 Early motor buses at the Bank crossing in about 1910. The Bank of England is on the left, the Royal Exchange in the centre, and the Liverpool & London & Globe insurance office on the right. It is clear that motor buses have not yet displaced horse buses entirely (reproduced from Howgego, 1977, plate 28)

Thus the major lines of approach to many of London's problems were set out – the *ad hoc* single-purpose body, the avoidance of governmental reform, the reliance on private endeavour, the shelving of the problems of central London, and the concentration on managing growth at the periphery.

(Garside, 1984, p.239)

Some measures were taken during the inter-war period in response to the continuing growth of motor transport. The first electric traffic-lights, in Piccadilly Circus, began to operate in 1925; and during the 1930s traffic lanes were marked by white lines, and a speed limit of thirty miles per hour was imposed. Taxation of motor transport – the Road Fund, introduced by Chancellor of the Exchequer David Lloyd George in 1909 – was to finance the new system of suburban bypasses and arterial roads, including the North Circular, the Western and Eastern Avenues, and the Great West Road (see Figure 5.13). The Restriction of Ribbon Development Act of 1935 was intended to control the speculative housing that these new roads also attracted.

Figure 5.13 Mixing concrete for the construction of the first few metres of the Great West Road, *c*.1925 (reproduced from Weightman and Humphries, 1984, p.58)

The Second World War and its aftermath caused a hiatus in the rise of the
motor car. Use of public transport in London (the buses, trams and
Underground had been taken into public ownership in 1933 under the London
Passenger Transport Board) reached a high point in 1949, equivalent to 479
rides per year per head of population (Barker and Robbins, 1974, p.12).
Thereafter it began its inexorable decline, as car ownership climbed, and
investment in roads was favoured over investment in rail. This bias also
affected London's public transport system, as buses were increasingly
preferred to trams in streets filling up with motor cars. From 1938, buses and
trolleybuses began to replace trams, and the last tram ran in 1952.

A sure sign of the pressure of private cars on London's inner roads was the
introduction of parking meters in 1958; underground car parks were another
remedy, and from 1973 traffic-lights were computer-controlled. The outer
suburbs and approaches to London were more susceptible to radical
restructuring, and space was made for the building of underpasses, elevated
roads and motorways An early example was the flyover road bridge at
Chiswick (1959); another at Hammersmith (1961) was built of prestressed
concrete, using precast sections, and incorporated electric cables to heat the
road surface. The central areas were not entirely immune: some of the most
drastic measures were taken in the City of London, where upper pedestrian
decks released space at ground-level for loading-bays, turning-circles and
parking-ramps (Hebbert, 1995, p.96).

The growing congestion of London's roads and public transport networks
contributed to the flight of manufacturing from inner London to the suburbs
and beyond. The gradual filling-in of the national motorway network
reinforced this trend, and high-technology enterprises in particular began to
locate in a 'western crescent' centred on the eastern end of the M4, and
stretching from Hampshire in the south-west, through Surrey and Berkshire, to
Hertfordshire in the north-west (Hall *et al.*, 1987, p.174). The M25 orbital
motorway, another transport link in this high-tech zone, was finally completed
in 1987; this was the main vestige of the system of ring and radial roads for
London proposed by Bressey and Lutyens but never fully implemented. In
1965, the GLC had attempted to resuscitate another of the proposed ring-roads
in the form of an 'Urban Motorway Box', some four to six miles (six to ten
kilometres) from the central area. The scheme was opposed by
environmentalists and conservationists, and scrapped when the Labour Party
regained control of the GLC in 1973 (Hart, 1976, pp.167–75). According to Sir
Peter Hall, one of the proponents of the Box Motorway:

> Our failure to handle the problem of the car in the city, I would argue, was the
> greatest single failure in the post-war planning of our cities and above all in the
> planning of London, where we did less road reconstruction than in Birmingham
> or Newcastle or Glasgow; and it simply reflected the sort of hopeless fudge, up
> the hill and down again, that has characterised successive policy lurches by
> governments of different complexions over far too long.
> (Hall, 1995, p.234)

Despite the GLC's renewed emphasis on London's public transport in the 1970s,
it has continued to decline through starvation of investment, though there have
been some new initiatives, notably the computer-controlled driverless trains of
the elevated Docklands Light Railway, which began to operate in 1987, and the
express rail link from Paddington to Heathrow, opened in 1998. In seeking
solutions to traffic congestion, planners now increasingly prefer not to build
more roads for motor vehicles; by the late 1980s, Hall, a former supporter of
urban motorways, saw them as 'hopelessly grandiose and expensive', and
looked instead to a Regional Express Rail network, along the lines of Paris's
Réseau Express Régional (RER) (Hall, 1989, pp.124, 147–52; see Section 5.3).

During the century from 1870, road improvements and the succession of transport innovations from the horse tram to the motor bus, along with the provision of services such as gas, electricity, water, sewerage and telephone lines, facilitated the extraordinary outward radial development of London. It is clearer in the case of services that technology *followed* suburban growth, except, arguably, in the case of the aggressive promotion of electricity in the inter-war years (Luckin, 1990). Much of the debate over the connection between transport and suburban development has centred on the London suburbs (Dyos, 1961; Thompson, 1974). There are instances where transport facilities preceded estate-building, and others where the reverse happened. But for most of the period under review, the outward spread of low-density housing was inexorable in response to seemingly unlimited demand. In consequence, it may be that housing development would have occurred whatever the state of the infrastructure.

Building construction

Though transport innovations helped empty the inner city of residents, they also brought in large numbers of day-trippers and tourists, stimulating the development of London as a centre for shopping and entertainment. The West End was increasingly, in the second half of the nineteenth century, given over to department stores, restaurants, hotels, pubs, clubs, theatres, music halls and concert halls. New, larger music halls, such as the Alhambra Palace in Leicester Square, opened in the later nineteenth century, with nightly audiences of about 2,000 people (Barker and Robbins, 1963, p.207). Some of the new buildings incorporated technological innovations – lifts were installed in hotels from 1859 and incandescent electric lighting from the 1890s – but for the shell of these buildings, traditional masonry construction was the order of the day. Many of London's major new buildings after 1870 reflected its traditional role as a political, cultural, educational and juridical centre, and were correspondingly traditional in appearance. Striking examples were the Royal Albert Hall (1867–71), with its iron-and-glass dome set upon a brick cylinder faced with terracotta; G.E. Street's neo-Gothic Royal Courts of Justice in the Strand (1868–82); and J.F. Bentley's eclectic Westminster Cathedral (1895–1903). The interest of much of London's late Victorian and Edwardian architecture lies primarily in its treatment of traditional designs and in the choice of materials. Increasing air pollution – more a result of domestic coal-burning grates than of noxious East End industries – lay behind the choice first of red bricks, and then, during the late nineteenth century, terracotta, instead of the now smoke-blackened yellow bricks of the Georgian terraces.

How did London respond to the opportunities presented by the reinforced-concrete or steel frame? Before the First World War, the London authorities constrained these new construction techniques; in 1894, for fire-security reasons, the LCC limited buildings to eighty feet (24.4 metres), the height that could be reached by a fireman's ladder. The US entrepreneur H. Gordon Selfridge had to overcome great resistance from the LCC before he could build his steel-framed department store (1909) to something like the dimensions he desired (Lawrence, 1990[3]). In the end, it was the Crown, exempt from the London building regulations, that was best-placed to experiment in the construction of civic buildings. The first large-scale use of reinforced concrete in London was Henry Tanner's King Edward Buildings (1907–11) for the new Post Office, constructed according to the patented prefabricated system of the French engineer François Hennebique (see Figure 5.14 overleaf). These were followed by the reinforced-concrete frame of the Stationery Office (1912); and steel

[3]   An edited version of the article by Lawrence can be found in Goodman (1999), the Reader associated with this volume.

framing lay beneath the unexceptional façade of the Public Trustee Office in Kingsway (1912–16). But LCC regulations approving the use of reinforced concrete only came into effect on 1 January 1916 (Port, 1995, pp.271–3, p.314, n.5). Thereafter, steel and concrete frames became more widely used, though they often remained imperceptible behind the heavy, often traditionally styled stone façades favoured for major buildings of the inter-war periods: for example, Sir Edwin Lutyens' Midland Bank in Poultry beside the Bank of England (1924–39); the BBC's Broadcasting House in Portland Place (1931); new hotels, such as the Dorchester on Park Lane (1930); and several large office-blocks, such as Adelaide House near London Bridge (1924–5) and Unilever House on the Victoria Embankment (1930–31). Stout masonry walls continued to bear the loads of some important inter-war structures, such as London University's Senate House, designed by Charles Holden (Ruddock, 1996). The structures least inhibited about modern materials and construction techniques were several new building types of the post-1870 period. These included transport facilities, such as bus depots (see Figure 5.15) and Underground stations (see Figure 5.16); power stations for electric lighting and traction; and the new electrically powered factories, many of them built by US firms, that proliferated on London's new arterial roads. In another important category were the structures of mass entertainment: sporting arenas, notably Wembley Stadium with its reinforced-concrete towers and terraces, and picture palaces such as the black granite Odeon in Leicester Square, which replaced the Alhambra Palace music hall.

During the Second World War, many parts of London suffered extensive damage, including some 800 hectares of Stepney and Poplar in the East End. This was mostly the result of raids by bombers, though from 1944 to 1945 the city fell prey to V1 flying bombs (doodlebugs) and V2 rockets. Out of this

Figure 5.14 The General Post Office extension, 1907, designed by Sir Henry Tanner, and engineered by Louis-Gustave Mouchel, the principal agent of Hennebique's reinforced-concrete system in Britain. Reinforced concrete was chosen to save costs in the construction of this otherwise unremarkable building in St Martin's-le-Grand, near St Paul's Cathedral; the use of the material was questioned in the House of Commons (Collins, 1959, p.79; photograph: reproduced from ibid., plate 19)

mayhem came an opportunity to realize the visions of a modern London foreshadowed in the plans of the 1930s and the wartime years. The Festival of Britain in 1951 on the South Bank, under its director Hugh Casson, gave a platform to Modernist architects. Although its structures, such as Ralph Tubbs' massive Dome of Discovery, were mainly transient, it left a permanent legacy, the Royal Festival Hall, which now sits alongside the controversial 'brutalist' concrete South Bank architecture of the early 1960s. Another landmark of Modernist post-war reconstruction was the New Barbican scheme, a complex mixture of commercial and residential towers and slabs. Situated in the most war-damaged area of the City of London, the Barbican is set on a raised deck among lawns, pools and terraces, with a pedestrian upper walkway. The scheme was proposed in 1954, though the first tower opened only in 1960; at forty-three storeys, the three residential towers were the tallest apartment buildings in Europe (see Figure 5.17, p.196). In general, the early 1960s marked the point at which the London landscape became dominated by tall buildings, as property developers made hay during the last term of the long Conservative administration from 1951 to 1964. The first of the new wave of concrete-and-glass office blocks was Castrol House (1958) in Marylebone

Figure 5.15 Stockwell Garage, built in 1952, photographed the following year. Between 1950 and 1953, two new garages were built, and several tram depots were reconstructed to accommodate the swelling bus fleet in South London. The construction of Stockwell Garage was so delayed by a shortage of steel that reinforced-concrete beams were used instead, giving a spectacularly large, unobstructed covered area (Barker and Robbins, 1974, p.340; photograph: reproduced from ibid., plate 144; © London Transport Museum)

Figure 5.16 Southgate Station, one of the show-piece tube stations designed by the architect Charles Holden for the Piccadilly line extensions to Cockfosters, Hounslow West and Uxbridge. These were built in 1932–3 with funding for public works provided by the Labour government in order to tackle unemployment. Holden turned the station building from a house into a hall, with Modernist features, including continuous fenestration and a horizontal canopy. Southgate, along with Arnos Grove, Northfields and Sudbury Town, was designated in 1971 as being 'of special architectural interest' (Barker and Robbins, 1974, pp.252–3; photograph: reproduced from Weightman and Humphries, 1984, p.91; © London Transport Museum)

Road; in 1962 the Shell Centre on the South Bank and the Vickers Tower on Millbank took office towers to unprecedented heights of more than 100 metres. The LCC sometimes agreed to developers' plans in return for land for traffic-improvement schemes; this was the case with Joe Levy's highly profitable Euston Centre, and Harry Hyams' Centrepoint office tower off Tottenham Court Road (Humphries and Taylor, 1986, pp.63–4). Other controversial intrusions on London's skyline were the Hilton Hotel (1963) in Park Lane, Eric Bedford's General Post Office Tower (1965), and Roebuck House (1963) in Victoria Street, an upmarket residential skyscraper that dared to overlook Buckingham Palace Gardens. Some of the towers of the 1960s

Figure 5.17 The Corporation of London's high-rental Barbican redevelopment, seen under construction in June 1968. It was designed to bring 6,500 residents back into central London (reproduced from Glendinning and Muthesius, 1994, plate XV; photograph: Miles Glendinning)

attained their heights with a technique that offered an alternative to the metal or reinforced-concrete frame (see Figure 5.18).

A temporary halt to the office boom was called in 1964 by the incoming Labour government, alarmed at the congestion caused by an increasingly car-commuting army of office-workers. Another sign of growing hostility to Modernist tower blocks was the battle in the 1970s over the GLC's plans to redevelop the run-down Covent Garden market area; conservationists succeeded in getting most of the historic buildings listed, and office-building has since been less wanton with the existing urban fabric. More latitude was found in the area of the docks, which started to run down after the decline of trade with the Commonwealth in the 1960s, and were killed off when Britain joined the European Common Market, and the Port of London Authority decided to invest in deep-water container facilities at Tilbury in Essex, along the Thames Estuary. The opportunity to redevelop the area was missed by a hamstrung GLC in the 1970s. In the 1980s and 1990s, this task was

Figure 5.18 The Commercial Union (the taller of the two) and P&O Buildings under construction, at the junction of Leadenhall Street and St Mary Axe in the City of London. Built in 1968–9, they deployed an innovatory high-rise structure that dispensed with peripheral load-bearing columns; instead, the floors were suspended on steel cantilevers set into a reinforced-concrete core that transmitted all the loads to the earth. The squat P&O tower is almost complete, structurally, but the core of the Commercial Union tower is clearly visible, as is the upper set of horizontal cantilevers and vertical hangers from which the upper twelve floors were to be suspended (reproduced from Mainstone, 1975, p.280; photograph: R. Mainstone)

entrusted to the unelected London Docklands Development Corporation, which in the very different unregulated planning climate of Thatcherite Britain was empowered to override the objections of local authorities. During this period, there was a boom in the financial services sector, as London took advantage of the globalization of banking and securities markets. London's ascendancy in the time-zone between Tokyo and New York prompted massive investment in office facilities in the City of London, and nearby London Bridge, Broadgate and Canary Wharf (Hebbert, 1995, p.94).

But the great bulk of London's built environment was residential. It is notable that from 1870 to 1939, London's new citizens were accommodated mainly by the outward rather than the upward development of the physical fabric. Up to 1870, London's building stock had grown mainly through a mix of countless market-led individual decisions and the separate planning of several aristocratic estates. Recent historical research, less confident perhaps in the superiority of the present-day city, has revised the 'Dickensian' view of a London dominated by slums, inefficient private water companies and venal local authorities. It is now thought that working-class housing was cheaper and more commodious in London than in other European capitals of the time (Davis, 1995, p.69). But whatever the motives of late Victorian middle-class reformers, educated public opinion was focusing on the 'housing question' during the last quarter of the nineteenth century, and the 1890 Housing Act empowered the fledgling LCC to clear slum areas, and purchase land compulsorily for working-class dwellings. This measure paved the way for the LCC and other local authorities to take the lead in the provision of low-cost housing. The 1900 Housing Act extended the LCC's powers of compulsory purchase to the suburban areas beyond the London county boundary.

Before the First World War, the LCC's involvement in municipal housing was limited to such out-of-county locations – notably cottage estates such as Totterdown Fields, Norbury, Tottenham (White Hart Lane) and Old Oak (Hammersmith). But there followed a vast increase in the suburban local authority housing stock, as a result of the 1919 Housing and Town Planning Act. It has been argued that fear of social unrest lay behind this measure, which now gave local authorities the main responsibility for low-rental housing (Swenarton, 1981). The LCC insisted on its unique statutory right to build outside its own area; one result was the great municipal housing development on over 1,000 hectares acquired in 1919 at Becontree in Essex. But there was growing resistance from the suburban boroughs to the LCC's plans to build public housing in their areas, and in the end the proportion of local authority housing built between the wars was much smaller in London than in other large British cities. Nearly 80 per cent of London's 175,000 new homes were built by small builders in the private sector in an unregulated fashion, using traditional building methods and materials; they were mostly of a vernacular, semi-detached design, in low-density neighbourhoods (Jackson, 1973). After the return of a Labour council in 1934, the LCC retreated into its own domain, and instituted a programme of slum clearance, resulting in a concentration of high-density flats in four boroughs south of the Thames: Lambeth, Lewisham, Camberwell and Wandsworth (Garside, 1984, pp.246, 249).

With the exception of some experiments in non-traditional housing in the years immediately following the First World War, house-building was pretty traditional fare, except that services were considerably improved. But technological innovation made an indirect contribution: the rapid spread of the suburbs into the Middlesex claylands was possible because the mechanization of urban transport meant that they were no longer needed for horse forage. Similarly, the Thames market-gardening plain could be built over, now that fruit and vegetables could be transported from further afield. Despite the semirural ideal exploited in the sale of these houses, new environmental problems arose in the unending task of housing London's millions. In order to build on clay, there was a pressing need for main-drainage schemes. It was the local claygrounds that had led to a brick-built city; but now the sulphates in some clays attacked the concrete foundations of the new suburban homes, and clay in general brought problems of settlement and subsidence. Rapid inter-war estate development led to flooding, and new sewage facilities to the pollution of wells. Sites were generally unplanned, with a monotonous layout, and no regard for existing vegetation (Jackson, 1973).

There were further experiments in non-traditional housing under the unusual conditions that prevailed in the aftermath of the Second World War (see Chapter 7). The acute housing shortage owing to bomb damage was compounded by a rise in London's population of over one million during the 1950s. As a result, the LCC now looked more favourably on the kind of high-rise residential developments pioneered on the Continent by the Bauhaus and Le Corbusier between the world wars. One of the first projects was Powell and Moya's Pimlico riverside housing estate of 1950, likened to an 'anglicized Gropian *Zeilenbau*' (Esher, 1981, p.105). A softer approach was the 'mixed development' at Alton West in Roehampton, incorporating eleven-storey point blocks, along with slab blocks and row housing (see Figure 5.19). From the

Figure 5.19 Mixed development by the LCC at Roehampton Lane (Alton West) in Victorian gardens in the south-western borough of Wandsworth. The estate was designed in 1953 by M.C.L. Powell, O.J. Cox, R. Stjernsdedt *et al.*; the designers decided to arrange the slab blocks in a staggered fashion, as advocated by Gropius in 1935 (photograph: Aerofilms)

Figure 5.20 The Brandon Estate, Southwark, designed in 1955 by E.E. Hollamby, and photographed during the course of construction in 1959. The six eighteen-storey point blocks in this development rose well beyond the previous LCC eleven-storey limit, following a relaxation in 1954 of the capital's fire safety requirements. In the left foreground are examples of the *Arcon* prefabricated house, discussed in Chapter 7 (photograph: Hulton Getty)

mid-1950s, the LCC sought to maximize its returns from the new building technologies, and especially from the lift, the most expensive component of the high-rise building. Tower blocks became thicker and higher, and were built using proprietary systems for the frame, claddings, windows and internal fittings (Glendinning and Muthesius, 1994[4]) (see Figure 5.20). The outcome in the war-damaged East End was far from Le Corbusier's vision of the Radiant City:

> Architecturally the result is a dusty museum of changing ideologies stretching all the way from the sub-Shaw tenements of Boundary Street (1895) via neo-Georgian King's Mead (1938) to the Festival/Swedish of the fifties and thereafter to all the systems of the sixties, culminating in the three mighty and much hated 26-storey monoliths of the Wellington estate in Hackney and the St George's towers that loom up south of Commercial Road.
>
> (Esher, 1981, p.120)

[4]   Edited excerpts from the book by Glendinning and Muthesius can be found in Goodman (1999), the Reader associated with this volume.

By the mid-1960s, residential slabs and towers had had their day. Their demise is often associated with the gas explosion that caused five deaths in the partial collapse of Ronan Point tower block in Canning Town in 1968 (see Figure 5.21). The architect Lionel Brett (Lord Esher) insists, however, that they were off the drawing boards at County Hall well before this event, as architectural fashion swung to high-density low-rise structures for public housing. It is important to keep London's slabs and towers in perspective; in 1971 only 6.4 per cent of local authority housing stock consisted of high flats (Esher, 1981, pp.130–31).

> By 1980 housing had ceased to be the major determinant of the look of London ... We were back with the 2-storey cottages from which the adventure had started a century ago.
> (Esher, 1981, pp.162–3)

Figure 5.21 The partial collapse of the system-built Ronan Point tower block, Canning Town in the East End, 1968. The block had no separate frame, but was a multicellular box consisting of prefabricated wall and floor panels.[5] When a gas explosion four storeys from the top of the building blew out a load-bearing wall panel, the unsupported wall and floor panels above fell through, causing a further collapse in the structure below, and killing five people (reproduced from Mainstone, 1975, p.91; photograph: R. Mainstone)

---

[5]   Details of alternative high-rise construction systems can be found in the edited excerpts from Glendinning and Muthesius in Goodman (1999), the Reader associated with this volume.

5.3 Paris

In 1850, the population of Paris was 1.05 million, less than half that of London; by 1968, the number had risen to 8.2 million, whereas the figure for London was less than 7.5 million by 1970/71 (Mitchell, 1998, pp.74–6). Although both populations had grown by broadly similar amounts, there were continuing contrasts between the two metropolises during this period. These were never so marked as at its beginning. In 1870, any comprehensive city plan for London was out of the question, as the autonomy of its local authorities was sacrosanct. By contrast, as David Goodman discussed in Chapter 3, the centre of Paris had already been ruthlessly reworked by Baron Haussmann, partly in response to its revolutionary past. Paris's vulnerability to both internal and external threats was underlined by the turmoil of 1870–71. The disastrous Franco-Prussian War brought down Napoleon III and the Second Empire. Paris, now under the Third Republic, was besieged and bombarded, and its citizens had to endure Prussian troops marching along the Avenue des Champs-Elysées and through the Arc de Triomphe. In the wake of the Prussians came civil war between the national government and the revolutionary Paris Commune, which culminated with the communards torching and blowing up some historic buildings, including the Hôtel de Ville (the town hall) and the Tuileries Palace. When order was restored, the authorities set about repairing the damage, though the gutted Tuileries was eventually demolished and the Tuileries Gardens extended. In the longer term, they sought to execute the rest of Haussmann's plan; Alphand, Haussmann's associate, was retained as director of works until 1891. There was, however, a greater reluctance to sweep aside historic buildings to make way for the straight boulevard. In fact, for most of the period up to 1958, when Charles de Gaulle founded the highly centralized and interventionist Fifth Republic, the Parisian authorities, much like those in London, adopted a *laissez-faire* approach to urban growth (Noin and White, 1997, p.55).

But the physical form and fabric of central Paris in 1870 had been shaped by the earlier French tradition of authoritarian intervention – in the persons of Henri IV, Louis XIV, Napoleon Bonaparte and Napoleon III – and by its ring of fortifications, the last of which, five kilometres from the city centre, were completed in 1845 (Noin and White, 1997) (see Figure 5.22 overleaf, and Chapter 3, pp.104–5). For these reasons Paris was a much more ordered and compact city than London; in 1913, it had an average population density of 370 people per hectare, compared with 161 in London and 265 in Berlin, a differential that continued to the end of the twentieth century (Evenson, 1979, p.271; Noin and White, 1997, pp.29–30). Paris's fortifications had acted as a tax-collecting point on goods entering the city – an arrangement that meant food was cheaper in the suburbs, so making these outer areas more attractive to working-class residents. Settlement was particularly dense near the gates and then followed the main axial routes. For much of the period after 1870 Paris continued to approximate to Sjoberg's model of a pre-industrial city, with the rich preferring to live near the centre, and the poor forced to live on the outskirts.

The beginnings of a change in the city's social geography became apparent during the 1920s: in 1921, the population of the city inside the fortifications reached its maximum of 2.9 million, while that of the suburbs stood at 1.5 million; ten years later, the city population had fallen slightly below 2.9 million, and that of the suburbs had risen above 2 million; and by 1970, the figures were 2.6 million and 5.6 million respectively (Evenson, 1979, pp.221, 238). Despite the familiar pattern of inner-city depopulation and suburban expansion indicated by these figures, the City of Paris remained sharply demarcated from the suburbs, and was loath to take on any financial responsibility for their development. According to a description in 1930, much of the working-class housing in the suburbs consisted of self-build shanties,

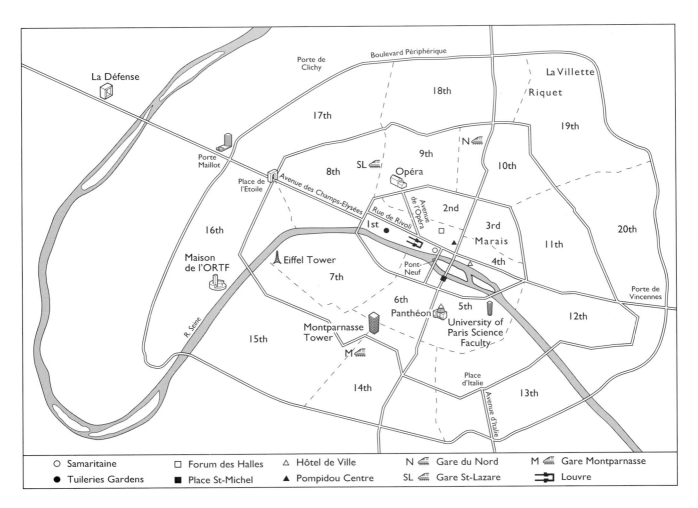

| ○ Samaritaine | □ Forum des Halles | △ Hôtel de Ville | N Gare du Nord | M Gare Montparnasse |
| ● Tuileries Gardens | ■ Place St-Michel | ▲ Pompidou Centre | SL Gare St-Lazare | Louvre |

Figure 5.22 The City of Paris, showing several buildings and other locations mentioned in the chapter. The three rings of the simplified street system (the Internal Boulevards, the Exterior Boulevards and the Boulevard Périphérique) largely follow the lines of successive city walls and ramparts; the Boulevard Périphérique was built on the zone immediately outside the fortifications of the 1840s. The sequence of numbers of the twenty *arrondissements* – the main administrative units of the City of Paris – takes the form of a clockwise spiral starting with the historic Right Bank central area

made of wood, plasterboard, tarpaper and corrugated metal, and dwellings adapted from old trucks and wagons (quoted in Evenson, 1979, p.207). There was no tradition of local government beyond the city boundaries, and in what was an area of mainly private development, urban amenities such as sewerage, water supply and even roads were completely lacking, and tuberculosis was rampant. There was growing support for the Communist Party in the suburbs in the 1920s, and it consequently became known as the 'red belt' (Evenson, 1979, p.231). Fear of the politicization of working-class residents was probably one of the reasons why planners and housing officials sought to control this diffuse suburban development. The fortifications of the 1840s (some 35 kilometres long and 130–35 kilometres wide) and their accompanying *glacis*, or field of fire (a further 250–300 metres in depth), which demarcated the suburbs from the city, were acquired by the municipal council in 1919, and redevelopment began in the 1920s and 1930s. The shanties of the 'zone', as the strip outside the fortifications was known to Parisians, had long presented a contrast with the city proper that shocked well-heeled visitors. In the end, no unified plan was implemented for the fortification area, and a mishmash of institutional, industrial, speculative and public uses prevailed (ibid., pp.272–6).

Concerns about public hygiene had taxed the government well before the 1920s. Eugène-René Poubelle, prefect of the Seine *département* between 1883 and 1896, made lidded rubbish bins compulsory (they are still named after him), and instituted surveys of tuberculosis mortality. These surveys were completed in 1904, and six *îlots insalubres* (unhealthy areas) were identified, a figure raised to seventeen after another survey in 1919. They were concentrated in the eastern half of the city, and despite the ambitions of planners to destroy them, remained largely intact during the inter-war years; even in 1925, nearly one-third of houses failed to comply with the law of 1894 requiring connection to the public sewer (Sutcliffe, 1970, p.106). This failure exemplified the tension in Paris between, on the one hand, technocratic visions of urban order and efficiency, and, on the other, property interests, which were increasingly buttressed by a conservationist respect for the historic fabric of the city. This post-Haussmannite sensibility was evinced in the foundation in 1884 of the Société des Amis des Monuments Parisiens – the first of several Parisian private middle-class historical societies – and reinforced by the creation in 1897 of the Commission du Vieux Paris, set up to advise the municipal council. Only the expropriation powers assumed during the Second World War finally enabled the authorities to set about vigorously demolishing what they considered the insanitary properties in the *îlots insalubres*.

In the intervening period, French planners began to look to the British Garden City movement for a solution to the morbidity of metropolitan life, though what resulted was even further from Howard's ideal than Letchworth, Welwyn or Hampstead Garden Suburb had proved. A Société des Cités-Jardins (Society of Garden Cities) had been founded in 1903, but it was not until the inter-war period that action was taken. Sixteen *cités-jardins* were laid out under the direction of Henri Sellier of the Seine *département*; they were conceived as working-class housing projects within commuting distance of the city, rather than the kind of self-sufficient settlement Howard had envisaged. For cost reasons, the British cottage-style single-family house was soon abandoned, and these developments took on the form characteristic of continental working-class housing, with the emphasis on three- to five-storey apartment blocks, and some tower blocks (see Figure 5.23).

Like London, Paris recognized the need for planning on a regional scale during the inter-war period. Responding to the haphazard growth of the city during the 1920s, the 'Prost plan' was completed in 1934. (It was named after Henri Prost, the chief government architect.) Like the London inter-war regional plans, it emphasized limitations on the growth of the metropolis, and reflected

Figure 5.23 *Cité-jardin* of Châtenay-Malabry, built in the 1930s on a site of fifty-six hectares south-west of Paris. This settlement included one eleven-storey tower, the first to be built in a *cité-jardin,* but three- to five-storey blocks were the norm (reproduced from Evenson, 1979, p.227; photograph: Norma Evenson)

Figure 5.24 Urban development and transport networks in the Paris region. The pattern of urban growth, including five new towns (Saint-Quentin-en-Yvelines, Cergy-Pontoise, Marne-la-Valleé, Melun-Sénart and Evry) is as it was projected in the revised Schéma Directeur of 1969. According to this regional plan, Paris was set to grow along two major east–west axes to the north and south of the City of Paris. The first motorway in the region was the A13, opened in 1946; it was followed by the A6 and the A1 in the 1960s, and by the A10, A4 and A15 in the 1970s, along with the Boulevard Périphérique. The A86 and the Francilienne, under construction at the time of writing, are intended to facilitate movements across the suburbs. The first portion of the suburban express railway network (the RER – see p.214) was opened in 1969 (based on maps in Noin and White, 1997, pp.64, 145, 153, and Evenson, 1979, p.345)

contemporary enthusiasm for the motor car in its call for a regional network of autoroutes. It also laid down standards for the provision of services in suburban developments, and stipulated that open spaces in the region should be protected. The City of Paris itself had much less open space than inner London; according to a study of 1975 it had 345 hectares, or 1.75 square metres per inhabitant, compared with nine square metres per inhabitant in London (Evenson, 1979, p.309). A version of the Prost plan was finally made law in 1941. There was, however, little new building in the inter-war period, and virtually none during the Second World War, though, as a result of the German occupation of Paris, the city escaped much of the damage to its building stock that London suffered. In the wartime years, however, legislation was enacted to provide for reconstruction of towns suffering war damage, and for a much more interventionist approach to town-planning. In Paris itself new powers of expropriation swept aside inter-war resistance to the redevelopment of the fortification zone, and many squatters (*zoniers*) were evicted.

The post-war era opened with concerns about the growing dominance of the Paris region (now known as the Ile-de-France; see Figure 5.24): in 1861, this region, consisting of the City of Paris and the inner and outer suburbs, contained 7.5 per cent of the total population of France, but by 1946 the proportion had risen to 16 per cent, a proportion that has been constant ever since (Evenson, 1979, p.336; Noin and White, 1997, p.22). In 1960, a regional plan was decreed, and the Plan d'Aménagement et d'Organisation Générale de la Région Parisienne (PADOG) outlined a vision of a constrained Paris with further expansion directed to suburban nodes and other metropolitan areas of France.

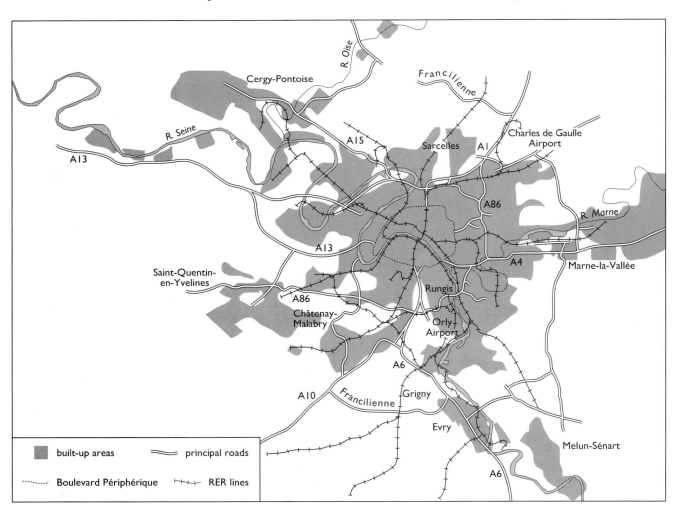

built-up areas   principal roads

Boulevard Périphérique   RER lines

But when a new plan for the capital – the Plan d'Urbanisme Directeur de Paris – was adopted in 1962, ideas about decentralization were put aside; President Charles de Gaulle reasserted the political centrality of Paris when he founded the Fifth Republic in 1958, and others saw Paris as 'the capital of Europe' in the new world of the European Common Market. The city plan of 1962 was superseded in 1968 by the Schéma Directeur d'Aménagement et d'Urbanisme de la Ville de Paris. Both plans involved a degree of zoning of the city, designating areas that were predominantly for business, for administration, for the university and for manufacturing. The second plan envisaged intervention to reverse the spontaneous march of commerce and administration into formerly residential western *arrondissements*, and the decline of industry in the east, by establishing poles of new employment, usually near railway stations (one such pole was the Gare Montparnasse in the south of the city, described later). In 1969 a revised version of a new regional plan (the Schéma Directeur d'Aménagement et d'Urbanisme de la Région de Paris) was published. This included provision for a series of five 'new towns', sometimes incorporating existing settlements and tied to the capital by motorways and railways (see Figure 5.24); construction began in 1969. Like Milton Keynes in the UK, they were relatively low-density settlements, presupposing widespread car ownership, but, unlike the British new towns, they were never intended to be independent of the metropolis. They had a mix of apartment blocks and single-family houses, and were also home to some bold architectural experimentation, especially in their shopping and leisure centres (see Figure 5.25).

Figure 5.25 Marne-la-Vallée: part of the centre of Noisy-le-Grand, one of the new town's separate nuclei of development. On the right is Le Palais, a reinforced-concrete apartment block designed by the Catalan postmodernist architect Ricardo Bofill; it forms part of an urban complex called Les Espaces d'Abraxas, completed in 1983. Marne-la-Vallée, founded in 1972, and now the location of the French Disneyland, was established to the east of Paris to balance population movement to the west (photograph: Greg Castillo)

Transport

During much of the nineteenth century, Paris was more innovative than London in the field of urban transport, thereby continuing the tradition of Pascal, who had pioneered urban public transport in Paris during the 1660s and 1670s.[6] In the late 1820s the omnibus, another French innovation, got off to a more successful start in Paris than in London, partly because of the French capital's greater population density: there were more tall buildings (five or more storeys) in Paris, and a more crowded urban fringe. And in 1854 Paris became the first European city to adopt the horse-drawn tram, which had originated in the USA (and was known in Paris as the *chemin de fer américain*). The grooved rails of this new tramway were laid flush with the street, an innovation that was the work of a French engineer, Alphonse Loubat. His system had at first been rejected by the Parisian authorities, and so had made its 1852 début in New York.

It was not until the early 1870s that a tramway network for the suburbs was built. Some of these tramways came under the wing of the Compagnie Générale des Omnibus (CGO), which in 1860 had been granted a fifty-year monopoly of Parisian omnibuses. In return, the imperial administration could influence fare policy, demand the operation of unprofitable routes, insist on the expansion of the network, and regulate all aspects of the enterprise, down to the dress and conduct of drivers (Larroque, 1988, p.43; Papayanis, 1993, 1996). But little changed in the first decades of the CGO's monopoly; their routes continued to be oriented to the west, where residents' higher incomes and shorter working hours enabled them to commute, and where the CGO could invest in upmarket property development. In the 1880s most factory workers walked to work from their crowded accommodation in the south and east of the city, and the factories themselves, marginalized in the Haussmannite grand scheme, continued to rely mainly on waterways for the movement of raw materials and products.

In contrast to the CGO monopoly of omnibuses, there were several concessions to private tramways. Although they were resisted in the centre of the city, trams became the main link between the city and the suburbs, and served to reinforce the pattern of working-class residence on the outskirts of the city, especially when mechanical traction was introduced – the congestion, noise, smell and animal suffering associated with horse-drawn urban transport was no less a cause for public concern in Paris than it was in London. Mechanization started with steam trams in the 1870s; other modes included Mekarski trams, which were powered by compressed air. But despite these innovations, transport was lagging behind rapid urban growth. This unsatisfactory state of affairs was tackled by the Paris municipal council, as it took a more radical turn in the 1890s. As the Exposition of 1900 approached, the council granted concessions to ten new tramway companies in order to link Paris with the suburbs; included in the concessions were requirements to introduce electric traction and to reduce fares. The new electric trams began to operate at the same time as industry was relocating to the suburbs, and this further reinforced the pattern of working-class housing in the outskirts of the city. Electric traction began with battery-powered tramcars in 1892; overhead wires were introduced in the suburbs in 1896 and underground conduits laid in the city centre from 1898 (see Figure 5.26). Although electrification might be seen as a victory for radical politicians over private entrepreneurs, according to Dominique Larroque the policy actually served the interests of two great new manufacturers of electric-traction equipment: the French Thomson Houston Company (set up in 1893 to operate the patents of the US Thomson Houston electrical engineering corporation) and the General Traction Company (Larroque, 1988, pp.58–62). Even so, electrification was by no means universal. By 1910 there were twelve companies operating more than 100 lines, some of

---

[6]   Urban transport in seventeenth-century Paris is discussed in Chant and Goodman (1999), the first volume in this series.

which still used steam and compressed air. Paris was 'virtually an open-air museum of traction equipment' (Evenson, 1979, p.82).

Unlike Londoners in the final quarter of the nineteenth century, Parisians made little use of suburban steam railways for the journey to work. By 1867, the six main railway termini were connected by a circular railway running just inside the fortification ring, and from 1877 to 1886 an outer circle was constructed through the suburbs. But the main point of these ring railways was to provide a route around Paris for the national network; only in the western suburbs did the big railway companies see an interest in providing stops for passengers travelling within the city. As in London, a reduced worker's fare operated from 1883 on some early-morning and late-evening trains, but this was still beyond the pocket of most of the working classes. By the end of the nineteenth century, the commuting habits of London and Paris were starting to converge: reductions in both fares and working hours prompted an increase in rail commuting, and led to a programme of electrification, beginning with the Invalides suburban line in 1900, and completed in 1967. The suburban lines of what is now the publicly owned national railway company, the Société Nationale des Chemins de Fer Français (SNCF), continue to make an essential contribution to Parisian mobility: in particular the services from the Gare Saint-Lazare to the western suburbs.

According to Larroque's interpretation, the new interest of capital in providing affordable urban transport opened the way for the municipal council to establish an underground railway network. This was a project long deemed necessary because of the inadequacy of trams and omnibuses to serve the central city; controversy had raged around it since the early 1870s. There were arguments about the relative merits of underground and elevated systems, and about the effects of depopulation on the inner city. There was a battle between the national government and the municipal council over the purpose of the network. The council wanted to confine it to journeys within the city. They attempted to

Figure 5.26 The Place Saint-Michel, *c.*1900, showing horse-drawn carriages and, on the left, a steam tram; on the right is an electric tram bound for the eastern suburbs. Though equipped with a trolley-pole for the suburban section of the line, here in the centre of the city the electric tram ran on batteries. Objections to overhead cables greatly restricted the contribution that electric tramways could make to transport in central Paris, and they mainly served the outer districts and the suburbs (photograph: Photographie Roger-Viollet)

Figure 5.27 Métro construction at the Place Saint-Michel in the early 1900s. Owing to the great depth of the line at this point, and the waterlogged subsoil, the Saint-Michel station was constructed on the surface and lowered into position. The vertical drum visible in the foreground is the bottom of the shaft for the lift and staircase (photograph: Photographie Roger-Viollet)

Table 5.3
Passenger-journeys in Paris, 1890 and 1913 (adapted from Larroque, 1988, p.58)

Mode of transport	1890 (millions)	1913 (millions)
Tram	c.137	c.489
Omnibus	c.115	c.246
Métro	–	c.467

exclude the big railway companies by introducing a narrow-gauge track of 1.30 metres rather than the standard network gauge of 1.44 metres, though this ploy was rejected by the war ministry for defence reasons; nevertheless, they succeeded in reducing the size of the tunnels, and in that way kept the interurban locomotives out. In 1898, with all minds concentrated by the approach of the 1900 Paris Exposition, the Métropolitain underground railway (popularly known as the 'Métro') was started; the first line, along the city's main east–west axis, from the Porte de Vincennes to the Porte Maillot, opened in 1900. The construction of the Métro had more in common with London's pioneering Metropolitan underground railway than with its deep-tube successors, though most of the railway lines were drilled out without disturbing the surface (see Figure 5.27 for an exception). The numerous cavities and galleries of the Paris subsoil made it unsuitable for deep tunnelling (Sutcliffe, 1970, p.85). All lines were to follow the path of wide streets, with the stone vaulting of the otherwise concrete tunnel usually about one metre below the road surface. In that way, the Métro's engineers avoided problems with private property above the ground, and with sewers, water-pipes and electric conduits, which were mostly beneath the tunnels' eight-metre depth (Evenson, 1979, p.106).

By 1914 an eighty-kilometre network was in place, with 400 million passenger-journeys per year; by 1930 this had risen to 888 million. In 1935, the number of passenger-journeys per kilometre of line (at 5.8 million) was much higher than in London (1.68 million) or Berlin (2.74 million); again, this reflected Paris's relatively high population density (Evenson, 1979, pp.109, 114). A dramatic rise in the mobility of Parisians resulted from the mechanization of urban transport (see Table 5.3): 'this jump reveals the extent of the distortion between need and demand maintained by a system centred only on a particular clientele' (Larroque, 1988, p.58).

The effect of the Métro, in the paradoxical fashion now widely recognized as a feature of many investments in transport infrastructure, was to increase surface congestion rather than reduce it, as the new system encouraged

Parisians to travel about the city more often than before (Sutcliffe, 1970, pp.86, 90). A sign of the increasing pressure of traffic on Paris's streets was a device invented by Eugène Hénard, a government architect and traffic expert, who as early as 1905 was predicting that Haussmann's thirty-metre-wide boulevards would be unable to cope with the motor car. Hénard's solution was that great metropolitan traffic innovation, the roundabout (*carrefour à giration*) (see Figure 5.28). The first of these, in 1907, was a twelve-branched intersection at the Place de l'Etoile, the site of the Arc de Triomphe. In 1909, the municipal council established a minimum width of fifteen metres for new streets, and a year later forced the CGO to replace its antiquated horse omnibuses with motor buses. In 1911 the Commission d'Extension de Paris was set up partly in response to the chaos caused by the mix of horse-drawn vehicles, motor vehicles, trams, bicycle-carts and handcarts. In its report of 1913, the commission recommended that existing boulevards be widened, partly by cutting back the characteristically wide pavements with their clutter of kiosks and street furniture, and also that Haussmann's street plan should be extended, though with more respect for the historic fabric, and less of his 'fetishism of the straight penetration' (quoted in Evenson, 1979, p.34). Much of this renovation had to wait until after the First World War.

A major administrative change came into effect in 1920, when the operation of all surface transport was leased for thirty years to the Société des Transports en Commun de la Région Parisienne (STCRP), the fixed assets coming into public ownership. This mix of state ownership and private operation was also applied to the railways and the Métro. Up to 1930, the Métro was essentially a system operating within the city boundaries, though during the 1930s a few short suburban extensions were added to the system. The inadequacy of the Métro in the context of continuing population growth was balanced by a great increase in commuting on the suburban surface railways, and a corresponding

Figure 5.28 Hénard's design for a traffic roundabout with underground pedestrian walkways converging on the centre. To eliminate the chaos and danger caused by drivers taking unpredictable paths at complex intersections, vehicles were required to approach the roundabout on the right-hand side of wedge-shaped traffic islands and to turn right on to the roundabout. They would then continue in the same direction until leaving, again on the right-hand side of one of the traffic islands (reproduced from Hénard, 1906, p.242; © Bibliothèque Nationale de France)

gravitation of shops and businesses to the main railway termini in the outer *arrondissements*, especially the Gare Saint-Lazare; in 1931, 363 million such passenger-journeys were made (Sutcliffe, 1970, p.178; Evenson, 1984, p.269). In 1929, despite the wholesale modernization of the tramway system during the 1920s, the authorities decided to replace electric trams with motor buses, which had been introduced in 1905 and were now felt to be less of an obstruction to the increasing numbers of motor cars. The measure was completed by 1937, and transport officials at the time were proud that Paris had become the first major European capital to free its streets from the lumbering, outdated trams (Evenson, 1979, p.83). Larroque, who considered the substitution completely unjustified in the case of the suburban lines, and lamented the 'anarchistic urbanization tendencies' motorization brought about, blamed the change on the supplanting of the Thomson Company's domination of the STCRP board by oil and motor car companies (Larroque, 1988, p.63); certainly Renault and Citroën were running private bus services successfully against state lines in the 1930s. Throughout this period of restricted nationalization, fares were kept stable, and the deficit resulting from the continuing modernization of the system was written into the public accounts.

During the Second World War the unavailability of petrol for motor vehicles meant that the Métro was used as never before: there were 1.32 billion passenger-journeys in 1943 (Evenson, 1979, p.114). In 1942, the surface and underground transport systems of Paris were integrated, and in 1948, the Régie Autonome des Transports Parisiens set up to direct the unified system. But at the same time as these communal forms of transport were being brought under firmer public control, Paris was leading the way in Western Europe in the embrace of the motor car; journeys by bus declined from 450 million in 1950 to 180 million in 1970 (ibid., p.86). The existing wide boulevards of Paris were much better suited to the car than were the streets of London; at one time Haussmann was credited with magical foresight in this regard (ibid., p.21), though he is now more likely to be blamed for the intolerable traffic congestion of the present-day metropolis. The car had already begun to dominate the streets of Paris during the inter-war years: the number in the Paris region had grown from 150,000 in the early 1920s to 500,000 on the eve of the Second World War. Although they all but disappeared from the streets during the German occupation, by 1970 their number had mushroomed to 2.5 million (ibid., pp.54–5); the urban fabric had somehow to adapt. From 1947, one-way streets, synchronized traffic lights and parking restrictions were introduced (Sutcliffe, 1970, p.236). During the 1950s and 1960s there was yet more widening of the boulevards, through a further reduction of the pavements, together with their characteristic lines of trees and street furniture. But such measures were hardly adequate to deal with the growth of traffic, and so planners turned their attention to an exterior orbital expressway intended to divert traffic from the inner city. The Boulevard Périphérique (see Figure 5.29), skirting the redeveloped land of the fortification ring and its zone of fire, was completed in 1973; it acted to worsen traffic congestion in the City of Paris, rather than improve it. The Périphérique was described by Norma Evenson as a 'concrete moat separating Paris from the suburbs':

> the circular motorway reinforced the image of Paris as a walled city, replacing the old line of defenses with an impenetrable barrier of high-speed vehicles.
>
> (Evenson, 1979, p.285)

In order to facilitate east–west movement through the city itself, without destroying historic roads and buildings, the city's tree-lined embankments (*quais*) were turned over to riverside expressways, connecting at either end with the Boulevard Périphérique. The Right Bank expressway, completed in 1967, includes a tunnel starting just before the Louvre and emerging at the Pont-Neuf.

Figure 5.29 The northern edge of Paris, looking east from the vicinity of the Porte de Clichy. The Boulevard Périphérique marks the city boundaries, within which lie the parallel bands of the zone, immediately to the right of the expressway, and next to it, dog-legging through the centre of the photograph, the redeveloped fortification area (photograph: Prefecture de Paris; © La documentation Française/Interphotothèque)

The motorization of Paris was becoming as controversial as it had been in London. President Georges Pompidou had announced that 'Paris must adapt itself to the automobile' (ibid., p.59); but after demonstrations by users of public transport in 1969 and 1970, and the first oil crisis of 1973, his successor, Valéry Giscard d'Estaing, halted the completion of the Left Bank expressway. There ensued, in the view of one Parisian technocrat, a brief 'golden age' of public transport, from 1974 to 1982. It was financed by a new tax on employers, the *Versement-Transport* (VT), and enacted alongside the pedestrianization of historic quarters and shopping precincts, and the detouring of through traffic. But although the 'manna' of the VT prompted an unprecedented expansion of public transport, it has been argued that this was more by the 'rehabilitation' of existing technologies (buses and trams) than by investment in transport innovation (Offner, 1993, p.99). Moreover, it resulted in greater use by existing consumers of public transport rather than by new ones, and so ownership and use of the car continued its inexorable rise. And since these existing consumers were paying the same as before, public transport got deeper into debt. The result of Parisians' embrace of the motor car and planners' attempts to accommodate it was, as might be expected, only a more diffuse pattern of regional settlement. In this respect, the British and French versions of the motorized society were converging: in 1967, nearly half of all French households possessed a car, and by 1987 it was more than 75 per cent, with 24 per cent owning two or more vehicles (ibid., p.99).

Building construction

As an emblem of the renewed commitment to the 'Haussmannization' of Paris under the Third Republic, and in anticipation of the Paris Exposition of 1878, the Avenue de l'Opéra (originally planned by Napoleon III in association with Haussmann) was opened by President Mac-Mahon in 1877 (see Figure 5.30). The new road afforded a vista from the Rue de Rivoli to the Paris Opéra, a confection of architectural styles designed by Charles Garnier and inaugurated two years earlier. This new boulevard marked something of a departure, however, in having no trees or street furniture; and its architectural unity was ensured only by building regulations limiting heights and window sizes, rather than by the government-imposed designs of the Haussmann era. It rapidly became a fashionable shopping street with the heaviest traffic of all the central boulevards (Sutcliffe, 1970, pp.53, 164–5).

The Opéra itself represented one branch of an institutionalized bifurcation in Parisian public building – the aesthetic approach of the Ecole des Beaux Arts. The other – the functional approach of the engineers of the Ecole Polytechnique – was represented by the metal structures of the 1889 Paris Exposition. The best-known is the controversial tower designed by Alexandre Gustave Eiffel, the survival of which beyond its planned twenty-year concession was partly because it was used for early radio transmissions; the first overseas broadcast from Paris was transmitted from the Eiffel Tower to Casablanca in 1907. There was also the Galerie des Machines, built to display the latest advances in machine technology; it boasted four and a half hectares of unbroken floor space, and an unprecedented 111-metre span (Evenson, 1979, pp.131, 136; Mainstone, 1975, pp.224–5; see Figure 5.31). Somewhere in between were the functional building types of the new commercial Paris, notably the department stores, or *grands magasins*, which combined decorative façades with the

Figure 5.30 Avenue de l'Opéra, *c.*1900 (photograph: Photographie Roger-Viollet)

Figure 5.31 Galerie des Machines, 1889 Paris Exposition. The great span achieved in this structure was due to the use of the three-pinned steel arch, as opposed to the continuous wrought-iron arches that made up the great train-sheds of railway termini such as St Pancras in London. Much larger sections could be prefabricated on the ground, and then raised and pinned together. The pins at the crowns of the arches can be seen in the illustration; the others were below floor level beneath the triangular feet of the arches (reproduced from Mainstone, 1975, p.225)

spacious internal dimensions permitted by metal framing. Perhaps the best-known was the Bon Marché on the Left Bank, designed by Eiffel and Louis Charles Boileau in 1876.

Since 1783, houses had been limited to a maximum height of twenty metres. A decree of 1859 sought to reduce heights further (though the upper twenty-metre limit was soon restored), the maxima allowed depending upon the street width; this was part of the regulatory framework giving the monotonous six- to seven-storey residential and commercial blocks of the Haussmannite scheme. The desire for greater architectural freedom, not unconnected with the desire to extract the maximum value from increasingly costly urban land, found expression in the municipal council's new building code of 1902, which among other things allowed for the addition of up to three storeys at attic level for buildings on major streets. This measure was in part a response to the desire of architects and builders to exploit the greater heights made possible by steel and concrete framing and lifts (Sutcliffe, 1970, p.211). Even this proved controversial, and the authorities continued to resist US-style skyscrapers in the inter-war period, despite the ambitions of Modernists such as Auguste Perret and Le Corbusier to show the Americans the proper way to enhance the urban built environment with tall buildings. Indeed in 1926, the municipal council forced the owners of the Hôtel Astoria to demolish the upper storeys which since 1907 had presumed to overshadow the Arc de Triomphe in the Place de l'Etoile (ibid., p.299). New building regulations issued in 1930 reaffirmed the existing limits on height. The continuing tight regime nevertheless allowed greater scope for individuality of building design, and sometimes this involved construction innovations. The façade of Frantz Jourdain's Samaritaine department store offered one of the first frank displays of a frame with an infill of metal-framed windows (see Figure 5.32); the building also had two giant cupolas, about which a '*scandale*' raged until, under pressure from the municipal council, the owner removed them in 1925 (Sutcliffe, 1993, pp.119, 155). The concrete frame was also fully expressed in Perret's apartment house on the Rue Franklin; at ten storeys this

Figure 5.32 La Samaritaine department store, in the Rue de la Monnaie near the Pont-Neuf, 1905. The metal frame, though prominent, was decorated with ornamental metalwork and inset faience panels (reproduced from Hitchcock, 1968, plate 133)

Figure 5.33 25 bis Rue Franklin, in the suburb of Passy in the 16th *arrondissement*, designed by Auguste Perret, 1903. It was Perret's first use of a reinforced-concrete frame; he selected it to make the most of a cramped site, and subcontracted the work to agents of the Hennebique system. The semi-octagonal inset of the façade provided additional window space, and served instead of an internal light court. Although the frame was boldly expressed in the façade, Perret faced it with ceramics in case it was not waterproof. In extending to ten storeys, the building fully exploited the freedom permitted by the 1902 regulations (Abram, 1987, pp.82–5; Collins, 1959, pp.178–84; photograph: Norma Evenson; reproduced from Evenson, 1979, p.269)

was one of the tallest functional structures in Paris (Abram, 1987; see Figure 5.33). More modest low-cost apartment blocks with concrete frames (known as *habitations à bon marché*) were built on the periphery during the inter-war years, often on land released by the demolition of the fortifications, though never in sufficient numbers to solve the housing shortage.

But it was not the until the most pressing infrastructure problems of post-war reconstruction had been faced that the focus returned to the chronic problem of the housing shortage in the Paris region. As far back as 1912, public housing offices had been set up to construct low-cost dwellings, but relatively little was achieved during the inter-war period, apart from the *habitations à bon marché* and the *cités-jardins*. The shortage worsened after the Second World War: shanty towns of immigrant workers (*bidonvilles*, after *bidon*, meaning 'oil drum') appeared on the outskirts of the city from the 1950s to the 1970s. Eventually modern construction technologies were brought to bear upon the problem, and several *grands ensembles* were built between 1953 and the suspension of the programme in the early 1980s (Noin and White, 1997, p.112). These were suburban complexes of high-rise towers and slabs, and blocks of five or six storeys, each planned to house between 30,000 and 40,000 people; by 1969 one-sixth of the population of the Paris region was reckoned to live in them (Evenson, 1979, p.238). They were connected to the city by motorways, the SNCF railway and more recently by a high-speed rail network – the Réseau Express Régional (RER), the first portion of which opened in 1969 (see Figure 5.24, p.204). Although they allowed for increasing commuting by car, they had none of the low-density characteristics associated with motorization in the UK and the USA:

> Although the Parisian concept of commuting is similar to that of many North American cities, the architectural components of the urban region seem virtually reversed. The North American suburbanite generally lives in a single-family house and travels to the city, where he is likely to work in a modern skyscraper. The Parisian suburbanite is likely to live in a modern high-rise building and may commute to the city to work in an eighteenth-century mansion or renovated nineteenth-century dwelling.
>
> (Evenson, 1979, p.240)

The *grands ensembles* varied in design and reputation. Sarcelles had the worst reputation among outsiders, being likened by its critics to a 'concentration camp' and a 'vertigo of technology' (quoted in ibid., p.246) (see Figure 5.34). An exception to the architectural monotony of the *grands ensembles* was Emile Aillard's La Grande Borne, which introduced colour, human scale and variety into the housing units, and separated pedestrians from motor cars (see Figure 5.35).

Between 1954 and 1974, there was a massive investment in new housing; 60,000 dwellings were demolished, and 270,000 new ones built; moreover, their average size grew from thirty-six to fifty-nine square metres. Nevertheless, in 1961, only 20.3 per cent of dwellings in the Paris region were single-family houses, compared with 78 per cent in Britain and 49 per cent in West Germany (Evenson, 1979, pp.236, 252). This striking contrast seems to have been the consequence of the high cost of land and building materials in the Paris region, and of the organization of the building industry, which was geared to the construction of apartment buildings. According to Jacques Riboud,

Figure 5.34 The *grand ensemble* of Sarcelles, in the northern suburbs of Paris, 1978. Ironically, in view of its present crime rate, the seeds of the settlement were sown in 1954 by some employees of the Paris Prefecture of Police; they formed a building society and acquired a tract of inexpensive building land in the vicinity of Paris that was well served by a main road and a railway terminating at the Gare du Nord (photograph: Sodel-Brigaud; © La documentation Française/Interphotothèque)

Figure 5.35 The *grand ensemble* of La Grande Borne, 1978. Built between 1967 and 1970, the settlement was situated twenty-five kilometres to the south of Paris at Grigny, and connected to the city by motorway. The presence of a radio station nearby limited building heights to five storeys. Its imaginative design was based on the use throughout of a standardized load-bearing panel, and three basic window types (Evenson, 1979, p.243; photograph: Service Régionale de l'Equipement)

the peculiar density of Parisian housing was not determined by new construction technologies; the high-rise architecture of the *grands ensembles* was 'less the inevitable result of financial, technical, or economic restraints than the expression of a fashion and the manifestation of the aesthetic conceptions of certain professionals' (quoted in Evenson, 1979, p.253).

The economic boom from the 1950s to the oil crises of the 1970s also created renewed pressure in the centre for office space, and the city authorities felt obliged to reconsider the embargo on tall buildings in central areas. A new building code was issued in 1961 and incorporated into the Plan d'Urbanisme Directeur of 1967. It allowed for increasing building heights the further out land was from the historic centre, giving the city a 'saucer-like shape' (Noin and White, 1997, p.72): the inverse of the economically driven US city with its high-rise commercial and administrative core, and low-density suburbs. Certain areas of the city were identified for urban renewal and freed from the usual restrictions; these districts, usually located in the outer *arrondissements*, were completely transformed, as dense masonry buildings devoted to working-class residence and industrial employment gave way to widely spaced curtain-walled

Figure 5.36
Redevelopment project in Riquet, in the 19th *arrondissement*, one of the Parisian districts designated for urban renewal in accordance with the 1967 building regulations (reproduced from Evenson, 1979, p.177; photograph: Service Régionale de l'Equipement)

Figure 5.37 Looking north from the Porte d'Italie, in the 13th *arrondissement*. The Avenue d'Italie is on the left, leading toward the Place d'Italie, from which redevelopment extends south and east over an area of eighty-seven hectares (© La documentation Française/Interphotothèque)

Figure 5.38 The comparatively low-rise skyline of Paris's Left Bank, as seen from the Cité Morland tower block in the Marais district on the Right Bank, 1986: the towers of the University of Paris science faculty (left foreground) and Montparnasse (to the right, further back) bracket the dome of J.-G. Soufflot's Panthéon, completed in 1791 (reproduced from Sutcliffe, 1993, p.173, figure 247; photograph: Anthony Sutcliffe)

slabs and towers for commercial, administrative and middle-class residential use (Evenson, 1979, p.295; see Figure 5.36). As a result, by 1974 two-thirds of all buildings over thirteen storeys had been put up in only four of the capital's outer ring of *arrondissements*; the greatest number were south of the Seine in the 13th *arrondissement*, in which the Place d'Italie was heavily redeveloped (ibid., pp.195–6; see Figure 5.37). In general, heights in the city remained restricted: the new code of 1974 kept two-thirds of the city more or less to existing levels (ibid., p.180) (see Figure 5.38). Nevertheless, the new developments aroused passionate opposition, and talk of the 'assassination of Paris' and its 'Manhattanization': 'the Paris skyline is now crenellated with towers, each more extravagant, higher and more profitable than its rivals' (Chevalier, 1994, p.264). No technocratic act was more lamented by Louis Chevalier than the demolition in 1971 of Victor Baltard's metal-and-glass pavilions at Les Halles, close to the Louvre in the centre of the city, following the removal of the meat market to La Villette on the northern edge of the city, and of the fruit and vegetable market to the southern suburb of Rungis; the old market gave way, though, to the Forum des Halles – a four-level underground complex including an RER railway station – rather than a high-rise structure.

Despite passionate resistance to the prospect of US-style high-rise cityscapes, Paris found a place for some striking examples of buildings exploiting the potentials of new building materials and construction technologies. Conspicuous among these was the circular structure and central tower of the

Figure 5.39 The Maison de l'ORTF, by Henry Bernard, built between 1956 and 1963 in the 16th *arrondissement*, overlooking the Seine. The building has the form of an electromagnet, clad in white metal, surrounding a central tower. The tower, which at seventy-five metres was allowed to breach the current building regulations, was the 'thin end of a very thick wedge' (Sutcliffe, 1993, p.164; photograph: © La documentation Française/ Interphotothèque)

Figure 5.40 Pompidou Centre, west front, 1986. Making a bold contrast with the historic environment into which it was thrust, the building flaunts its supporting structure and services, including a glass-enclosed escalator, on the outside – a feature Richard Rogers, one of its architects, repeated in his Lloyds Building in the City of London. It has been described as looking like 'the middle section of the hull of a beached Queen Mary … with all the superstructure and exterior plating removed' (Sutcliffe, 1993, p.182). Sponsored by President Pompidou in 1969, the building opened in 1976; it was sited on the Plateau Beaubourg, an open area near Les Halles (photograph: Anthony Sutcliffe)

Maison de l'ORTF, the headquarters of the French radio and television service (see Figure 5.39). An even more outspoken testament to post-war building technology, and in particular to the increasing importance of services in the design of tall buildings, was the Georges Pompidou National Centre of Art and Culture, dedicated in 1976 and designed by Renzo Piano and Richard Rogers (see Figure 5.40). Apart from these single structures, there was a hotel and exhibition complex, completed in 1968, at the Porte Maillot; this was formerly the gate where the Voie Triomphale, the main east–west axis, cut through the old fortifications, and is now a major junction with the Boulevard Périphérique (see Figure 5.41). In the south of the city, the old Montparnasse and Maine railway stations were amalgamated into a single Gare Montparnasse, and surrounded by massive slab blocks. The station complex was completed in 1964, and by 1973 was overshadowed by the Montparnasse Tower, a collaboration between French architects and US engineers, and at 210 metres the tallest building in Europe at the time (see Figure 5.42).

Figure 5.41   The Porte Maillot, viewed from the west. The Boulevard Périphérique is visible in the foregound, passing under the Avenue de Neuilly, which forms part of the Voie Triomphale, the great east–west axis of Paris extending from the Louvre to La Défense. Adjacent to the intersection is the Centre International de Paris, constructed between 1971 and 1974. It includes the high-rise Hôtel Concorde Lafayette and the Palais des Congrès, an ovoid structure containing an ensemble of meeting halls. The architects were Guillaume Gillet, Henri Guibout and Serge Maloletenkov (reproduced from Evenson, 1979, p.48; photograph: Préfecture de Police)

Figure 5.42   Montparnasse redevelopment scheme, c.1974, making dramatically clear the planned relationship between railway commuting and inner-city commercial and administrative development. Sited on a hill facing Montmartre across the historic city core, the tower slowly and all too visibly ascended at the very time that opinion was moving against the high-rise projects of the 1950s and 1960s (reproduced from Sutcliffe, 1993, p.17; © La documentation Française/Interphotothèque)

Figure 5.43   The Défense district in 1989, looking out of the city towards the Grande Arche, one of the 'great projects' of the presidency of François Mitterand (1981–95). The office complex was located on a pedestrian deck spanning the Avenue de Neuilly, with several levels of underground services. After the lid was taken off building heights in the late 1960s, the Groupement d'Assurances Nationales (GAN) building rose to 200 metres, and another controversy raged as it drew attention from the Arc de Triomphe in the vista from the Louvre in the historic centre (© La documentation Française/Interphotothèque; photograph: A. Guyomard)

The most dramatic intrusion of the steel-and-glass tower on the post-war suburban skyline of Paris has been the development of a government-controlled commercial and residential complex at La Défense (see Figure 5.43). Situated outside the western city boundary at the end of the Voie Triomphale, it was seen in the mid-1950s as a way to provide office accommodation without redeveloping the historic centre of the city. The government undertook to provide the transport infrastructure, including a multi-level subterranean RER railway station, underground roads and parking, and a surface loop-road connecting with the Boulevard Périphérique. The buildings were put up by private concerns, with building heights strictly controlled. But in order to recover their infrastructure costs, the government decided in the late 1960s to allow taller structures, thereby increasing their revenues. Formerly restricted to twenty-five storeys, towers now rose to forty-five storeys and beyond. The highly controversial result was a 'fashion show of curtain-wall cladding materials' (Evenson, 1979, p.187).

The complex was defended by President Pompidou, who lamented the Parisians' 'retrograde' prejudice against height (ibid., p.190); but in 1974, France entered a new political and economic climate, and Pompidou's successor, Giscard d'Estaing, cancelled not only the Left Bank expressway, but also a planned 179-metre tower in the Place d'Italie district.

5.4 *Conclusion*

London and Paris present some intriguing parallels and contrasts in their deployment of new technologies in response to the two sets of challenges considered in this chapter: the years of rapid expansion leading up to the First World War, and the time of reconstruction and development from the end of the Second World War to the economic crises of the 1970s. The history of the urban fabric in these periods clearly reflects the potentials and limitations of prevalent technologies: in both capitals, the growth pattern associated with a mix of steam railway and horse-drawn carriages was quite distinct from the urban morphology connected with a later mix of electric traction and motor transport. But differing economic, social and ideological constraints clearly influenced the implementation of the available technologies. For most of the period after 1870, London continued to spread outward, using mainly traditional building methods and materials, rather than upward, using new technology. Although feverish investment in railways during the middle decades of the nineteenth century endowed the capital with a well-developed suburban system, its pluralist system of local government thereafter made it generally slow to adopt new modes of transport, or to adapt to their consequences, despite appalling congestion and housing conditions. Though it was quicker to replace electric trams with buses than many other European cities, London was still some way behind Paris in this aspect of motorization. The French capital remained much more compact and more wedded to apartment living. In both capitals, it is clear that the building of high-rise slabs and towers, and the balance struck between private and public transport, was not the inevitable outcome of technological progress, but heavily dependent upon the political hue of the national and civic administrations.

But the necessarily broad-brush treatment in this chapter reveals that there was much in common in the ways that the two cities exploited the new technologies of transport and construction. Both cities appear to have embraced the same transport and construction technologies in much the same sequence. Both experienced similar tensions between the individual's love of private transport and low-density housing, and the planner's goal of an efficient integrated transport system and a cost-effective means of housing millions. Both cities had historic buildings that resisted the demands of the motorized society and were increasingly valuable as tourist attractions. Similarities in the measures adopted stem partly from the sharing of a common technological and planning culture; each city increasingly looked to the other for solutions to its acute metropolitan problems. First, the Paris of Haussmann was a model of decisive action to be aspired to; then, British attempts to decentralize London through garden-city satellites found favour in Paris. But there remained key differences, partly because of the intractability of two urban fabrics that were quite distinct historical palimpsests, and also because of differing traditions of government. Finer focusing than has been possible here would surely reveal an even more distinctive historical grain. There was, even in an age of the growing internationalization of transport and building technologies, some local technological colour: London's deep-tube railways were necessary because the city lacked the wide boulevards that Haussmann had carved through Paris, and which enabled the Métro to be built just below their surface. Perhaps the differences between the two cities are symbolized in London's attempt to outstrip Paris's Eiffel Tower; London's proposed equivalent, on the site of the future Wembley Stadium, foundered for lack of investment and hardly got off the ground (Kostof, 1991, p.321).

References

ABERCROMBIE, P. (1944) *The Greater London Plan 1944*, London, HMSO.

ABRAM, J. (1987) 'An unusual organization of production: the building firm of the Perret Brothers, 1897–1954', *Construction History*, vol.3, pp.75–93.

ARMSTRONG, J. (1996) 'The development of the Park Royal industrial estate in the interwar period: a re-examination of the Aldcroft/Richardson thesis', *London Journal*, vol.21, pp.64–79.

ATKINS, P.J. (1993a) 'Freeing the streets of Victorian London', *History Today*, March, pp.5–8.

ATKINS, P.J. (1993b) 'How the West End was won: the struggle to remove street barriers in Victorian London', *Journal of Historical Geography*, vol.19, pp.265–77.

BARKER, T.C. and ROBBINS, M. (1963) *A History of London Transport: passenger travel and the development of the metropolis*, London, George Allen and Unwin, vol.1.

BARKER, T.C. and ROBBINS, M. (1974) *A History of London Transport: passenger travel and the development of the metropolis*, London, George Allen and Unwin, vol.2.

BROOKS, P.W. (1957) 'A short history of London's airports', *Journal of Transport History*, 1st series, vol.3. pp.12–22.

CHANT, C. and GOODMAN, D. (eds) (1999) *Pre-industrial Cities and Technology,* London, Routledge, in association with The Open University.

CHEVALIER, L. (1994) *The Assassination of Paris* (trans. D.P. Jordan), Chicago, University of Chicago Press.

COLLINS, P. (1959) *Concrete – the vision of a new architecture: a study of Auguste Perret and his precursors*, London, Faber and Faber.

DAVIS, J. (1995) 'Modern London 1850–1939', *London Journal*, vol.20, no.2, pp.56–90.

DYOS, H.J. (1961) *The Victorian Suburb: a study of the growth of Camberwell*, Leicester, Leicester University Press.

ESHER, L. (1981) *A Broken Wave: the rebuilding of England 1940–1980*, London, Allen Lane.

EVENSON, N. (1979) *Paris: a century of change, 1878–1978*, New Haven, Yale University Press.

EVENSON, N. (1984) 'Paris, 1890–1940' in A. Sutcliffe (ed.) *Metropolis 1890–1940*, London, Mansell, pp.259–87.

FIRTH, J.B. (1906) 'The ruin of Middlesex', *Fortnightly Review*, vol.79, new series, pp.1068–78.

FORSHAW, J.H. and ABERCROMBIE, P. (1943) *County of London Plan 1943*, London, Macmillan.

GARSIDE, P.L. (1984) 'West End, East End: London, 1890–1940' in A. Sutcliffe (ed.) *Metropolis 1890–1940*, London, Mansell, pp.221–58.

GLENDINNING, M. and MUTHESIUS, S. (1994) *Tower Block: modern public housing in England, Scotland, Wales and Northern Ireland*, New Haven, Yale University Press.

GOLD, J.R. (1995) 'The MARS plans for London, 1933–1942: plurality and experimentation in the city plans of the early British Modern Movement', *Town Planning Review*, vol.66, no.3, pp.243–67.

GOLD, J.R. (1997) *The Experience of Modernism: modern architects and the future city, 1928–53*, London, E. and F.N. Spon.

GOODMAN, D. (ed.) (1999) *The European Cities and Technology Reader: industrial to post-industrial city*, London, Routledge, in association with The Open University.

HALL, P. (1977, 2nd edn) *The World Cities*, London, Weidenfeld and Nicolson.

HALL, P. (1989) *London 2001*, London, Unwin Hyman.

HALL, P. (1995) 'Bringing Abercrombie back from the shades: a look forward and back', *Town Planning Review*, vol.66, no.3, pp.227–41.

HALL, P., BREHENY, M., MACQUAID, R. and HART, D. (1987) *Western Sunrise: the genesis and growth of Britain's major high tech corridor*, London, Allen and Unwin.

HART, D.A. (1976) *Strategic Planning in London: the rise and fall of the primary road network*, Oxford, Pergamon.

HAYWOOD, R. (1997) 'Railways, urban form and town planning in London: 1900–1947', *Planning Perspectives*, vol.12, pp.37–69.

HEBBERT, M. (1995) 'London recent and present', *London Journal*, vol.20, no.2, pp.91–101.

HENARD, E. (1906) *Etudes sur les transformations de Paris, Fascicule 7. Les voitures et les passants*, Paris, Librairies-Imprimeries.

HIGGINSON, M. (ed.) (1993) *Tramway London: background to the abandonment of London's trams, 1931–1952*, London, Light Rail Transit Association.

HITCHCOCK, H.-R. (1968) *Architecture: nineteenth and twentieth centuries*, Harmondsworth, Penguin Books.

HOWGEGO, J.L. (1977) *The Victorian and Edwardian City of London from Old Photographs*, London, B.T. Batsford.

HUMPHRIES, S. and TAYLOR, J. (1986) *The Making of Modern London, 1945–85*, London, Sidgwick and Jackson.

JACKSON, A.A. (1973) *Semi-detached London: suburban development, life and transport, 1900–1939*, London, George Allen and Unwin.

KOSTOF, S. (1991) *The City Shaped: urban patterns and meanings through history*, London, Thames and Hudson.

KOSTOF, S. (1992) *The City Assembled: the elements of urban form through history*, London, Thames and Hudson.

LARROQUE, D. (1988) 'Economic aspects of public transit in the Parisian area, 1855–1939' in J.A. Tarr and G. Dupuy (eds) *Technology and the Rise of the Networked City in Europe and America*, Philadelphia, Temple University Press, pp.40–66.

LAWRENCE, J.C. (1990) 'Steel frame architecture versus the London building regulations: Selfridges, the Ritz, and American technology', *Construction History*, vol.6, pp.23–46.

LUCKIN, B. (1990) *Questions of Power: electricity and environment in inter-war Britain*, Manchester, Manchester University Press.

MAINSTONE, R. (1975) *Developments in Structural Form*, London, Allen Lane.

MARMARAS, E. and SUTCLIFFE, A. (1994) 'Planning for post-war London: the three independent plans, 1942–3', *Planning Perspectives*, vol.9, no.4, pp.431–53.

MITCHELL, B.R. (1998) *International Historical Statistics: Europe 1750–1993*, London and Basingstoke/New York, Macmillan/Stockton Press.

MUMFORD, L. (1948, 2nd edn) *The Culture of Cities*, London, Secker and Warburg (first published 1938).

NOIN, D. and WHITE, P. (1997) *Paris*, Chichester, John Wiley.

OFFNER, J.-M. (1993) 'Twenty-five years (1967–1992) of urban transport planning in France', *Planning Perspectives*, vol.8, no.1, pp.92–105.

OLSEN, D.J. (1976) *The Growth of Victorian London*, London, B.T. Batsford.

PAPAYANIS, N. (1993) *The Coachmen of Nineteenth-century Paris: service workers and class consciousness*, Baton Rouge, Louisiana State University Press.

PAPAYANIS, N. (1996) *Horse-drawn Cabs and Omnibuses in Paris: the idea of circulation and the business of public transit*, Baton Rouge, Louisiana State University Press.

PORT, M.H. (1995) *Imperial London: civil government building in London, 1851–1915*, New Haven, Yale University Press.

ROBERTS, G.K and STEADMAN, J.P (1999) *American Cities and Technology: wilderness to wired city*, London, Routledge, in association with The Open University.

RUDDOCK, T. (1996) 'Charles Holden and the issue of high buildings in London, 1927–47', *Construction History*, vol.12, pp.83–99.

SAINT, A. (ed.) (1989) *Politics and the People of London: the London County Council, 1889–1965*, London, Hambledon Press.

SCHUBERT, D. and SUTCLIFFE, A. (1996) 'The "Haussmannization" of London?: the planning and construction of Kingsway-Aldwych, 1889–1935', *Planning Perspectives*, vol.11, pp.115–44.

SIMS, G.R. (ed.) (1903) *Living London: its work and its play, its humour and its pathos, its sights and its scenes*, London, Cassell, vol.3.

SUTCLIFFE, A. (1970) *The Autumn of Central Paris: the defeat of town planning, 1850–1970*, London, Edward Arnold.

SUTCLIFFE, A. (1984) 'Introduction: urbanization, planning, and the giant city' in A. Sutcliffe (ed.) *Metropolis 1890–1940*, London, Mansell, pp.1–18.

SUTCLIFFE, A. (1993) *Paris: an architectural history*, New Haven, Yale University Press.

SWENARTON, M. (1981) *Homes Fit for Heroes: the politics and administration of early state housing in Britain*, London, Heinemann.

THOMPSON, F.M.L. (1974) *Hampstead: building a borough, 1650–1964*, London, Routledge and Kegan Paul.

WEIGHTMAN, G. and HUMPHRIES, S. (1984) *The Making of Modern London 1914–1939*, London, Sidgwick and Jackson.

Chapter 6:
A SHORT HISTORY OF EVERYDAY BERLIN, 1871–1989

by Nicholas Bullock

6.1 Introduction

Berlin figures prominently in the history of the last hundred years or so. The city has seen victory, revolution, defeat and armed confrontation: in 1871 the triumphant Prussian armies marched through the Brandenburg Gate, fresh from the defeat of the French; in 1918 the Weimar Republic was proclaimed from the balcony of the Reichstag; in 1932 Hitler seized power on the streets of Berlin before becoming Chancellor; in May 1945 the city saw the final hopeless skirmishes as the Red Army fought its way through the city centre's last ragged defences to end the war in Europe; in June 1948 the Russians blockaded the city, signalling the decisive confrontation between East and West and forcing the West to supply all the city's needs, from coal to medicines, by air; in 1989 the crowds at the Brandenburg Gate breached the Wall to mark the end of the Cold War. For over a century Berlin has been at the heart of world events.[1]

But alongside this grand history of emperors and revolutionaries, battles won and empires lost, there is another history of Berlin, a history of the city's everyday life. This tells of where people go to work in the morning and where they return home in the evening, of the pattern of family life, of weekends, birthday parties and summer outings. This history has its own rhythms. It records the march of industrialization and factory production, the slow extension of tramways and the suburban railway system, and the growth of representative local government.

This everyday history of Berlin follows a number of themes. Like other cities across Europe, Berlin was changed by the new technologies that were also transforming London and Paris. Drainage and fresh water supplies, omnibuses and electric trams are the hard technology of these changes. Technology, understood in its widest sense to include intellectual innovation, must also cover the financial, scientific and other developments that change the city: the new legislation that made town-planning possible, the growth of scientific medicine and its impact on public health, and the development of new ways of assembling the capital necessary for building the factories, the public works and, above all, the densely packed workers' housing that make up the greater part of the fabric of the city.

---

[1] There is a comprehensive literature on nineteenth- and twentieth-century Berlin in German. This is summarized in the decennial surveys on Berlin published by the Historische Kommission zu Berlin. Of particular interest for the development of the city are the following. Hegemann (1930) provides a spirited, if partisan, account of developments up to the 1920s; Haus (1992), Dietrich (1960), Engelmann (1986) and Schinz (1964) offer general summaries of the city's development. There is no authoritative history of Berlin in English though Ladd (1997) may convey the complexity of the city's history and the way it shapes the city's form. Read and Fischer (1994) and Masur (1970) offer vapid accounts for the general reader. Friedrich (1974), Jackson (1988), Gelb (1986) and Wyden (1989) focus on particular periods in the history of the city. Ladd (1990) links the form of the city to cultural and political processes; Balfour (1990) focuses only on the area around the Potsdamer Platz, equivalent perhaps to examining the area around Marble Arch as a starting-point for a history of London.

Central to the way these themes of everyday life interact with Berlin's grander history is the direction given to these developments by government, both local and central. Germany has a long tradition of interventionist government. While British prime ministers of the nineteenth century dwelt on the inability of the government to regulate the economy, Bismarck saw no reason to limit the state's powers to shape society: if Germany could establish a system of state welfare in 1880s, could it not also regulate the growth and the form of the city? Government has certainly influenced the development of Berlin. The ambition of the government during the Second Reich to create a modern *Weltstadt* (world city) to rival Paris or London was as critical for the way the city developed before 1914 as the social policies pursued by socialist Berlin were for the rapid outward expansion of the city during the late 1920s.

Change is most easily remembered in the events of Berlin's world history: defeat, revolution, tyranny, war and occupation touch most families in one way or another and are easily grasped as markers for the city's history. Yet this leaves out the slow but sweeping transformations that change people's everyday lives: the benefits of electricity in the home, the ready availability of the bicycle, the coming of convenience foods and television. To tell the history of Berlin it is necessary to bind these two orders of history together, to convey the city's intimate history along with Berlin's imperial role. To do this, this essay focuses on four composite snapshots of key stages in the physical development of Berlin to document the way in which certain innovations have changed the form of the city between 1871 and 1989. The first snapshot introduces Berlin during the turbulent years of the founding of the empire in the early 1870s, a period known as the *Gründerjahre* (the years of the founding of the German empire); the second examines Berlin at the turn of the century, after the first great wave of industrialization had already transformed the city of the *Gründerzeit* (the founding time of the German empire); the third is of the city in the late 1920s as the brief prosperity of the Weimar Republic was beginning to touch the city that had emerged from the First World War; and the final snapshot reveals the divided city of the late 1960s and early 1970s, when Berlin was defended by the West as the 'shop window of capitalism' while the material benefits of post-war reconstruction, the new housing and slum clearances were already being questioned by an impatient younger generation.

6.2 *The boom of the* Gründerjahre

The Berlin to which Bismarck's victorious armies returned in 1871 was already a city of 774,000 people, comparable in size to Vienna and larger than any other city in the German-speaking world (Bullock and Read, 1985; see Figure 6.1). The capital of the newly-formed Reich might have been less than a third the size of London, less than half the size of Paris, but it was already the leading centre for manufacturing industry in the country. Firms such as Borsig, Schwartzkopf and Hoppe had already given Berlin the greatest concentration of heavy engineering on the continent. Yet despite a rapid increase in population during the 1860s, when the number of people living in Berlin rose by fifty-seven per cent from just under half a million, the city was still recognizable as the Berlin of the 1830s, the city of Stein and Humboldt, of Lenne and Schinkel. During the 1860s the density of building within the city walls had increased but, apart from new building to the north around the Hamburg Tore and Rosenthal Tore (Hamburg and Rosenthal Gates), and to the south-east in inner Luisenstadt, outward growth was still contained within the line of the old city walls.

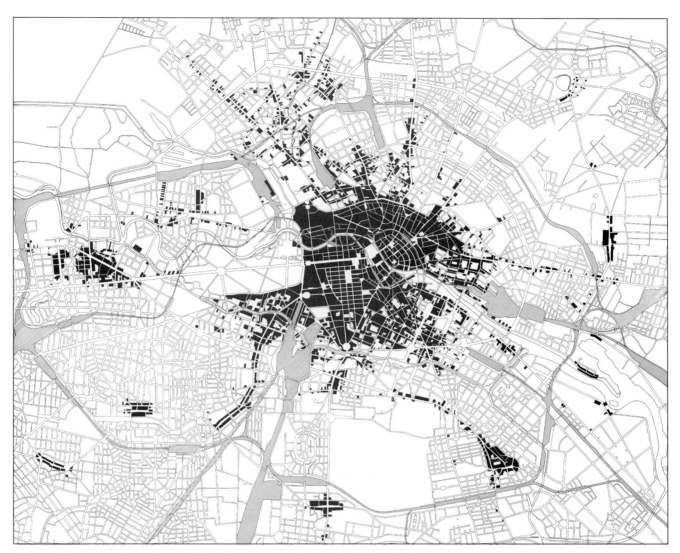

Figure 6.1   Berlin after the boom of the *Gründerjahre*. The area shaded in black is the built-up area in 1868. The areas developed between 1869 and 1881 are shaded in grey (reproduced from Geist and Kürvers, 1984, p.344)

Between 1871 and 1875 this was to change dramatically. In 1871 Berlin stood poised at the beginning of a period of furious growth. The boom of the *Gründerjahre*, the early 1870s, was to see the city expand way beyond the line of the old walls as nearly a quarter of a million immigrants flocked to seize the opportunities created by the explosive growth of the economy. How and where were the 20,000 people arriving in the city every year during the first half of the 1870s to be accommodated?

The strain on housing was acute. At the height of the boom many families simply pressed into any form of shelter that they could find. Contemporaries describe the spread of *Barrackenstädte* (shanty towns) across areas of common land, such as the Tempelhofer Feld, as being like the land invasions of South America (Geist and Kürvers, 1984). At the peak of growth in 1873–4 the proportion of empty dwellings was down to 0.6 per cent, the lowest figure recorded at any time before the First World War (Bullock, 1990). Overcrowding ran at record levels with a surge in the number of lodgers, particularly *Schlafleute* (transient lodgers), who rented a share of a room or even a share of a bed for the night, in the working-class districts to the south-east of the city, where many dwellings doubled as workshops during the day.

To meet the demand for housing, the whole process of city development was transformed.[2] During the *Gründerjahre*, Berliners were astonished and troubled to see the speculative house-building industry that had started in the 1860s to grow on a scale that was without precedent in the German-speaking world. The starting-point of the typical speculative development was the preparation of agricultural land for residential building by the *Terraingesellschaft* (land companies) who in the 1860s and early 1870s were still buying land directly from farmers in the areas around the city. Before, the boom prices generally reflected agricultural use, but as farmers came to realize the value of their land for housing, prices began to escalate. The price of land held by a land company and ready for development was largely determined by the developers' estimate of the return to be made on the sale of a building plot, in turn based on the probable level of rents and the maximum number of dwellings that might be crammed into the maximum permissible envelope. Having acquired a site, the land company would divide it into building plots and, at around half the cost of the eventual sale, would lay out and arrange drainage and street-paving.

To reduce costs, the land company would wish to sell its plots as quickly as possible, either to a *Baustellenmackler* (dealer), who would in turn hope to sell it on at a profit, or direct to a builder. As builders generally operated with very little or no capital of their own, the land company would need to offer a mortgage as part of the terms of sale. It could do so in the knowledge that its investment was protected by the system of registering in order of priority all charges on a property in a *Grundbuch* (register of interests), and the value of the site; while the builder would be highly exposed. Inevitably the first (and the frequent second or third) mortgage arranged by the land company would not meet the costs of construction and the builder would have to borrow further finance (known as the *Baugeld*), at higher rates of interest, in order to finish the building.

Like the land company, the builder, too, would seek to reduce costs by completing the building as rapidly as possible and by delaying payment to the work-force and suppliers, sometimes not paying until the building was sold to a landlord. For the builder who could sell at the peak of the boom, these risks were justified by the huge returns that could be made with virtually no capital, but for those selling on a falling market – a frequent occurrence given the counter-cyclical nature of financing speculative building – the risk of bankruptcy was high. The landlord, the final link in a chain of speculative development, was by the early 1870s generally very heavily mortgaged already, and often looking for a quick return through a sale of a fully-rented building. Failing that, the landlord might look to a long-term but modest return by letting the building. Like builders, landlords flourished in boom times when they might be able to secure tenants for a building before its completion, but in a slump, when demand fell away and property remained empty, the risk of bankruptcy was ever present, as the statistics of the late 1870s show only too clearly (Reich, 1912).

The results of this speculatively-driven development were clearly visible by the late 1870s. For the expanding middle classes, the industry built the new *Villen Colonien* (villa estates) that graced the suburbs, offering seclusion and fashionable comfort away from the noise and dust of the city (see Figure 6.2). Carstenn, the 'Napoleon of developers' and the most powerful figure in the industry at the time, with over 1619 hectares of land ready for sale – a fifth of the area available for development at the time – developed extensive areas to

---

[2]   There is an extensive literature on the way in which Berlin was developed before 1914: see Reich (1912), Carstenn (1917) and Voigt (1901); for accounts that are more sympathetic to the interests of property, see Ascher (1914) and Meinardus (1913). This literature is summarized in English in Bullock and Read (1985) and Bullock (1990).

Figure 6.2 Prospectus for a villa development of the 1870s (courtesy of Bildarchiv Preussischer Kulturbesitz)

the south of the city (Voigt, 1901). He commanded the resources to build for the most affluent. At Lichterfelde, for example, about twelve kilometres south of the city centre, he followed the example of English developers with a suburban halt to enable the owners of his villas to travel by rail to the Potsdamer Bahnhof (Potsdamer Station) in the heart of the city. But Carstenn was the exception rather than the rule. For the most part, the land companies, such as the Thiergarten-Bau-Gesellschaft, which developed the area around the western suburbs (the present-day Breidtscheidplatz), were much smaller and less stable, capable of developing small plots with a handful of villas and seeking to sell at the crest of the wave before interest charges or, worse still, a chill in the market, threatened their very existence.

For the working classes, the building industry was throwing up housing as rapidly as it could in the new suburbs, within walking distance of the major centres of employment. To the north-west in new suburbs such as Moabit and Wedding, employees in the engineering and metal-working industry needed housing. So, too, did those working in Luisenstadt and the Stralau Viertel (Stralau Quarter) to the south-east of the centre, where the range of industries and trades was wider. Here, major trades such as clothing manufacture, building and wood-working jostled amongst the back courts alongside other forms of employment such as chemical production and printing. In these areas, workers still walked to work, reserving the expense of a ride on the horse-drawn trams and buses for outings at the weekend to escape the city.

The housing in these suburbs was predominantly in the form of five-storey blocks along the street, with side buildings and back buildings enclosing small courts to take advantage of the deeper building plots that developers were now laying out. This was possible because the only limitations on the overall building form were the building regulations of 1853, drawn up when the typical residential development was very different (Bullock and Read, 1985; Frick, 1970). These regulations specified a minimum court size determined by the dimensions of a hand-operated fire pump, and related the maximum heights of the building to the width of the street (see Figure 6.3 overleaf).

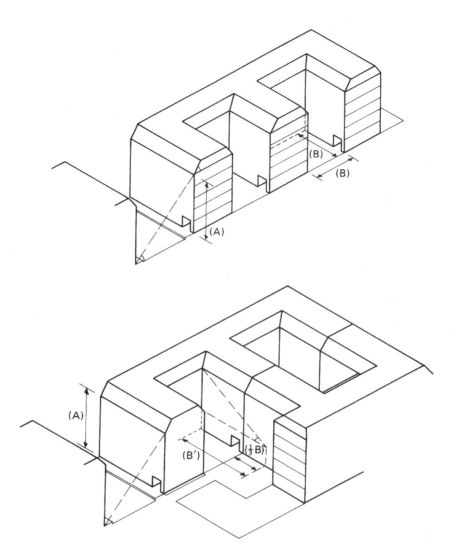

Figure 6.3 Simplified representation of building bulk controls in the Berlin building regulations of (top) 1853 and 1887 and (bottom) 1897 (reproduced from Bullock and Read, 1985, p.92)

Top 1853 The maximum height of the building (A) was determined by the street width (12 m minimum to 22 m maximum) measured to the opposing building line; the minimum court dimensions (B) were fixed at 5.60 x 5.60 m giving a minimum court area of 31.60 m².

1887 The maximum height of the building (A) was determined as before and the number of habitable storeys limited to 5. The roof storey was generally set back at an angle of 45°, but this might be increased to 60° if the resulting increase in overall height was equal to or less than half the height of the roof storey set back at 45°. Court size was set at a minimum of 60 m², although for corner sites a minimum of 40 m² was permitted: the minimum court dimensions (B) were fixed as the height of the building less 6 m.

Bottom 1897 The maximum height of the building (A) was determined as before, but provisions were made to allow a more elaborate treatment of the façade with turrets, balconies, gables and bays. Minimum court area (in building zone 5) was increased to 80 m², but corner sites were still permitted a minimum court size of 40 m². Minimum court dimensions could now take advantage of combining courts on adjacent sites: the minimum dimension for combined courts (B') was fixed as the minimum court dimension for a single court (B) plus $\frac{1}{3}$B to take account of the advantage conferred by the adjacent court

Naturally the concentration of dwellings on the fifth floor and in court buildings was much higher in these new working-class districts than in the city as a whole, but even aggregate figures for the whole city reveal these changes: in 1861, twenty per cent of all dwellings were on the third and fourth floors; by 1880 this category accounted for about thirty per cent of all dwellings, with twelve per cent of all dwellings on the fifth floor. As plots increased in size, the proportion of dwellings giving onto the minimal courts increased: in 1861, only just a quarter of all dwellings were located in court buildings, but by 1880 this proportion stood at just over a third (Bullock, 1990). The general reaction against this relentlessly dense form of tenement building was expressed by the derogatory term *Mietskaserne* (rent barracks). What contemporaries found so oppressive was the monotonous effect of block after five-storey block, completely shutting out any impression of nature, and the ruthlessness with which the process of speculation ensured that every square metre of space was put to profitable use (see Figure 6.4).

Figure 6.4 A view through the entrance into the courts of a Berlin *Mietskaserne* at Meyer's-Hof, Wedding (courtesy of Ullstein Bilderdienst)

But while the process of development was being transformed, the process of house-building remained much as it had been. There were changes in the building industry but they did not affect speculative house-building: contractors involved in railway-building and major public-works were becoming larger and technically more sophisticated, as were many firms in the building materials industry. By the end of the boom years, there were fewer but larger and more efficient brickworks than before, and various building components (joinery items such as doors and windows, and external plaster ornaments such as pre-cast columns, pediments, sections of cornice and other decorative devices) were being mass-produced off-site by the mid-1870s (Geist and Kürvers, 1984). By contrast, the house-builders, derisively nicknamed *Baulöwen* (literally, building lions), were rarely more than adventurous contractors who made up in daring for what they lacked in capital (Reich, 1912). The labour-force they commanded had few skills; at the peak of the boom it was said that contractors were hiring unskilled migrants as they stepped off the trains that had brought them to Berlin. A gang assembled in this way would often live rough in the building during construction, in part to meet the difficulties of irregular and uncertain wages. These were generally settled up when the carcass of the building was completed but sometimes payment might have to wait until the sale (or even the letting) of the finished building. But if this labour-force had few skills, construction remained simple and technical progress slow. Mechanization was minimal and horses, not steam power, were used when sheer human effort was not enough. Piped water supply had been introduced to Berlin by an English company in 1855 but was not universally available until the city took the company in 1873. Nor were water-closets or water-borne household drainage more widespread until the introduction of a comprehensive city-wide system in the mid-1870s (Geist and Kürvers, 1984).

The constraints imposed on these market forces by local administration and central government were minimal. True, there was for Berlin, in a way that there was not for London or indeed for most contemporary cities, a plan intended to govern the overall character of new development. James Hobrecht's plan of 1862 was an attempt to channel the forces of growth in order to secure for Berlin the advantages of the wide streets and handsome squares that Haussmann was producing in Paris, using the extraordinary powers granted to him by Napoleon III (Geist and Kürvers, 1984; Sutcliffe, 1981). Hobrecht's plan for Berlin looked back to an earlier tradition of city layout, that of Schinkel and Lenne, that was dependent on the powers of the monarch and the old eighteenth-century tradition of absolutism. Yet even with these powers, Hobrecht's plan set only the overall schema of development; the main roads and ornamental squares. Within this outline, the activities of the developer, in relation to the lines of the streets and building density, were subject to only minimal control.

The boom of the early 1870s could not last. Furious economic expansion was followed by a sudden crash. In the recession that followed, employment slackened, immigration fell, and with it, the demand for housing. In 1873, the first failures of banks heavily involved in lending to land companies and builders began the progressive collapse that was to bring the whole house-building industry to a halt. Barely a year after the peak of the boom, accommodation across the city was impossible to let: in 1876, for the first time for eight years, the percentage of empty dwellings passed three per cent, and it increased steadily until 1879. By 1879 more than five per cent of all plots in the city were the subject of forced sales by those going bankrupt (Reich, 1912).

The surge of growth prompted debate on how to regulate the development of the city. Legislation put forward by the reformers (the loose alliance of doctors, public health campaigners, architects and city administrators, associated with organizations such as the VFÖG – Verein fur öffentliche Gesundheitspflege, i.e. the German Association for the Promotion of Public Health – and the VDAIV – Verband deutscher Architekten- und Ingenieurvereine, i.e. the Confederation of German Associations of Architects and Engineers), giving the Prussian cities the power to control urban development, now won widespread support. Hailed as the equivalent of the Magna Carta for the German cities, the Prussian Enteignunsgesetz (the Expropriation Act, 1874) and the Fluchtliniengesetz (the Building Alignment Act, 1875), which gave cities the power to buy land by compulsory purchase and to lay out, drain and pave new streets, represented a first attempt to control urban growth (Fehl and Rodriguez-Lores, 1983; Sutcliffe, 1981). In the face of protests from the house-building industry and the property lobby, reformers protested that these measures did not go far enough: the form of city development was still too dense and housing conditions were still not adequately policed. The battle to extend control of the growth of the city was to be keenly fought over the next thirty years.

6.3 Berlin at the turn of the century

Between 1870 and 1900, despite the succession of bust and boom, Berlin had become a true *Weltstadt* and the size of its population now rivalled that of Paris. The five years of depression that followed the *Gründerjahre* saw a slower rate of growth, but in the second half of the 1880s, a new flood of immigrants entered the city and by 1890 the population stood at over 1.5 million. This growth was to continue until just before the First World War. Despite the severity of the slump of the early 1890s and another sharp depression in the early 1900s, the population of the city had passed two million by 1905. By 1914, the population of the city and its suburbs together totalled around four million, making it the fifth largest city in the world (Bullock and Read, 1985; see Figure 6.5 opposite).

These thirty years again saw a radical transformation of the form of the city as Berliners exploited advances in the technology of manufacture and transportation. Old Berlin, clearly recognizable in 1871, was by 1900 no more

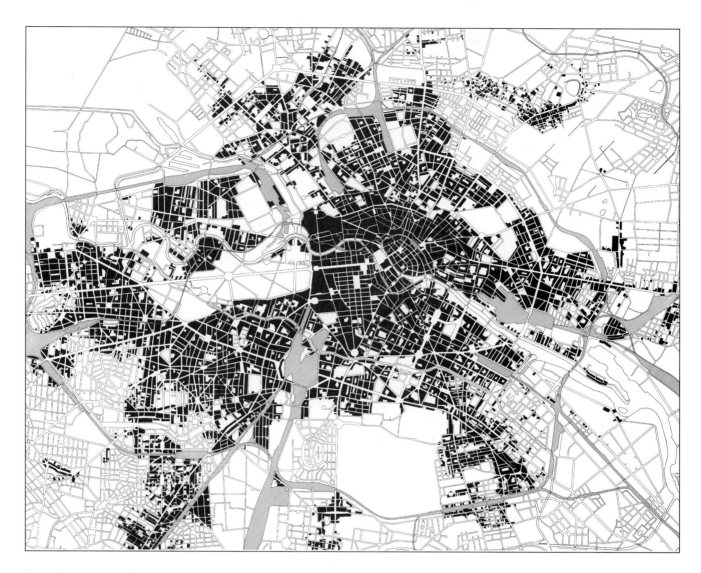

Figure 6.5 Berlin at the turn of the century. The area shaded in black is the built-up area in 1895. The areas developed between 1896 and 1908 are shaded in grey (reproduced from Geist and Kürvers, 1984, p.368)

than the centre of a built-up area eight times larger, a built-up area that was already spilling out well beyond the boundaries set in the 1880s by the *Ringbahn* (ring railway), Berlin's peripheral suburban railway. As non-residential land uses forced housing out of the central districts, the relocation of industry further out, the so-called *Randwanderung* (outward movement of industry) to the *Ringbahn*, made the outer districts increasingly attractive for working-class housing. In 1895, the population of the suburbs formed only seventeen per cent of the total population of Greater Berlin, but by 1900 this had risen to twenty-nine per cent, and by 1910 to forty-five per cent. The basic forces behind this radical transformation were to be found in the new pattern of accessibility created by the expansion of public transport, and the relocation of industry.

The development of the railway system, in particular the completion of the *Ringbahn* in 1877 and the opening of the east–west *Stadtbahn* (city railway) link across Berlin in 1882, had done much to improve accessibility throughout the city in the 1880s. During the next two decades, these improvements, and continuing expansion of public transport, were to have important consequences for the way in which the city was to develop. But the immediate impact of these changes was on the location of industry, not housing (Thienel, 1973; Zimm, 1959).

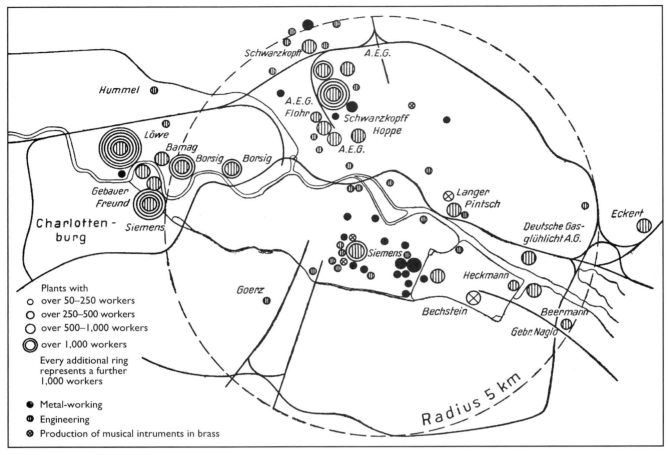

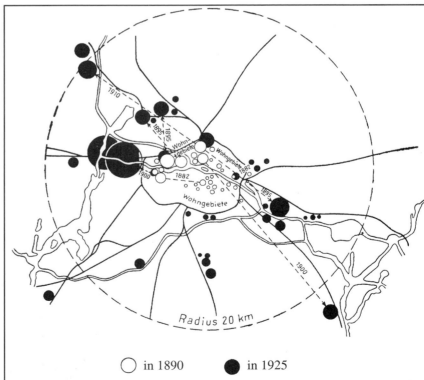

Figure 6.6 The first and second *Randwanderungen*: the location of mechanical engineering plants employing more than 50 persons in (top) 1895 and in (bottom) 1890 and 1925. ('*Wohngebiete*' means 'residential areas) (reproduced from Zimm, 1959, pp.66, 89)

Seeking larger sites, industries such as metal-working, engineering and electrical engineering were drawn beyond the built-up area to sites around the *Ringbahn* which offered cheap land and good access to waterways and railways (see Figure 6.6). Berlin companies such as AEG, Borsig and Siemens had become national, even international, leaders, and had to expand to maintain their position. This outward movement of industry, the first *Randwanderung*, to the *Ringbahn* and areas about five kilometres from the city centre, increased the attraction for the work-force of areas such as Moabit, Wedding, Charlottenburg and other areas where the growth of working-class housing could follow the growth of industry. Borsig had closed his works near the Oranienburg Tor (Oranienburg Gate) in 1878 to move his factory, which already employed over 5,000 people, to a larger site in Moabit that was well served by both rail and water. Siemens and Halske acquired a large site in Charlottenburg in 1883 which, greatly expanded four years later, was to form the basis of their operations in Berlin for the next thirty years. AEG, which had traded from the Schlegelstrasse for over twenty years, moved in 1887 to a new factory in Wedding, and in 1895 to a 90,000 square metres site on the Voltastrasse, from which it could gain access to the *Ringbahn* on a specially constructed branch line. By the late 1880s and the early 1890s, the population in the outer districts and the suburbs of the city was growing apace. This outward growth – the second *Randwanderung* – continued and even accelerated during the twenty years before the First World War, creating new concentrations of industry between fifteen and twenty kilometres from the city centre. Again, the example of Borsig illustrates the way in which heavy industry continued to search even further out for more and cheaper land with direct links to water and rail. In 1898 the firm moved its operations from Moabit, five kilometres from the centre, to a huge twenty-eight hectare site on the Tegler See, more than twelve kilometres to the north-west of Berlin. As a necessary adjunct to expansion on this scale, Borsig built housing, schools, and, in effect, a complete settlement to provide for the day-to-day needs of his work-force.

By 1914, the location of manufacturing industry was very different from the situation in the 1870s. To the north, manufacturers such as AEG were following Borsig, developing along the *Nordbahn* (north-bound railway line) and its branch lines, and seeking access to the system of waterways on that side of the city. To the south and south-east, a new concentration of industries, including Konrad Lorenz and AEG (again), was expanding between the Frankfurter and the Goerlitzer railways out into Mariendorf, Buckow, Lichterfelde and other areas, some fifteen kilometres from the centre. These areas were well served, too, by waterways such as the Teltow canal. To the west, what was later to be Siemensstadt was well connected to the Havel (part of the waterways system), and was already starting to grow along the Hamburger and the Lehrter railways.

This major relocation of employment led to a substantial increase in the demand for housing in the outlying districts of the city and in the suburbs, though the expansion of the suburbs was not just a product of new employment opportunities. The dramatic increase in the availability of public transport during the 1890s also stimulated the centrifugal pattern of growth. The decade witnessed an enormous increase in the overall volume of travel in the city. In the period 1890–95, trips for all types of transport increased by thirty-five per cent and this increase continued yet more rapidly during the second half of the decade: between 1895 and 1900 the number of trips rose by a further sixty-eight per cent. In the decade after 1900 the volume of travel continued to increase, but the rate of growth slowed to fifty per cent in 1900–05 and to thirty per cent in 1905–10 (Radicke, 1979; see Figure 6.7 overleaf).

The greater part of this growth was made up of commuter travel, and between 1890 and 1914 the proportions of different forms of transport remained broadly stable. Half of all trips were by tram, a form of transport which, with

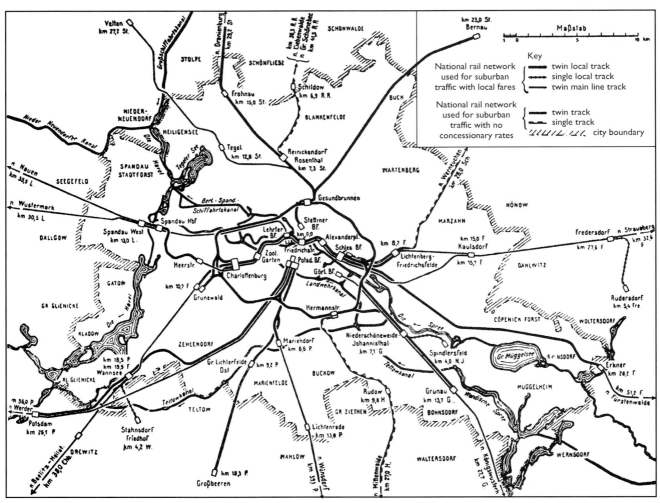

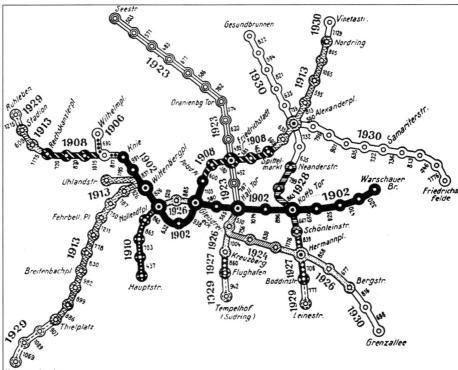

Figure 6.7 *Top* The suburban and local rail system in Berlin and the surrounding areas *Bottom* The underground railway system in Berlin, showing the date of construction and the distance between stations for each section of track (reproduced with the permission from the publisher from Weber *et al.*, 1979, pp.12–13)

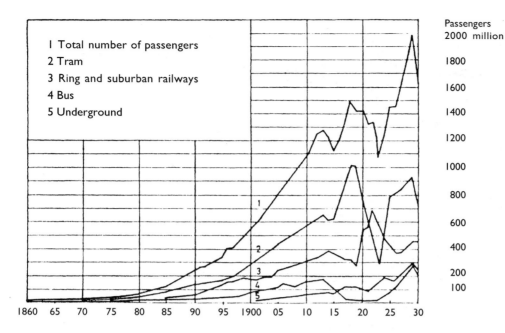

1 Total number of passengers
2 Tram
3 Ring and suburban railways
4 Bus
5 Underground

Passengers
2000 million
1800
1600
1400
1200
1000
800
600
400
200
100

1860 65 70 75 80 85 90 95 1900 05 10 15 20 25 30

Figure 6.8 The total number of passengers and number of passengers using different forms of transport (reproduced from Weber et al., 1979, p.14)

electrification (progressively from 1895 onwards) and a new city-wide structure of administration, became the most efficient way of travelling to work within the city. More important for the growth of the outer districts of the city and the suburbs was the expansion of the services provided by the underground and the various suburban rail systems known as the *Ring-*, *Stadt-* and *Vorort-bahnen* (the peripheral, city and suburban rail services). These maintained their share of the greatly increased volume of travel, emphasizing the value of suburban rail transport both for trips between the city and the outer districts and suburbs, and between areas of the outer districts and suburbs themselves. By 1914 the underground, first opened in 1902, was providing an important east–west service and carrying 6.5 per cent of all traffic (see Figure 6.8).

During the late 1880s and the early 1890s, the volume of trains that could be run on the *Ring* and on the key radial suburban lines to the south-west and the south-east was greatly increased by doubling the number of tracks. Not only did trains run more frequently; they also served more stations, thus extending and consolidating the impact of public transport in opening up these outer areas of the city for new housing. Equally important was the effect of the new fares introduced at the beginning of October 1891. In place of the standard state-wide tariff, the authorities now introduced for suburban travel a system of concessionary fares, priced by zones, to complement that already available for travel on the *Ring*. This considerably reduced the cost of suburban rail travel and made it possible for the first time for workers to live in the suburbs and commute to work by train.

These increases in accessibility and the changes in the pattern of employment during the 1890s transformed the population of the suburbs. In the graphs plotted by Leyden recording population growth, the impact of these developments can be seen clearly in the surge of growth in the new municipalities such as Charlottenburg, Schöneberg, Lichtenberg and Neukölln/Rixdorf, and in the smaller suburban communities such as Weisensee and Reinickendorf (Leyden, 1933). This rapid expansion in numbers represented a massive increase in the demand for new housing, and, in response, there was a wave of new building. By 1905, for example, average occupancy rates in many of the working-class suburbs were higher than those in the city of Berlin, even though dwelling sizes were, if anything, smaller than those in the city (Bullock, 1990; see Figure 6.9 overleaf).

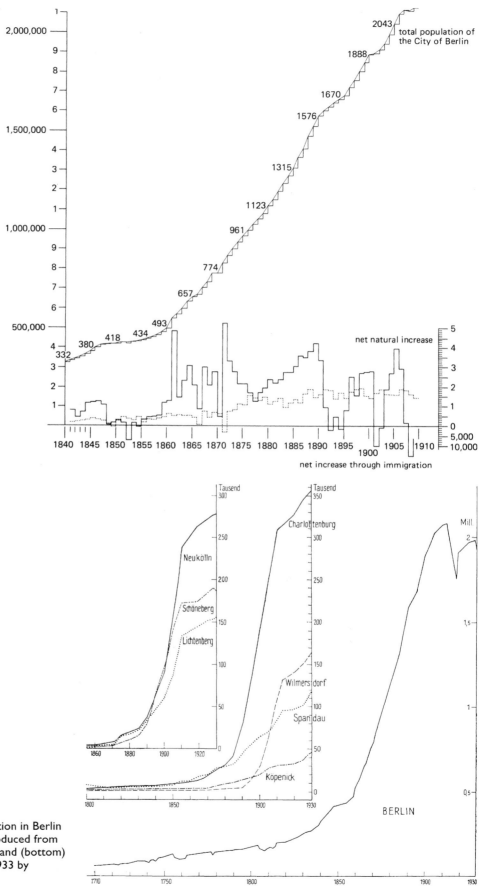

Figure 6.9 The growth of population in Berlin
and the surrounding districts (reproduced from
(top) Bullock and Read, 1985, p.19; and (bottom)
Leyden, 1933, pp.86–7; copyright 1933 by
Ferdinand Hirt in Breslau)

While these changes were taking place in the suburbs, the process of economic growth was also transforming the centre of the city where housing was being forced out by commercial and other non-residential land uses (Leyden, 1933). Between 1880 and 1914, districts lying at the very heart of old Berlin, within one and a half kilometres of the Hausvogteiplatz, had experienced a fall in population from 256,000 to just over half that figure, 137,000. With intense competition for land from the new department stores, shops, offices and various forms of government buildings, the number of dwellings in the very centre of the city was being reduced by what contemporaries were already calling 'city-building'. But unlike London, where working-class housing was destroyed to make way for uses of this kind, working-class areas such as the Stralauer Viertel and the Rosenthal Vorstadt, which lay just to the east and within easy walking distance of the centre, seem to have remained largely unaffected by these pressures. While owners of more expensive housing bemoaned the flight of their tenants to the new suburbs, large numbers of workers, many of them semi-skilled or unskilled, remained tied to the central areas by the location of their employment. For the skilled worker, with the promise of secure and often well-paid employment in one of the new industries, the suburbs might be attractive. But for the semi-skilled and the unskilled, access to the variety of job opportunities in the centre, in trades such as building and clothing manufacture, was generally dependent on finding somewhere affordable to live within walking distance, or at most a short tram ride away from work, even if this did mean crowding into older housing in the centre of the city.

By the turn of the century, new building, both in the suburbs and in the centre, was controlled more actively than had been possible in the 1870s. After more than thirty years of active campaigning for better housing and the control of urban growth, organizations such as the VFÖG, the VFSP (Verein für Sozial Politik; the Association for Social Policy) and VDAIV had created the expectation that the state should play a role in this (Bullock and Read, 1985). Armed with the powers granted by the Fluchtliniengesetz of 1875, Prussian cities could determine the way in which new streets and squares were laid out, paved, planted and drained, and Stübben's redevelopment during the 1880s of the area left vacant by the fortifications in Cologne was a widely-quoted example of what could be done with the new powers, given the necessary political will (Fehl and Rodriguez-Lores, 1983; Sutcliffe, 1981). The building regulations now demanded higher standards of construction, water supply to the individual dwelling and internal (but shared) water-closets on every landing. These regulations also limited the density of developments by capping the maximum height of residential buildings to five floors and by increasing the minimum size of internal courts. Finally, the introduction of the new 'stepped' density regulations in a number of cities in the 1890s, and in Berlin in 1892, made it possible to establish different densities of building for different areas and thus to ensure a more open form of development in suburban districts where new building was taking place.

By 1900, Berlin was beginning to follow the lead of Frankfurt in attempting to regulate the use of land within the city both through the introduction of new taxes on trading in land, and through some form of *Bodenpolitik* (a policy on land ownership and land use; Bullock, 1990). In answer to the increasingly shrill demands of national pressure groups led by Damaschke's Bund Deutsche Bodenreformer (Association of German Land Reformers), the Prussian government was prepared to countenance various taxes on property and land in an attempt to secure for the community the increase in value that the community itself had created, rather than allowing this to fall into the hands of the land speculator as unearned income. In the Kommunalabgabengesetz (Communal Taxation Act) of 1893, cities were given general powers to introduce the Umsatzsteuer – a tax on unused land and on the conveyancing of land and property. Frankfurt introduced the Wertzuwachssteuer (a tax on unearned increment) in 1904, the first tax on the unearned increase in the value of land between sales. Though not in the vanguard of progress, the city authorities of Berlin had instituted a form of the Umstazsteuer in 1895 and the Wertzuwachssteur in 1910, though a number of the suburbs had done so earlier, despite the furious opposition of property interests (Haberland, 1931). Again following the lead of Frankfurt, the city authorities in Berlin and in the surrounding suburban municipalities were buying land and making it available on the newly-established leasehold basis for non-profit housing and communal uses: by the turn of the century, the authorities owned almost thirty-five per cent of land within the city boundaries (Bullock, 1990).

But hopes of developing a systematic approach to the co-ordination of new building, of expanding the suburban and underground railways, and of preserving areas of open space, woodland and access to areas of open water, remained unsuccessful, frustrated by competition between the different local authorities (and national authorities such as the War Office, which owned large areas of suburban land). The establishment of the Zweckverband Gross-Berlin (1911) as a metropolitan planning authority was a measure both of these hopes and of the continuing strength of the opposition to granting the necessary political powers for planning on this scale (Hegemann, 1930; Sutcliffe, 1981).

By the turn of the century, not only could housing reformers point to the way in which the state was beginning to regulate the activities of private enterprise, but they could also take comfort in the support of the state for the non-profit housing sector. In some areas, such as the Rheinland and in cities such as Frankfurt and Düsseldorf, the expansion of the non-profit sector made an important contribution to the supply of working-class housing. In Berlin, the overall impact was less, but even here the increase in non-profit house-building was substantial: before 1890 the various non-profit housing organizations in the city had built fewer than 600 dwellings, but by 1914 the non-profit sector, including the *Beamtenwohnungsvereine* (the housing associations for white-collar workers) had built in the order of 11,000 dwellings (Bullock and Read, 1985).

The non-profit sector set new standards for housing: new standards for the design of the individual dwelling and new standards of housing layout. The housing built for artisans by the Berliner Spar- und Bauverein (The Berlin Savings and Building Association) on the Proskauerstrasse showed what could be achieved even in a central working-class district such as Friedrichshain (see Figure 6.10). Messel's design is built around a generous courtyard with a kindergarten and communal laundry, and these shared facilities were matched by well-designed individual dwellings with windows on both sides, which offered not only the benefits of cross-ventilation for the hot summer months, but a view for most over both the courtyard and the surrounding streets.

Figure 6.10 The Berliner Spar- und Bauverein block on the Proskauerstrasse (reproduced with the permission of the publisher from Rentschler and Schirmer, 1974, p.149)

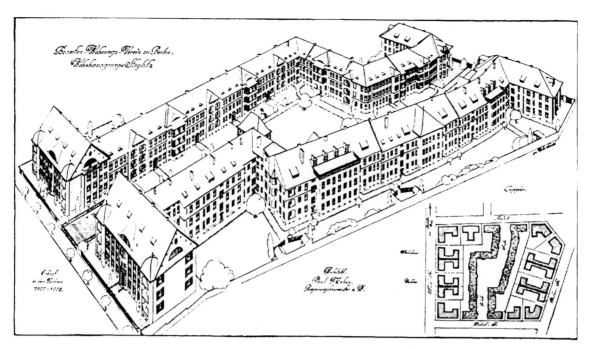

Figure 6.11 Housing built by the housing associations for white-collar workers (reproduced with the permission of the publisher from Frank and Rentschler, 1970, p.121)

Further out to the west in areas such as Schöneberg and Steglitz, the associations were building housing that was even more open in layout and more generous in the design of the individual dwelling. Developments such as Sophie-Charlotte Park, built in Charlottenburg by Albert Gessner for the Berlin Administration Housing Association, showed how the new non-profit organizations could take advantage of the zoning regulations to build an open form of housing in the suburbs, which was at the same time linked directly to work and the centre of the city by suburban railway and, just before the First World War, by underground (Wedepohl, 1970; see Figure 6.11). Private enterprise, too, was happy to build in these areas for an affluent middle class who could afford the rents for housing built to high standards and with an open layout. Georg Haberland's Berlinische Boden- und Baugesellschaft (the Land and Building Society) was responsible for developing the area around the Bayrischer Platz as one of the most fashionable areas in Berlin's fast-growing western suburbs. Built in the low-density manner with large open courts and generous planting, the layout of the district was centred on large public squares, planted formally and informally, which offered facilities such as children's playgrounds and paddling-pools. Here was a stark contrast to the middle-class housing of the 1890s built in areas such as Charlottenburg, where the individual dwelling might be opulently equipped but densities remained high and the layout enclosed.

Even further out still, along the *S-Bahn* (*Stadtbahn*) line to Wannsee and served, from 1913 onwards, by the *U-Bahn* (underground) line to Thielplatz, new suburbs such as Dahlem, Zehlendorf and Nikolassee were being slowly and elegantly developed from the turn of the century onwards on the old royal estates (Machule and Seiberlich, 1970). The population of Dahlem, which stood at just 182 in 1900, had risen to 5,500 by 1914; in Nikolassee, still further out and still more exclusive, the population had not reached 2,000 by 1914. However, in Zehlendorf, which was less smart, development had started before 1871 and the population had by this time reached just under 20,000. With density limited by the zoning regulations of 1892 and 1910 to a maximum of two storeys, these suburbs were developed with villas, many designed by well-known architects such as Gessner, Mebes and Muthesius, to an overall plan, such as Jansen's plan for Dahlem, embodying the formal ideas set out by Camillo Sitte and Theodor Goecke in the new planning journal, *Der Städtebau* ('City-building'; Posener, 1979; Sutcliffe, 1981). Those living in the new low-density outer suburbs could, like the Wertheim family, take advantage of the improved accessibility to get to work (in the family department store, for the Wertheims) in the heart of Berlin on the Potsdamer Platz with a minimum of delay: just before the First World War, an express suburban railway service, the so-called *Bankierzug* (bankers' train), ran at rush hours from Wannsee station to the Potsdamer Bahnhof, covering the nineteen kilometres in barely half an hour.

6.4   Berlin at the end of the 1920s

If someone who had not seen Berlin since the turn of the century had wandered around it in the summer of 1929, they would have had little difficulty in recognizing the city they had last seen. The war and the revolution that had swept away the monarchy had had little effect on the fabric of the city. During the months before the allied blockade had been lifted and again during the inflation crisis of 1923, life for the mass of Berliners had been hard enough, but with the stabilization of the mark and the reconstruction of the German economy, city life had quickly become quite as bearable as it had been at the turn of the century. With lower rents, higher wages and consumer goods in the shops, it was easy to believe that after the turmoil of the post-war years, things were looking up. For the working classes, the routine of the working week, the tram ride to work and the narrow confines of the tenement might be balanced, as they had been before the war, by the easy pleasures of summer weekends on the *Schrebergarden* (the suburban allotment). This way of life was characterized with affection in the well-known cartoons of Heinrich Zille, in which the men play skittles over a few beers while the women, at ease without their corsets, make jam, pickle gherkins, swap gossip, and eat and drink away the heat of the evening with their neighbours.

But for more affluent Berliners the rhythms of everyday life were beginning to change. Growing numbers of white-collar workers commuted to work, employed perhaps in the administration of a large company or in government service, and for them prosperity offered choices and advantages that had not been available before 1914. To them the city could offer a new apartment, perhaps in the new Modern manner, furnished by one of the down-town department stores, and with plenty of space outside for the children to play. At the weekends, unlike their working-class relatives, office workers and their families might see a film in one of the new UFA cinemas, or if they were fortunate enough to own one of the 40,000 private cars in the city, they might take a day out in the country with Sunday lunch at a restaurant in a favoured spot such as Birkenwirder.

The forces that were changing the city and feeding the rapid growth of the suburbs were not new, but were a continuation of the forces already at work before the war (Leyden, 1933). By 1919, over half of all Berliners lived outside what had historically been Berlin (1,971,000 in Old Berlin and 3,804,000 in Greater Berlin). In 1920, in recognition of the real extent of the city and its economic activities, the administrative boundaries were redrawn to merge the former city districts of Berlin with the eight neighbouring municipalities, fifty-nine rural districts and seven estate-districts, to form Gross-Berlin (Greater Berlin; see Figure 6.12). By 1925, the population of Greater Berlin was over four million; by 1929 it had increased by a further 250,000 to 4.29 million. But these aggregate figures conceal the extent of the redistribution of population that was taking place within the city and the rate of growth of the suburbs. While administration, commerce and other land uses were reducing housing in the centre, suburban areas such as Reinickendorf, Britz and Steglitz were growing even more rapidly than had been the case before the war.

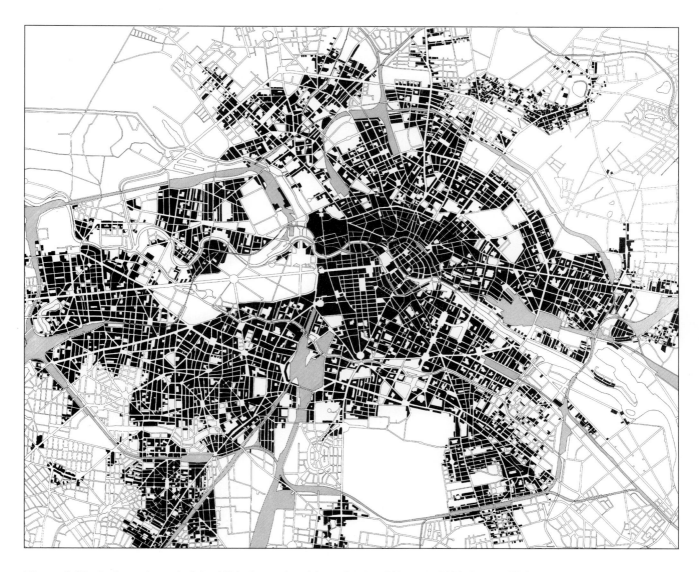

Figure 6.12 Berlin at the end of the 1920s (reproduced from Geist and Kürvers, 1984, facing p.376)

Again, the underlying reasons for this were essentially the same as they had been before 1914. Although industry in Berlin was badly disrupted by the economic difficulties of the early 1920s, and it was not until 1928–9 that production was to exceed (by ten per cent) the 1913 level, the 1920s were a period of restructuring and rationalization, not of radical change (Zimm, 1959; Kreidt, 1971). The large-scale employers, such as AEG and Borsig, who had moved to the periphery of the city around the turn of the century, continued to consolidate and expand their production in areas that were as far as fifteen kilometres from the centre. Siemens, for example, was to continue extending its cable-works on the Gut Gartenfeld (Garden Field Estate; first developed in 1911) through until the beginning of World War II. Similarly, in Siemensstadt, the company continued to expand by building a new switch-gear factory between 1926 and 1928 and the huge new Wernerwerk xv complex, started in 1928 but not finished until 1942 (Czada, 1969; Kreidt, 1971). While existing firms expanded, other large-scale firms engaged in everything from the mass production of bread to artificial rubber, and generating services, such as electricity, moved out to join them.

These centrifugal forces were matched by the complementary attraction of the city centre for other forms of employment, again a continuation of the

city-building process that had attracted comment before 1914. Here, the
growth of white-collar employment, for commerce, finance and for
government administration, was attracting large numbers of men and women
to work in the business and government districts in the heart of Berlin. In a
study of the city's economic and social geography, Leyden identifies the
banking district around Unter den Linden, the government quarter around the
Wilhemstrasse and the newspaper and commercial centre at Friedrichstrasse,
as particular concentrations of white-collar employment (Leyden, 1933).

With industrial employment moving to the suburbs and white-collar
employment increasing in the centre, the need to link the different sectors of
the suburbs with each other and with the centre became more important than
ever. While Berliners were naturally tempted to seek employment within the
sector where they lived, the number of those commuting and the length of
time that they spent travelling increased rapidly, continuing a pre-war trend.
Siemens again provides an illustration: the company employed 23,000 people
in 1913 in its works in north Charlottenburg but provided housing for only
6,100, leaving the others to commute into work, many from other sectors of
the city. By the late 1920s, the number of commuters had grown still higher,
despite the best efforts of the company to build housing for its employees in
the new Siemensstadt development (Czada, 1969; see Figure 6.13).

Figure 6.13 Siemensstadt in 1930 (reproduced with permission of the publishers from Frank and Rentschler, 1970, p.25;
photograph: Siemens)

What was happening at Siemens was happening across the city, and by the late 1920s there were three times as many trips on public transport as there had been at the turn of the century (1,600 million in 1927, compared with 545 million in 1900) though the proportion of trips by each form of transport remained broadly constant, apart from in the immediate post-war years (Radicke, 1979). Thus not only did the number of trips by tram, generally used for shorter trips within the city, increase threefold, but so also did the number of trips and the length of trips on the suburban and underground railways, which provided the links between employment and the new housing developments at the edge of the city. To cope with these new demands, the underground network was greatly extended during the 1920s, with the addition of two north–south lines and important extensions of existing lines down to the east and south-east providing much-improved access to employment and housing in these important areas of industrial growth.

The demand for suburban housing could be met only by a new form of housing production to replace the old system of speculative development with its reliance on private capital. This old system had been killed off by the war and showed no signs of reviving in the troubled economic climate of the post-war years when rent levels continued to be restricted under the 'command economy' as they had been during the war. Its place was to be taken by a new role for the state in housing production. The replacement of the monarchy by the republic had encouraged a very different interpretation of the powers of the state. Socialists now commanded a majority in both the Prussian state government and in Berlin. In place of their long-standing indifference to social and economic problems such as housing, the new administrations of both state and city had shown willingness to wrestle with the intractable difficulties created by the war and the inflation years. Already before the end of the war, the city authorities in Berlin had shown themselves willing to protect tenancies, regulate rents and increase the supply of housing. During the 1920s, the government was to intervene directly in the financing of non-profit housing. In 1924, with the stabilization of the mark and the start of the reconstruction of the German economy, a new system of finance was introduced which drew resources partly from the open capital market and partly from the proceeds of the Hauszinssteuer (or HZS, a tax on property built before 1914). Here inflation had eroded the value of the landlords' mortgage debts, and so rents, which had risen with inflation, were now largely unencumbered profit.

With funding from these sources, non-profit housing was to become the most important means of house-building in Berlin during the 1920s, accounting for 82,914 dwellings, or sixty-one per cent of the 135,000 dwellings built between 1924 and 1930 (Doerre, 1970; Schallenberg and Krassert, 1926). Rents for this form of housing were lower than a straight economic rent but were still higher, perhaps by as much as fifty per cent, than the rents for housing built before 1914 in which rent control was still in force. Rents in the pre-war unfurnished accommodation were, in the late 1920s, much lower than they had been before 1914. But although rents for new housing were higher, the accommodation still seemed good value for money to those who wished to escape from the traditional working-class quarters of the city to something better. It was generally situated in the suburbs, where the regulations ensured lower, more open development, and was built to high standards of space and equipment: new apartments offered hot and cold running water, a separate bathroom with a fixed bath, and electricity to power the range of smart new electrical goods now coming into the shops. Though much of this housing was built in the architectural manner of the pre-war years, the uncompromisingly modern appearance of large estates, such as Siemensstadt in Charlottenburg and the Weissestadt in Reinsickendorf, served as a very visible reminder of the way in which government wished to modernize the city (Bullock, 1993; Schinz, 1974).

Figure 6.14   The GEHAG housing estate at Berlin-Britz by Bruno Taut (reproduced from Junghanns, 1983, fig. 156)

Among the best known of these modern estates is the Hufeisensiedlung, built by the GEHAG housing association in Berlin-Britz, to the south-east of the centre, for members of the Berlin trade union association (see Figure 6.14). Designed by Bruno Taut (one of the leaders of the avant-garde architects in Berlin) in collaboration with Martin Wagner (later *Stadtbaurat* – city architect – for Berlin), the estate of more than 1,000 dwellings was grouped around a distinctive horseshoe-shaped area of green space with a pond at its centre (Junghanns, 1983; Rossow, 1986). It was conceived by its designers as a polemical demonstration of the benefits of co-operation as an alternative form of production to capitalism. Here was housing built to high standards and let at affordable rents because it had been built by the co-operative firm owned by the building trades who did not make a speculative profit. No longer caught in the old battle between labour and management, the building co-operative had introduced Taylorist ideas on the scientific management of construction. New methods of construction, new construction equipment, and the erosion of traditional barriers between skilled and unskilled workers and between trades, all helped to increase productivity and reduce costs.

But for Taut and Wagner, modernization meant more than just an enthusiasm for the machine and the obvious benefits of new technology. They wanted to redraw the balance between Man and Nature. Thus the Hufeisensiedlung was also a demonstration of the way in which working-class housing, developed on land owned by the city and not bought on a speculative land market, could be planned as part of a programme to 'green' Berlin (Schindler, 1972). Moreover, the individual tenants were to benefit not only from lower rents but also from the allotment gardens laid out around the central space where they could grow their own vegetables and enjoy working with nature. All this was to be had within easy reach of employment and the city centre, accessible by underground from a station that was only a ten-minute walk away. For Taut, Wagner and the GEHAG, here was an attempt to bridge the divide between the utopian ideals of the Garden City movement and the hard practice of building social housing.

Large-scale planning provides another example of the way in which the Berlin authorities could now do what had been impossible before 1914.

Appointed as *Stadtbaurat* in 1926, Martin Wagner was able to use the planning powers granted by the Prussian state in 1925 to co-ordinate the development of public transport, employment, housing and open space. With the building regulations of 1925, the city restricted still further the permissible density of building and set out for the first time the way in which land was to be zoned, differentiating between industrial, mixed and residential uses. The progress Wagner was able to make in the short time available to him before 1933 fell far short of the hopes he had entertained since 1917 when he had first drawn up a plan for a green Berlin with pathways of green space and parkland running from the countryside right into the centre of the city. But his battle to preserve the special qualities of the Berlin landscape, such as the combination of water and pine forest around the Havel, to establish municipal swimming beaches such as that at Wannsee, the largest inland beach in the country, and to preserve areas of open space such as the Jungfernheide as large-scale people's parks, did make a real contribution to the character and texture of the city. In place of the dense brick-built canyons of the *Gründerjahre*, the city was changing. Berliners could now enjoy more open space than ever before, they could take advantage of the surrounding countryside, and the city's new streets were generously wide and lined with trees. Gradually, Berlin was turning green.

6.5   Berlin in the early 1970s

The continuities in the everyday history of Berlin between 1914 and 1918 are not hard to trace: a war was lost, the Wilhelmenian monarchy was replaced by the Weimar Republic, but the physical fabric of the city remained virtually unaltered. The impact of the Second World War and its political aftermath was very different. By 1970 the Berlin of the late 1920s had been transformed (see Figure 6.15).

The heavy raids started in 1943 and had destroyed whole tracts of building in the centre of the city before the spring of 1945. To this was added the destruction caused by the battle for Berlin which was to bring the European war to a close. This began in earnest on 20 April 1945. By 24 April, the Russians had reached the periphery of the city proper and there was fighting in Neukölln, Zehlendorf and Tempelhof. Between 28 April and 1 May, the front closed ever more tightly around the innermost defensive ring, the government centre and the old pre-industrial core of the city. When battle ended on 2 May, forty per cent of the city lay in ruins (Diefendorf, 1993; Geist and Kürvers, 1989).

After defeat, the painful process of reconstruction with scarce resources proceeded slowly, complicated by the partitioning of the city between the four occupying powers. The closing of the borders of the three Allied sectors by the Russians in June 1948 marked the freezing of the divisions between East and West, and with it, the fragmentation of the old Berlin. There were attempts to consider the future of the city as a whole: in 1946, in an act of calculated optimism, a team under the leadership of Hans Scharoun had drawn up plans for the future of a united city (Geist and Kürvers, 1989). But in the face of fundamental political divisions, reconstruction was inevitably focused on the creation of two separate cities: the one based on

Figure 6.15 The districts of East and West Berlin since the division of the city in December 1948 (reproduced with the permission of the publisher from Heinrich and Mielke, 1964, p.64)

the us, British and French sectors, the other based on the Russian sector and designated as the capital of the Deutsche Demokratische Republik (DDR). Though formally separated, the two Berlins remained economically and functionally linked in a variety of *ad hoc* ways: the *S-Bahn*, run by East Berlin, continued to serve the city as a whole; West Berlin relied for rebuilding on the bulk building materials of the East; and throughout the 1950s large numbers of East Berliners, perhaps as many as 70,000, crossed the border between the two cities every working day to jobs in the West (Elkins, 1988). The informal arrangements between the two cities were never easy, and always subject to revocation at the first hint of political pressure, national or international. With the building of the Wall in August 1961, the border between the two cities was sealed and contacts were frozen for the next decade. Only after the four-power agreement in 1971 on the special status of Berlin were they to resume.

To a visitor who knew the Berlin of the late 1920s, the divided city of the early 1970s would have been confusing. The key monuments would still have been recognizable, though the rupture of once-familiar connections between different parts of the city would have been disorienting. So, too, would have been the reconstitution of the city as two separate Berlins, each with its own centre. The centre of West Berlin had now moved west to Charlottenburg, along the glitter of Kurfürstendamm and around the Breitscheidplatz Platz. In East Berlin, too, there was a new centre. The hub of this Berlin had moved eastwards to re-occupy the site of the older but less prestigious city at Alexanderplatz. Meanwhile, what had once been the most central, the smartest and the most vital areas of Berlin – Friedrichstrasse, Wilhelmstrasse and the Tiergarten edge of Friedrichstadt – were now deserted wastelands of self-seeded birch trees, buddleia and broken paving, with only the burnt-out shells of the Japanese and Italian embassies to speak of the area's pre-war glories.

In the Western sectors, even with the resources provided by the Marshall plan, reconstruction was slow. Much of the industry in these sectors had suffered heavily during the war and much of what had survived had been removed by the Russians in the first post-war months before the establishment of the separate sectors. Now cut off from its natural market and the range of supporting companies, industry in West Berlin was bound to look to the Bundesrepublik Deutschland (BRD) for raw materials, for skills to make up for those lost, and for a market. However, notwithstanding these difficulties, manufacturing industry did revive: the index of production, which had fallen as low as nineteen per cent of pre-war production at the end of the blockade in 1949, bounced back to pass 100 per cent of pre-war levels by as early as 1955, and by 1961 it had risen to 167 per cent of its 1939 levels (Elkins, 1988; Kreidt, 1971). But this revival was based on a different pattern of employment from that of pre-war Berlin. Many of the large-scale manufacturers, such as Siemens and AEG Telefunken, who were typical of Berlin's early industrial history, abandoned the city and moved west to the BRD. Gone, too, by the early 1970s were industries such as clothing manufacture which had survived on the cheap labour of the 53,000 *Grenzgänger* (cross-border commuters), who, like a further 20,000 women engaged in domestic service and cleaning work, lived in East Berlin and worked in the West. Notwithstanding this general exodus of manufacturing industry, the assistance offered by the BRD in subsidizing transport costs to and from Berlin and in meeting the taxes imposed by the DDR on goods entering and leaving the city, encouraged a number of large firms, such as Schering and Eternit, to remain in Berlin throughout the period of partition. But such firms were not sufficient to make up for the employment opportunities lost with the departure of large-scale industry. For the most part this was made up by smaller companies engaged in the reconstruction of the city itself or in providing services, for example milk production, that were no longer available from the surrounding area. Additional employment was created by the opening in the city of a number of federal government departments, in defiance of the city's special status. More important was the expansion through the 1960s of specialized employment in research and highly-skilled employment which took advantage of the high level of training and education available to the Berlin work-force. By the early 1970s, electronic engineering and electronics had become the largest single source of employment in the city, followed by the production of food and drink and, in third place, by machine-building (Elkins, 1988).

Though the city's economy might adapt relatively painlessly to partition, West Berlin's isolation set an agenda for planners that differed inescapably in a number of critical respects from those of other German cities. With partition, water supply and sewerage – hardly top priorities in most other northern European cities – became issues demanding urgent attention. Cut off from the established water supply to the East, at Müggel See, but fortunately retaining the plant on the Tegeler See, West Berlin was unable to come to an accommodation with the East Berlin authorities and had to make good this loss by extracting more ground water by bank-filtration methods. Despite the reduction in industrial use of water, demand continued to rise (by nearly twenty-five per cent in the twenty years from 1963), progressively reducing the city's water table and prompting fears of a water shortage that were relieved only with reunification in 1990. By contrast, sewage disposal illustrates the kind of collaboration that was possible between the two cities in areas where there was a clear benefit to the DDR. After partition, most of the city's sewage farms were in the DDR, and West Berlin had to negotiate an agreement under which the DDR treated the city's sewage and used it for agricultural purposes. But with the construction of more modern treatment facilities at Ruhleben, West Berlin came to depend on the DDR for only a quarter of its sewage treatment capacity. As with the provision of both electricity and gas, it was possible to find a technical solution to the problems created by the city's political status.

The consequences of the city's isolation were not always damaging. West Berlin has suffered less than most major European cities from the disruption, the congestion and the pollution caused by the motor car. By the early 1970s, not only was the level of car ownership lower than the average for West Germany, but Berliners also used their cars less. Given the relative location of the main residential areas and the major centres of employment, and already possessed of an efficient and comprehensive system of public transport, West Berlin experienced nothing like the commuter traffic that was clogging the roads in most large German cities: even in the early 1980s, it was possible to find kerbside parking in the centre of the city. As a result, West Berlin was mostly spared the large-scale road surgery forced on major West German cities such as Hamburg and Frankfurt (Stephan, 1964). True, the strategic traffic plan – drawn up in 1956 on the then optimistic assumption that Berlin would at some time be reunited – envisaged an inner motorway ring and a series of links through to the *Autobahnring* (the existing motorway ring road) that had been built right around the city in the 1930s. But of this grand vision only a portion had been built by 1970, much of it following the route of the *Ringbahn* to minimize disruption to the fabric of the city, and the idea of building what would have been most disruptive section, through Kreuzberg, was already abandoned by the mid-1970s.

If by 1970 the city was more fortunate than many German cities in escaping the most damaging effects of rapidly increasing car ownership, Berlin could not escape the need felt acutely across the BRD to modernize key elements of the physical fabric of the city. One of the most critical was housing, an issue now causing concern across the BRD, as the contrast became sharper and more visible between new suburban developments with growing prosperity, and the dense, run-down, nineteenth-century housing. Like Frankfurt and Hamburg, Berlin had both to meet the demands for new and better housing made by the more affluent, and to address the problems of providing for the growing number of so-called guest workers, students and other low-income groups living in run-down inner-city areas. Moreover, for Berlin these difficulties were further complicated by the need to address these issues within the narrow compass of the land available to the city. Those needing new housing could not simply be exported beyond the city's boundaries to a new town; even the potential for suburban expansion was tightly constrained.

The demand for housing was directly related to the size and structure of the city's population which, in turn, reflected the changing political and economic fortunes of the city. The population of West Berlin had risen from 1.73 million in 1945 to 2.14 million in 1950 (Doerre, 1970). Thereafter the size of the population had drifted upwards to a peak of 2.28 million in 1957 as refugees from the DDR streamed into Berlin, most of them on the *S-Bahn*: by 1961 about 1.5 million refugees had arrived. The majority of these moved on almost immediately to the BRD but a sizeable portion chose to remain, more than compensating for the very small number of those going from the West to the East. The building of the Wall ended this westward flow overnight, accentuating the fall in West Berlin's population that had first been detectable in 1958. By 1968 the city's population was lower than it had been in 1950. Underpinning this fall was an imbalance in the age and sex structure of the population: many more women than men had survived the war and the proportion of people aged over 65 rose from 9.4 per cent in 1950 to 13.2 per cent in 1970, and this despite the presence of around 100,000 students and other young people, many drawn to the city by the freedom from military service that resulted from the city's special status.

However, as the economy of the city gathered pace after a brief slowing of growth in 1966–7, the city again started to attract immigrants, but from much further afield (Elkins, 1988). As late as 1968, Berlin still had only 18,754 registered foreign workers, less than one per cent of the city's population, but from the late 1960s onwards, the numbers of guest workers arriving annually from Turkey and Yugoslavia increased rapidly, rising from under 4,000 in the mid-1960s to a peak of 25,000 in 1971. By the early 1980s, the immigrant population made up around twelve per cent of the city's population, or nearly 250,000 people. These people were a welcome addition to the labour-force in a city which could offer employment to unskilled men and women. But they also represented an important increase in the demand for housing and services such as education and health, especially after 1970 when the first wave of male immigrants were joined by women and children, and whole families began to arrive in the city. The arrival of these guest workers was to play an important part in sharpening contemporary perceptions of the housing shortage in the city.

By the mid-1950s, the extraordinary shortage of housing of the immediate post-war years had been made good through rebuilding war-damaged buildings and infilling in areas of existing housing, and the city was ready to start building afresh. In 1957, with substantial assistance from the federal government, the city was able to embark on its first large-scale reconstruction project, in the form of the redevelopment of the Hansaviertel, which had been a smart area before the war, just to the north of the Tiergarten (Kleihues, 1989; Weitz and Friedenberg, 1957; see Figure 6.16). Intended as a demonstration of the West's commitment to modernization and to continuing investment in West Berlin, the project brought together a group of international stars, including Walter Gropius, Le Corbusier, Aalvar Aalto and Oscar Niemeyer, to create a mixed development of tower and slab blocks of apartments, combined with groups of individual family houses, in a carefully landscaped demonstration of the promise of modern architecture made real.

Figure 6.16 The Hansaviertel (reproduced with the permission of the publisher from Heinrich and Mielke, 1964; photograph: Landesbildstelle, Berlin)

This was followed by a number of major initiatives taken by the Berlin senate to modernize the city's housing stock. A large housing project was started to the designs of Hans Scharoun in Charlottenburg Nord, and in the early 1960s this was followed by two even larger developments, the first stage of Gropiusstadt, or BBR, linking the south Berlin suburbs of Britz, Buckow and Rudow, and the Märkisches Viertel (Rave, 1970; see Figure 6.17). It was billed from the start as another demonstration of Berlin's transformation from the city of nineteenth-century rent barracks into a modern city offering all the facilities to be found in any other West German city. It was conceived as a new community for 45,000 people to be built in an area to the north of the city that was then covered with shabby *Schrebergarden* (allotment) plots, dilapidated weekend shacks and occasional cottages. The few buildings judged fit to remain were to be

Figure 6.17 The Märkisches Viertel (photograph: Reinhard Friedrich)

'enclosed' or 'gathered' together between the 'arms' of the new project. Laid out at relatively low density, the arms of the new development were to take the form of apartment buildings varying in height between four and sixteen storeys and were also to enclose a new centre providing shops, schools and other facilities for the whole community, including the original cottages. Any danger that the forms of the development would appear too uniform was to be avoided by involving a number of architects to design different elements of the whole.

The complex was to be equally diverse in its construction: different elements of the complex were to be built using different methods, including a large-panel system similar to those being used for the huge multi-storey housing development at Marzahn in the DDR. Architecturally ambitious, the Märkisches Viertel appeared at first sight to answer the expectations of those looking for an escape from the drab and run-down housing in the city centre.

As families started to move into the first of these developments, the failure to co-ordinate the construction of housing with the provision of schools, shops and other communal facilities, and above all the failure to extend the underground to the new community, attracted widespread criticism and growing disillusionment with developments of this kind. This in turn sparked a critical debate, parallel to the discussions in London, Paris and West German cities, of the rival merits of suburban 'greenfield' developments against the rehabilitation of run-down areas in the inner city. In Berlin, critics of the senate's policies questioned whether modernization in these terms was sustainable; whether a city so short of open space could continue to extend outwards, stretching links with the centre, increasing the burden of commuting and requiring the building of new infrastructure and services (Kleihues, 1989; Siedler and Niggemeyer, 1964). Such critics called for rebuilding in the inner city as a way of maintaining the vitality of the heart of the city and the continuity of established working-class areas such as Kreuzberg.

This line of criticism was intimately linked to two key changes already taking place in the inner city: first, the programme of slum clearance principally linked to plans for the construction of the southern sector of the east–west urban motorway; and second, the increasing pressure on inner-city housing caused by the growing number of immigrant families and by others such as students searching for cheap housing. Plans for large-scale slum clearance in six inner-city areas were announced by the senate as a high priority in 1964 as soon as the first major reconstruction projects at BBR and the Märkisches Viertel were under way. The most sensitive and one of the largest of these projects was to be in Kreuzberg, where 1,600 dwellings housing 37,022 people were to be cleared (Stephan, 1964; Suhr and Enke, 1991). The grounds for clearance were essentially the constructional and sanitary disrepair of the housing: over ninety-one per cent of all dwellings in the clearance area had been built before 1918 and over seventy-five per cent dated from before 1875. As was to be expected given the date of these developments, over forty per cent of dwellings faced onto courts and most were equipped with only the most minimal sanitary facilities: only fifteen per cent were equipped with both a WC and a bath or shower, a further nineteen per cent had only a WC and a further sixty per cent only had access to a shared WC on the landing, while six per cent still had only the use of a shared external WC in an internal court. Not only were these courts minimal in size, but many were occupied by small businesses, many of which were in poor economic health.

The plans for the Kreuzberg/Kottbusser Tor area, first put forward by the senate, envisaged the wholesale clearance of the area and the building of a key southern section of the east–west motorway loop which the senate still envisaged as part of a future unified motorway ring around the whole city of

Berlin. But revised plans for redevelopment on a reduced scale, retaining the existing street pattern and refurbishing rather than demolishing property, were soon submitted. Yet even this scale of clearance involved sweeping away back buildings and gutting remaining buildings to replace services, and clearing courts of workshops and other mixed uses. Those living in the area were inevitably faced with higher rents, while those renting workshops generally faced a five-fold increase in rent. The result was fewer and more expensive dwellings, which forced out tenants. Those who could afford to do so, moved to new housing in the suburbs; and those who could not, tried to find alternative accommodation in the reduced housing stock in the area or elsewhere in the inner city.

The effect of these first clearances was to create an organized opposition among the new arrivals in the area; immigrants, students and other groups who were attracted by low rents and were less critical of housing conditions than were the German families leaving for new housing in the suburbs. By the early 1970s, the groups were united in an increasingly politicized alliance in opposition to the city's slum-clearance programme. What had started in Kreuzberg spread across the city. The protesters turned the tactics of confirmation that had been used against us involvement in Vietnam and national political issues, to local use, fighting an aggressive campaign against the senate's plans. The immediate result was to frustrate these plans; the building of the motorway was finally abandoned in 1976. More important was the impetus that it gave for a change in policy away from large-scale suburban development and towards modernization through redevelopment, refurbishment and conservation in the inner city. This redirection of policy, signalled by the foundation in 1977 of the IBA (Internationale Bau-Austellung; International Building Exhibition) to revitalize the city through architecture, was to point the way forward into the coming decade (Kleihues, 1989; *Architectural Review*, 1984). By the mid-1980s, not only were a substantial number of new projects, many of them by internationally-known architects, being built across the inner city, but, under the aegis of the IBA's Altbau section, which dealt with the refurbishment of existing buildings, run-down blocks in areas such as Kreuzberg were being refurbished and modernized in a variety of ways ranging from state-of-the-art technology to the most basic forms of self-help (see Figure 6.18).

Figure 6.18   Redevelopment in the inner city. Flats, shops and old people's day centre at the junction of Ickensteinstrasse and Sculesiscaestrasse, block 121: SO36; architects: Alvard Siza and Peter Brinkert (photograph: *Architectural Review*)

6.6   Conclusion: Berlin beyond the 1990s

Looking back over the last hundred years of Berlin's history, what force has done most to shape the form of the city? Has it been its place in world events or has it been the slow but cumulatively sweeping changes in Berlin's everyday affairs, brought about by innovations such as the expansion of the railway system and the introduction of new building regulations? More than most European cities, Berlin still bears the scars of its recent past. Defeat of the Nazis' imperial ambitions led to the destruction of nearly half of the central area of the city. The division of Germany after the war shaped the city in a way that is without parallel: which other major European city has endured partition as long and as complete as that of Berlin? The development of the two separate centres from 1948 to 1989 is the most enduring mark left by the city's imperial past on its current form.

But with the breaching of the Wall, the forces of everyday development which shape other European cities, reinforcing the commercial pull of the centre and draining residential growth out to the suburbs, have returned to Berlin. The small plots on what had become West Berlin's eastern edge, forgotten and undervalued, one used as an artist's sculpture garden, another as a training-yard for guard dogs, have been transformed. Shortly after the fall of the Wall they were snapped up cheaply by rich dealers, the descendants of the *Terraingesellschaften* and the *Baustellenmackler*, and sold on for redevelopment to major corporations such as Mercedes and Sony to become a new and prestigious Potsdamer Platz at the heart of a unified Berlin. Now that the Wall is no longer there to contain the movement outwards, West Berliners are again building new suburbs. Meanwhile in the East, increasing affluence, rising car ownership and an influx of people from the West are strengthening and democratizing the forces of suburbanization that had been regulated by the state. In the East, too, the suburbs are growing: in place of the stark contrast between concentrations of high-rise blocks (*Hohenshoenhausen*) at Marzahn and the elegantly landscaped houses for the elite of the DDR around the Müggel See, there are private houses for those who can afford them.

What does this return to 'business as normal' mean for the balance of the two orders of history that we have been considering? The call to retain a section of the Wall as a reminder of the city's recent past, and the proposals by Daniel Liebeskind for a rebuilding of the Potsdamer Platz in a form that would serve to stress the continuity of the present with Berlin's history, are attempts to defend the past against the relentless pressure of the everyday forces of development. Many Berliners fear that these forces will gradually overpower and devour the past. Even in Berlin. By the middle of the twenty-first century, will we be able to detect the scars of the Second World War or the line of the Wall? Tyrants, kings and generals may leave their monuments, but they cannot stay forever the forces that shape the workaday city.

References

ASCHER, S. (1914) *Die Wohnungsmieten in Berlin von 1880–1910*, Berlin, Bodenpolitische Zeitfragen Heft 7.

Architectural Review (1984) 'Berlin as model' (special issue), vol.CLXXVI.

BALFOUR, A. (1990) *Berlin: the politics of order 1737–1989*, New York, Rizzoli.

BULLOCK, N. and READ, J. (1985) *The Movement for Housing Reform in Germany and France 1840–1914*, Cambridge, Cambridge University Press.

BULLOCK, N. (1990) 'Berlin' in M. Daunton (ed.) *Housing the Workers 1850–1914*, London and New York, Leicester University Press.

BULLOCK, N. (1993) 'Searching for the new dimensions of the city' in J.P. Kleihues and C. Rathgeber (eds) *Berlin–New York: like and unlike*, New York, Rizzoli.

CARSTENN, V. (1917) *Zur Geschichte und Theorie der Grundstückskrisen in deutschen Grosstädten, mit besonderer Berücksichtigung von Gross-Berlin*, Jena, Fischer Verlag.

CZADA, P. (1969) *Die Berliner Elektroindustrie in der Weimarer Zeit, eine regionalstatistische-wirtschaftshistorische Untersuchung*, Berlin, Historische Kommission zu Berlin.

DIEFENDORF, J. (1993) *In the Wake of War: the reconstruction of German cities after World War II*, New York and Oxford, Oxford University Press.

DIETRICH, R. (ed.) (1960) *Berlin: Neun Kapitel seiner Geschichte*, Berlin, Walter de Gruyter Verlag.

DOERRE, A. (1970) 'Entwicklung und Ergebnisse des sozialen Wohnungsbaues' in D.R. Frank and D. Rentschler (eds) *Wohnungsbau, Berlin und seine Bauten*, vol.IVa, Berlin, Verlag Wilhelm Ernst und Sohn.

ELKINS, T.H. (1988) *Berlin: the spatial structure of a divided city*, London, Methuen.

ENGELMANN, B. (1986) *Berlin, eine Stadt wie keine andere*, Munich, Bertelsmann.

FEHL, G. and RODRIGUEZ-LORES, J. (eds) (1983) *Stadterweiterungen 1800–1875, von den Anfängen des modernen Städtebaues in Deutschland*, Hamburg, Christians Verlag.

FRANK, D.R. and RENTSCHLER, D. (eds) (1970) *Berlin und seine Bauten, Wohnungsbau*, vol.IVa, Berlin, Verlag Wilhelm Ernst und Sohn.

FRICK, D. (1970) 'Einfluss der Baugesetze und Bauordnungen auf das Stadtbild' in D.R. Frank and D. Rentschler (eds) *Berlin und seine Bauten, Wohnungsbau*, vol.IVa, Berlin, Verlag Wilhelm Ernst und Sohn.

FRIEDRICH, O. (1974) *Before the Deluge*, London, Michael Joseph.

GEIST, J.F. and KÜRVERS, K. (1984) *Das Berliner Mietshaus 1862–1945*, Munich, Prestel Verlag.

GEIST, J.F. and KÜRVERS, K. (1989) *Das Berliner Mietshaus 1945–1989*, Munich, Prestel Verlag.

GELB, N. (1986) *The Berlin Wall*, London, Michael Joseph.

HABERLAND, G. (1931) *Aus meinem Leben*, Berlin, H.S. Hermann.

HAUS, W. (1992) *Geschichte der Stadt Berlin*, Mannheim, BI Taschenbuch Verlag.

HEGEMANN, W. (1930) *Das steinerne Berlin, die grösste Mietskasernestadt der Welt*, Berlin, Wasmuth.

HEINRICH, E. and MIELKE, F. (eds) (1964) *Berlin und seine Bauten, Rechtsgrundlagen und Stadtentwicklung*, vol.II, Berlin, Verlag Wilhelm Ernst und Sohn.

JACKSON, R. (1988) *The Berlin Airlift*, Wellingborough, Patrick Stephen Press.

JUNGHANNS, K. (1983) *Bruno Taut 1880–1938*, East Berlin, Elefanten Verlag.

KLEIHUES, J.P. (1989) 'From the destruction to the critical reconstruction of the city: urban design in Berlin after 1945' in J.P. Kleihues and C. Rathgeber (eds) *Berlin – New York: like and unlike*, New York, Rizzoli.

KREIDT, H. (1971) 'Industriebauten' in K.K. Weber (ed.) *Berlin und seine Bauten, Industriebauten und Bürohäuser*, vol.IX, Berlin, Verlag Wilhelm Ernst und Sohn.

LADD, B. (1990) *Urban Planning and Civic Order in Germany 1860–1914*, Cambridge, Mass., Harvard University Press.

LADD, B. (1997) *The Ghosts of Berlin: confronting German history in the urban landscape*, Chicago and London, University of Chicago Press.

LEYDEN, F. (1933) *Gross-Berlin, Geographie der Weltstadt*, Breslau, F. Hirt.

MACHULE, D. and SEIBERLICH, L. (1970) 'Die Berliner Villenvororte' in D.R. Frank and D. Rentschler (eds) *Berlin und seine Bauten, Wohnungsbau*, vol.IVa, Berlin, Verlag Wilhelm Ernst und Sohn.

MASUR, G. (1970) *Imperial Berlin*, New York, Dorset Press.

MEINARDUS, L. (1913) *Die Technik des Terraingewerbes*, Berlin.

POSENER, J. (1979) *Berlin auf dem Wege zu einer neuen Architektur, das Zeitalter Wilhelms II*, Munich, Prestel Verlag.

RADICKE, D. (1979) 'Die Entwicklung des öffentlichen Personennennahverkehrs in Berlin bis zur Gründung der BVG' in K.K. Weber, P. Güttler and D. Ahmadi (eds) *Berlin und seine Bauten, Anlagen und Bauten für den Verkehr (1) Städtischer Nahverkehr*, vol.Xb, Berlin, Verlag Wilhelm Ernst und Sohn.

RAVE, J. (1970) 'Die Wohngebiete 1945–67' in D.R. Frank and D. Rentschler (eds) *Berlin und seine Bauten, Wohnungsbau*, vol.IVa, Berlin, Verlag Wilhelm Ernst und Sohn.

READ, A. and FISCHER, D. (1994) *Berlin: the biography of a city*, London, Pimlico Press.

REICH, E. (1912) *Der Wohnungsmarkt in Berlin 1840–1910*, Munich and Leipzig.

RENTSCHLER, D. and SCHIRMER, W. (eds) (1974) *Berlin und seine Bauten, Wohnungsbau*, vol.IVa, Berlin, Verlag Wilhelm Ernst und Sohn.

ROSSOW, W. (ed) (1986) *Martin Wagner 1885–1957, Wohnungsbau und Weltstadtplanung: die Rationalisierung des Glückes*, Berlin, Akademie der Künste.

SCHALLENBERGER, J. and KRASSERT, H. (1926) *Berliner Wohnungsbauten aus öffentlichen Mitteln, die Verwendung der Hauszinssteuer-Hypotheken*, Berlin, Bauwelt Verlag.

SCHINDLER, N. (1972) 'Gartenwesen und Grünordnung in Berlin' in K.K. Weber (ed.) *Berlin und seine Bauten, Gartenwesen*, vol.XI, Berlin, Verlag Wilhelm Ernst und Sohn.

SCHINZ, A. (1964) *Berlin: Stadtschicksal and Städtebau*, Braunschweig, G. Westerman.

SCHINZ, A. (1974) 'Das mehrgeschossige Mietshaus von 1896–1945' in D. Rentschler and W. Schirmer (eds) *Berlin und seine Bauten, Wohnungsbau*, vol.IVb, Berlin, Verlag Wilhelm Ernst und Sohn.

SIEDLER, W.J. and NIGGEMEYER, E. (1964) *Die gemordete Stadt*, Munich and Berlin, Herbig Verlagsbuchhandel.

STEPHAN, H. (1964) 'Städtebau und Verkehrsentwicklung' in E. Heinrich and F. Mielke (eds) *Berlin und seine Bauten, Rechtsgrundlagen und Stadtentwicklung*, vol.II, Berlin, Verlag Wilhelm Ernst und Sohn.

SUHR, H. and ENKE, D. (1991) 'The 60s Phase' in M. Suhr (ed.) *Urban Renewal, Berlin: experiments, examples, projects*, Berlin, Senator für Bauwesen.

SUTCLIFFE, A. (1981) *Towards the Planned City: Germany, Britain, the United States and France, 1780–1914*, Oxford, Blackwell.

THIENEL, I. (1973) *Städtewachstum im Industrializierungsprozess des 19ten Jahrhunderts, das Berliner Beispiel, Veröffentlichungen der Historischen Kommission zu Berlin*, vol.39, Berlin.

VOIGT, P. (1901) *Grundrente und Wohnungsfrage in Berlin und den Vororten*, Jena, G. Fischer.

WEBER, K.K., GÜTTLER, P. and AHMADI, D. (1979) *Berlin und seine Bauten, Anlagen und Bauten für den Verkehr (1) Städtischer Nahverkehr*, vol.Xb, Berlin, Verlag Wilhelm Ernst und Sohn.

WEDEPOHL, E. (1970) 'Die Wohngebiete 1896–1918' in D.R. Frank and D. Rentschler (eds) *Berlin und seine Bauten, Wohnungsbau*, vol.IVa, Berlin, Verlag Wilhelm Ernst und Sohn.

WEITZ, E. and FRIEDENBERG, J. (1957) *Interbau Berlin 1957*, Berlin, catalogue to International Bauaustellung Berlin.

WYDEN, P. (1989) *Wall: the inside story of divided Berlin*, New York, Simon and Schuster.

ZIMM, A. (1959) *Die Entwicklung des Industriestandortes Berlin, Tendenzen der geographsichen Lokalisation bei den Berliner Industriezweigen von überörtlicher Bedeutung sowie die territoriale Stadtentwicklung bis 1945*, East Berlin, Deutscher Verlag der Wissenschaften.

Chapter 7: RECONSTRUCTION: NEW WAYS OF BUILDING AND GOVERNMENT SPONSORSHIP

by Nicholas Bullock

7.1 Introduction

On 15 June 1945, morning shoppers and passers-by in Oxford Street turned to watch in excitement as the first section of an AIROH prefabricated bungalow[1] was delivered, ready painted and glazed, for erection on a specially prepared site and settled into position by a four-man crew (see Figure 7.1). This section was followed shortly afterwards by a second containing the bathroom and the

Figure 7.1 A slice of house: one of the four sections of the temporary aluminium house being transported to the site next to Selfridges on Oxford Street (reproduced with permission from *Architects' Journal*, 1945, p.452)

[1]   The AIROH (Aircraft Industry Research Organization on Housing) or Aluminium was one of the types of bungalow accepted for production as part of the government's Temporary Housing Programme. For an account of the Temporary Housing Programme see White (1965) and Vale (1995).

kitchen, complete with electric cooker and refrigerator. By 4 p.m. the crew had joined the four sections together, connected the mains and made the house ready for inspection by the public. The people liked what they saw. Here in the very heart of London was a demonstration of how the government was planning to meet the needs of those who had been bombed out, those who were sharing with neighbours, in-laws or another family and those whose families were split up because of the shortage of housing. Technology that had been developed and applied during the war was now to be used to solve the most pressing social problem of reconstruction.

7.2 Government plans for reconstruction

The installation of the AIROH bungalow on Oxford Street was an illustration of only one of the ways in which the government had been preparing since the early years of the war to meet the challenge of post-war reconstruction. The starting point for these preparations is to be found in the debate on reconstruction which began with the ending of the phoney war in the summer of 1940. Then, as London, Liverpool, Bristol and other cities suffered assault from the air, it was natural, if only as an act of defiance against a seemingly hostile fate or as a way of bolstering morale, to talk of the way in which Britain should be rebuilt after the war.

Popular interest ran well ahead of government action. The heavy raids of October 1940 spurred discussion in the press. With this came the recognition that a massive programme of rebuilding lay ahead after the war and that here was an extraordinary opportunity not just to rebuild, but to improve on what had been. During the next five years reconstruction was a word on everybody's lips (see Figure 7.2). It was to become the subject of innumerable newspaper articles, pamphlets, books, exhibitions and films. It also provided a focus for the activities of reforming groups keen to champion their own blueprint for the future and to shape the future of post-war Britain. Besides the predictable caution of the professions, there were a variety of campaigning bodies such as the Modern Architectural Research Society (MARS) and the Royal Academy's Reconstruction Committee who pressed very different, if equally radical (and implausible), alternatives as the best way to rebuild the country (Gold, 1997). Central to both the popular debate and to the position of the campaigning groups was the assumption, sometimes implicit, more often explicit, that only the government could direct the programme of reconstruction. In contrast to the prevailing *laissez-faire* wisdom of the 1930s, it was generally accepted that the state would need to maintain in peacetime a role very similar to that which it was playing during the war.

How was the state to set about organizing the massive task of reconstruction? As early as April 1940, the Cabinet established a committee to consider the issue, and in January 1941, Greenwood as Minister without Portfolio was put in charge of government thinking on reconstruction.[2] But it was not until the establishment of an Official Committee on Post-war Economic Affairs, staffed by senior civil servants from the main departments, and the reorganization of the Reconstruction Committee in a more powerful form under Jowitt in March 1942, that the administrative machinery necessary to consider a problem of this scale was put in place. With the civil servants who understood the nuts and bolts of government, backed by the executive power of a Cabinet committee, progress was at last possible.

At the heart of the government's discussion of reconstruction was the central difficulty of balancing the country's expectations of reconstruction with the

---

[2]   There are a number of general studies of reconstruction; for example, Addison (1975). But there is little on government plans for physical reconstruction. See White (1965), Cullingworth (1975) and Gold (1997, pp.157–85).

Figure 7.2 Poster for a wartime exhibition of proposals for rebuilding Britain, arranged by the Royal Institute of British Architects for the building industry at the National Gallery, February 1943 (reproduced from RIBA, 1943, cover)

resources that it believed might be available when the war ended. The memory of the boom after the First World War and the accompanying inflation was still vivid and served as a clear warning of what would inevitably happen if demand and supply were not carefully controlled in what would in effect be an empty economy after the war. During the spring of 1942 the Official Committee had reviewed the painful experience of boom and slump between 1918 and 1921 when the huge purchasing power built up during the war, combined with the need to replace material and stocks run down during the years of hostilities, had fuelled a surge of demand. The sudden removal of the wartime controls and subsidies and a return to an open economy had only increased the effect of the boom, which had been followed in the summer of 1920 by a slump of corresponding ferocity. Given the damage sustained across the economy since 1939, there was every reason to believe that the same pattern would be repeated on an even greater scale unless the government could act to prevent it.

Government plans for the post-war housing programme (see Figure 7.3) illustrate the scale and complexity of the task involved.[3] The housing drive was conceived as three separate campaigns. The first was to provide mass-produced temporary bungalows (of the kind erected in Oxford Street in June 1945) to meet the immediate needs of those without housing, the 'razor's edge of demand', while the second campaign got under way. This second phase was to make good the loss of housing caused by enemy action and was set a target of 220,000 permanent houses to be completed two years after the end of the war. Finally, in answer to widespread hopes for a better future, the third campaign was to continue through into the long term – at least five years after the war – an expanded version of the policy of slum clearance and improvement that had been cut short by the war. Having identified the housing programme as the largest and the most urgent priority for the post-war construction industry, the government had to find a way of accommodating the programme's needs to the capacity of the building industry and the demand for construction in other fields, all of which had to be reconciled with the demands of other sectors of the economy.

To meet this challenge, the government needed to plan on a number of fronts. A first requirement for the housing programme was to get the building industry back into action as soon as possible after the war by securing an adequate work-force for both the building and the building materials industries. This was addressed by officials from a number of ministries – the Ministries of Works, Labour, Health and the Board of Trade who explored how skilled labour might best be released from the services as part of the general process of demobilization after the war. Complementary to these initiatives were attempts to increase recruitment into the industry. This in turn required delicate negotiations with the building unions and a long-term commitment to training in the building industry, which resulted in the White Paper, *Training in the Building Industry*, published in 1944 (Ministry of Labour, 1944). A second major area of planning was the concern to encourage the building industry to explore ways of saving labour, particularly skilled labour, and economizing on the use of materials that would be in short supply after the war. This was to be achieved most effectively, not by the lowering of standards, which the government hoped would in fact be raised by the various reconstruction programmes, but by developing new forms of construction for a variety of different building types.

This programme of innovation was to be the responsibility of the Directorate of Post-war Building, which had been established with Hugh Beaver as director in the autumn of 1941 to oversee the preparation of the building industry for post-war reconstruction and had quickly assembled an awesome array of committees to shape government policy on the design of certain building types, on the design of structures and on building standards and codes of practice.[4] Within a year, the Directorate had set up a special Interdepartmental Committee on House Construction, under Sir George Burt, a director of Mowlems and a friend of Hugh Beaver, to examine how permanent housing might be built using non-traditional materials. The task of this committee was to vet the proposals sent in by sponsors, principally by individual building firms and local authorities, and to recommend the most promising ideas for further development back to the Directorate of Post-war Building which would then issue a licence for the construction of a demonstration house either by the sponsor or by the Ministry of Works on their test site at Northolt. This demonstration house would then be assessed by a team

[3]   Despite the importance of the housing programme for government plans for reconstruction, there is still no comprehensive account of wartime plans for post-war housing. The subject falls between the different volumes of the official History of the Second World War, though there is some discussion of the subject in Kohan (1952). The documentation of wartime planning for the housing programme is to be found in the minutes and papers of a number of key committees.

[4]   The work of the Directorate of Post-war building, and wartime plans for post-war building, are described in Kohan (1952), White (1965) and Finnimore (1989).

from the Building Research Station and, if judged successful, would be designated as one of the limited number of systems of construction officially adopted for use in the housing programme. Initially the committee's brief had been to investigate non-traditional design for permanent housing only. But with the recognition by the end of 1943 that some form of emergency housing would be necessary for the homeless while the Permanent Housing Programme got under way, the committee was asked to consider designs for temporary housing as well.

7.3 Designs for non-traditional housing

By its last meeting, the committee had received over 500 submissions.[5] Many were trivial or simply irrelevant; some were extravagantly silly: for example, the torrent of correspondence from an inventor, working under the motto 'Troutstream and Thatch', who urged the advantages of Goodyear tyre canvas and thatch sprayed with cement as a form of cheap housing. But there was no shortage of serious proposals. Of the 101 firms who were listed in the committee's final report as being granted licences, most were from the building industry: twenty were builders, five were engineering firms with strong links to the building industry and fourteen had a major interest in building materials, components or fittings. Many of these, like Wates who built the pre-cast concrete Mulberry Harbours for use in the 1944 Normandy landings, had wartime experience of new forms of construction and were keen to secure a slice of the action in the post-war building programme. Others were attracted by keeping production going during the difficult period of transition from war to peace: proposals by the British Power Boat Company for the Jicwood plywood house were submitted as a way of using manufacturing capacity previously used to make motor-torpedo boats and other light craft; the AIROH temporary aluminium bungalow was developed for the same reasons by firms in the aircraft industry and the Ministry of Aircraft Production.

There was also an active interest in non-traditional construction in architectural circles and architects were closely involved in the design and development of most of the successful systems. A number of progressive architects may have remembered the interest in new forms of house-building shown by leaders of the pre-war avant-garde: Le Corbusier had collaborated briefly and inconclusively with the Voisin Aircraft Company on plans for mass-produced housing to be erected in war-damaged areas of northern France after 1918; Walter Gropius had referred in his book, *The New Architecture and the Bauhaus* (1935), to his own work on prefabricated housing and included a few suggestive if cryptic photographs of the houses developed with a copper manufacturer in the early 1930s. This interest was not confined to Modernists alone. Staff of the Building Research Station were also certainly aware of experiments by continental Modernists with permanent non-traditional construction: Ernst May's large-panel construction system, used on Frankfurt estates such as the Römerstadt Siedlung in the late 1920s, was well documented, and the French Mopin system of construction was even being used for the construction of the Quarry Hill estate in Leeds in 1939.[6] Independent architectural interest in the field is exemplified by the work of the

[5]   The findings of the Burt Committee are summarized in the three reports, Post-war Building Studies nos. 1, 23 and 25; the papers of the Burt Committee are held in the Public Record Office (HLG94/1–16), and the work of the committee is briefly described in White (1965).

[6]   The Building Research Station received the reports of their German equivalent, the Reichsforschungsgesellschaft (Rfg), which published, *inter alia*, detailed reports on May's Frankfurt large-panel construction system (Rfg 1929). For British knowledge of the Mopin system see White, 1965, pp.97–103.

Committee for the Industrial and Scientific Provision of Housing (CISPH), founded in November 1941 to look at ways of speeding up the production of housing (CISPH, 1943). Meeting irregularly in London over the next eighteen months, the thirty members of the committee investigated ways both of extending mechanized production of building components and materials, and of speeding up and systematizing the process of their assembly by looking to the example of motor-car and caravan production. CISPH published two reports but did not establish formal links with such groups as the Burt Committee working under the Directorate of Post-war Building. Though there is a possibility that their ideas influenced the design of the AIROH aluminium bungalow, CISPH's work is principally of value as evidence of the interest of young and often politically radical architects who were keen to explore new ways of solving current social problems such as housing, and were willing to accept the new architectural forms that would result.

More important as a source of ideas for new forms of construction was the extraordinary achievement of the Americans in providing a number of what in effect were mass-produced prefabricated new towns built at great speed to house the work-force for the rapidly expanding munitions industry. This feat was known to some of those working at the Building Research Station and in 1943 members of the Burt Committee visited the USA to see the American achievement at first hand (Casson, 1945).[7] What they saw impressed them. From the mid-1930s onwards American industry had, with the encouragement of a variety of federal agencies, been experimenting with ways of speeding up the production of low-cost housing, generally by building on the American tradition of lightweight timber construction. Many New Deal public works programmes called for the creation of new communities, often in rural areas where it was difficult to build conventional housing, and where prefabricated units were the simplest way to house a new labour-force. The lightweight demountable units designed and built for the Tennessee Valley Authority (TVA) were typical of this kind of housing, designed for delivery by road with three sections to a house. The TVA housing was widely illustrated both in the USA and in wartime Britain (Huxley, 1943). Other US government agencies had an equally distinguished record of achievement in this field. Before the war, under the aegis of the Farm Security Administration (FSA), architects with a national reputation such as Vernon de Mars had been engaged to design minimal-cost mass-produced housing for low-income farm workers. The results, for example at the small community of Yuba City, California, attracted national attention as a successful marriage of modern design with constructional innovation (*Architectural Review*, 1944; Albrecht, 1995).

With the beginning of the war in Europe and the need for a rapid expansion of the defence industries, these ideas were developed still further. At Vallejo, de Mars began the construction of hundreds of houses for those working in the navy yards. These houses were factory built on an assembly line using stressed skin and glued plywood panels, techniques borrowed from aircraft production. When the Federal Works Agency (FWA) set about increasing the volume of wartime house-building for the defence industry, it turned to modern architects such as de Mars, William Wurster, Walter Gropius, Marcel Breuer, George Howe, Louis Kahn, Hugh Stubbins, Antonin Raymond and even Frank Lloyd Wright to encourage the construction industry to experiment with new ways of building. The achievements – from Wurster's housing in Vallejo in California to the housing built in Pennsylvania by Gropius and Breuer and by Howe, Stonorov and Kahn – were widely illustrated in the States, and impressed the team from the Burt Committee as an inspiration for British practice after the war (see Figure 7.3).

---

[7]   For the Burt Committee's responses to American practice see PRO HLG94/7 and 9.

Figure 7.3 Wartime housing at Coatesville, Pennsylvania, by Walter Gropius and Marcel Breuer (reproduced courtesy of *The Architectural Review*, 1944, p.72, plate 3)

American housing exhibition at the Royal Institute of British Architects

(1 and 2). *Family units and small community house in Yuba City, California—a housing project for farm workers.*

(3). *A two-bedroom unit at Windsor Locks, New England. Inside, there is a zig-zag relationship between the dining corner, kitchen and utility space to give an illusion of privacy and spaciousness.*

(4). *Detail of window treatment in houses at Windsor Locks. The outswinging casements are fixed panes of glass set directly into the structural members.*

Figure 7.4 Wartime exhibition of American housing, September 1944, showing housing at Yuba City by Vernon de Mars and at Windsor Locks by Hugh Stubbins (*Architectural Design and Construction*, 1944, p.220; copyright John Wiley and Sons Limited. Reproduced with permission.)

Just as impressive as the quality and the novelty of the design and construction of the individual house was the speed at which the Americans were developing whole settlements, many of them the size of large towns. These varied from small rural communities to new towns. Vanport near Vancouver was built to house the 45,000 workers in the Kaiser ship-building plants engaged in the mass production of Liberty ships (Hise, 1995). Within three years, using a mix of permanent and temporary housing, 4,452 hectares of dairy and garlic farms at McLoughlin Heights had been transformed to provide a new town with city-wide water, sewage and electricity systems, four schools, recreation and day-care centres, a branch library, a medical clinic, two shopping centres (the larger of which boasted thirty-three shops covering 4,645 square metres) and the largest food store in the region. Nor was Vanport exceptional: new towns of similar size were being developed to accommodate the furious growth of the defence industries in California, Michigan, Oklahoma and along the East Coast.

The American experience was seen at first hand by members of the Burt Committee, and was soon broadcast to a wider public, through publications, films and the Museum of Modern Art's exhibition, 'American Housing in War and Peace' (see Figure 7.4). The exhibition opened in London in July 1944 and attracted widespread interest and an enthusiastic response from the British architectural press. Commenting on the scope and the quality of the work on display, the *RIBA Journal* declared much of the work to be of 'quite superlative excellence' and urged British architects to learn from American example (*RIBA Journal*, 1944a).

British architects were keen to take on the challenge. When RIBA had convened a debate in April 1944 to determine the profession's views on the subject, the majority expressed themselves wholeheartedly in favour of a more innovative approach to solving the post-war housing problem (*RIBA Journal*, 1944b). The issue for architects was now to decide what buildings built in a non-traditional way would look like. Already architects of all kinds, from establishment figures such as Grey Wornum to rebels such as Anthony Chitty, were engaged in designing non-traditionally built housing for both the Temporary and the Permanent Housing Programmes. This kind of design was generously covered in the professional press and came to acquire a certain gritty glamour, an exciting sense of drawing real plans for the post-war world.[8]

7.4   Non-traditional forms of construction for permanent housing

By the end of 1944, government plans were beginning to bear fruit. In the summer of 1944 the Ministries of Works and Health published jointly a memorandum for local authorities on government plans for temporary housing (Ministry of Health and Ministry of Works, 1944), the Building Research Station published a report on the demonstration houses it had built at Northolt (Ministry of Works, 1944) and the Burt Committee published its first report on new ways of building permanent housing (Ministry of Health, Secretary of State for Scotland and Ministry of Works, 1944). These permanent houses were to be built to the improved standards recommended by the Dudley Committee which had been established to consider the detailed design and layout of post-war housing (Ministry of Health, 1944; Bullock, 1987). For permanent housing the Burt Committee eventually recommended 101 different types of construction: steel, timber and concrete, both pre-cast and *in situ*, submitted

---

[8]   See, for example, the coverage given to both the Temporary and the Permanent non-traditional Housing Programmes in *Architectural Design and Construction*, which ran a regular series of articles under the title 'Housing Forum' from 1943 until 1947.

by sixty-one sponsors (Ministry of Health, Secretary of State for Scotland and Ministry of Works, 1948). Because of the post-war rise in the cost of timber, however, steel and concrete systems of construction were more popular than any form of timber construction before 1950 (Bowley, 1966; see Table 7.1).

Table 7.1 Numbers of non-traditional houses of different types in England and Wales, April 1945 to 31 March 1955 (from Bowley, 1966, based on Housing Returns for England and Wales)

Type of house	Up to 30 Sept. 1950	Oct. 1950 to 31 March 1955	Total
1 Aluminium (special scheme)	13,856	3,041	16,897
Steel frames*			
2 BISF†	31,076	440	31,516
3 Howard†	1,404	–	1,404
4 Unity Structures	2,314	11,387	13,701
5 All other steel frames	3,211	1,679	4,890
Totals (2)–(5)	38,005	13,506	51,511
Timber frames*			
6 Spooner†	1,308	2,199	3,507
7 Scottswood	–	1,067	1,067
Totals (6)–(7)	1,308	3,266	4,574
Concrete*			
Precast (a) Pier-and-panel			
8 Airey	20,920	4,647	25,567
9 Cornish Unit	4,234	18,939	23,173
10 Orlit	7,126	1,323	8,449
11 Woolaway	975	3,870	4,845
Totals (8)–(11)	33,255	28,779	62,034
Precast (b) Large slab			
12 Newland and Kingston§	2,307	374	2,681
13 Reema	738	6,329	7,067
14 Smith's Building Systems	1,124	2,753	3,877
15 Stent	–	1,253	1,253
16 Wates	5,320	13,456	18,776
Totals (12)–(16)	9,489	24,165	33,654
In situ			
17 Laing's 'Easiform'	14,690	28,516	43,206
18 Wimpey's 'no fines'	6,488	46,883	53,371
19 Other in situ	268	3,515	3,783
Totals (17)–(19)	21,446	78,914	100,360
Totals (8)–(19) all concrete	64,190	131,858	196,048
Grand total	117,359	151,671	269,030

* as far as separately recorded and listed in PWBS, no.25
† wholly or partly clad with aluminium, steel or asbestos cement
‡ i.e. Cussins, Steane, Trusteel and Hills' Presweld
§ built by Tarran Industries Ltd and its successor Myton Ltd.

The most widely used type of steel-framed house was the British Iron and Steel Federation house (BISF) designed by Frederick Gibberd in a number of variants (Ministry of Health, Secretary of State for Scotland and Ministry of Works, 1946; see Figures 7.5 and 7.6). The 'B' version was the most common. The basic structure was a steel frame of cold-rolled sections assembled in units which were light enough to be handled by two people. The uprights of these units were placed just over a metre apart and spot-and-ridge welded to a transverse beam at ground-floor, first-floor and eaves level. The roof structure was formed of a lightweight steel truss and the floors of cold-rolled steel joists; external walls were of concrete render on steel sheet to first-floor level, with horizontally ribbed steel sheeting, galvanized and painted above. Party walls were of two independent 7.5 cm skins of foamed slag and internal walls were of plasterboard panels bonded to fibreboard skimmed *in situ*. Heating was by open fire in a masonry hearth with an encased flue above.

The most popular of the pre-cast concrete systems was the Airey system, and its rural version the Airey 'rural type', which used single-storey pre-cast concrete posts and small concrete panels, which were small enough to be easily handled by two people (Ministry of Health, Secretary of State for Scotland and Ministry of Works, 1946; see Figures 7.7 and 7.8). The concrete posts were cast around a steel tube and connected by a steel lattice joist at first-floor level, and the resulting goalpost-like frames were then erected at 45 cm centres with the upper storey frames located by dowels onto the frames beneath. The cladding panels, 90 cm long and 24 cm high, were attached like concrete weather boarding to the posts with copper wire twisted around a softwood fillet cast into the back of each post; these were laid without mortar and the vertical joints staggered. The roof was of traditional timber construction with tiles or slate. Internally the walls were lined with sheets of plasterboard or cellular plywood backed by additional thermal insulation and fixed to the softwood fillet on the posts. Easily erected by unskilled labour without special equipment, the Airey 'rural type' houses were well suited to rural or scattered sites and, today, they are still a common sight on the edge of villages as far apart as the Fens and the Borders.

The most common form of *in situ* concrete construction, the 'no fines', was developed by sponsors such as Wimpeys (Ministry of Health, Secretary

Figure 7.5 'A' type house. Designed for the British Iron and Steel Federation by Frederick Gibberd (reproduced from *Post-war Building Studies*, 1946, plate 22)

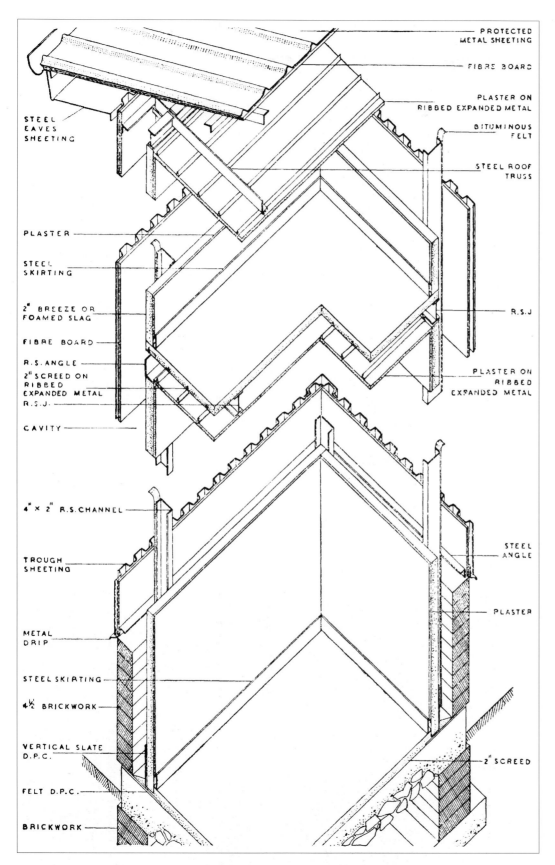

Figure 7.6 The construction of the BISF 'A' type house (reproduced from Ministry of Works, 1946, figure 8, p.48)

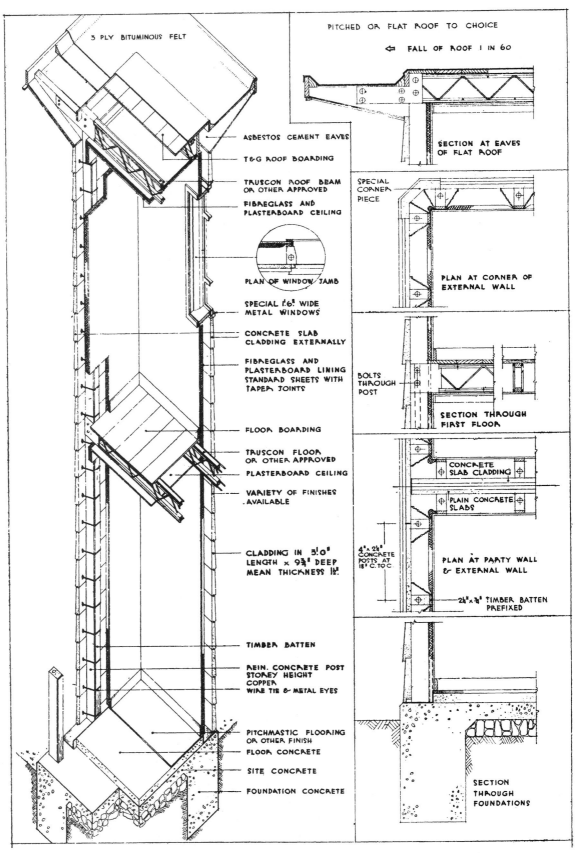

Figure 7.7 The construction of the Airey house (*Architectural Design and Construction*, 1945, p.300; copyright: John Wiley and Sons Limited. Reproduced with permission)

Figure 7.8 A pair of Airey 'rural type' houses (reproduced from White, 1965, plate 3.19)

of State for Scotland and Ministry of Works, 1948; see Figure 7.9). The common characteristic of this approach was the construction of the walls of a house from cement and large aggregate only, omitting the fine aggregate of normal concrete – hence the name – to leave small voids uniformly distributed throughout the structure. The rough concrete finish that resulted was rendered to improve appearance and to reduce water penetration. This form of construction offered excellent thermal properties: a 20 cm-thick wall of 'no fines' concrete rendered externally had a standard of insulation equal to that offered by a 27.5 cm cavity wall. Unlike the large panel systems which made use of expensive reusable shuttering, 'no fines' concrete used cheap and reusable shuttering consisting – in the form developed by the Ministry of Works and used by Laings – of small frames covered in wire mesh to give three 'lifts' to a storey height and making it relatively easy to accommodate window and door openings in the shuttering process (White, 1965; see Figure 7.10).

Figure 7.9 The shuttering for a Wimpey 'no fines' house in the process of erection (reproduced from Ministry of Works 1948c, plate 9)

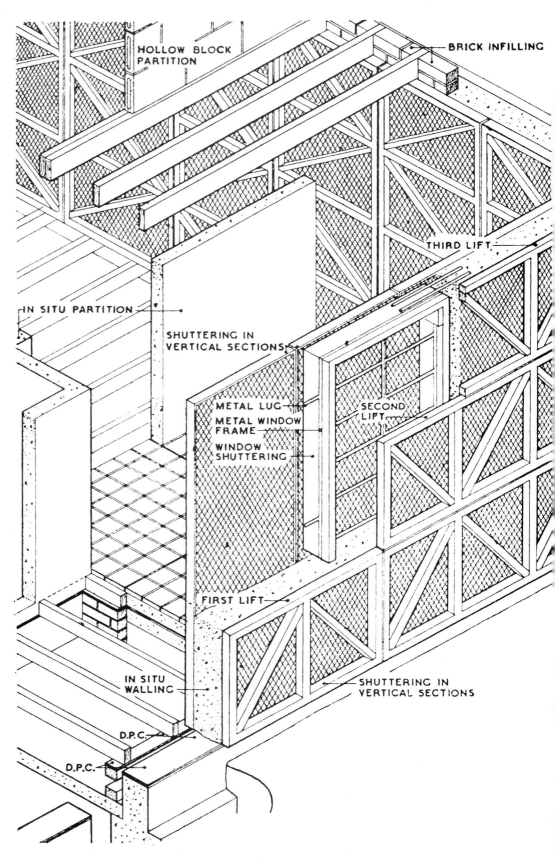

Figure 7.10 The 'no fines' construction system and its steel-mesh shuttering system (reproduced from White, 1965, p.164, fig.3.10)

7.5 *Bungalows for the Temporary Housing Programme*

Complementary to the non-traditional design for the Permanent Housing Programme were the proposals for the temporary housing that was to be erected immediately after the war, before the Permanent Programme had got under way (Ministry of Health and Ministry of Works, 1944; Vale, 1995). These houses were to be built to lower standards than those for permanent houses and were to be built for an explicitly limited life without considering housing layout and town-planning. The Committee recommended a number of types constructed of different materials – timber, aluminium, steel frame – each using very different proportions of factory-based and on-site labour. The AIROH aluminium bungalow, for instance, prefabricated in sections like the TVA housing, is an example of factory-based manufacture, while the *Arcon* house demonstrates the advantages of an 'open' approach based on the on-site assembly of standardized and widely available building components.

The structure of the AIROH house (see Figure 7.11) consisted of four sections whose width, at 2.25 metres, was set by the maximum transportable by road.

Figure 7.11 The aluminium or AIROH bungalow of the Temporary Housing Programme (reproduced from White, 1965, plate 3.2)

Figure 7.12 Production line for sections of the aluminium bungalow (reproduced from White, 1965, plate 3.7)

These sections were structurally discrete and joined on site in the manner of the wings and the fuselage of an aircraft (White, 1965). Like a motor car, each section was built up on a chassis to which were attached wall panels of aluminium sheet finished internally with bitumen-bonded insulation. The roof consisted of a trussed assembly attached to both external and internal walls so that the whole section would act like a deep trussed beam when lifted by crane for transport or assembly. Internal wall linings and partitions were of plasterboard, the ceilings of fibreboard, and the floors of tongue-and-groove boarding on joists fixed to the section's chassis. All the services, including a slow-combustion stove for central heating, were provided by the Ministry of Works standard kitchen/bathroom unit which was built into one of the house's four sections. All glazing, wiring and finishing work, including painting, was completed in the factory (see Figures 7.12 and 7.13). Finished sections were delivered to the site by low-loader ready for erection on a prepared site slab, similar to those used for other temporary bungalows, which contained all mains services. The four sections were levelled on this slab by jacking points positioned at three of the section's four corners, the sections joined together and the joints sealed. With site labour kept to a minimum, it was claimed that

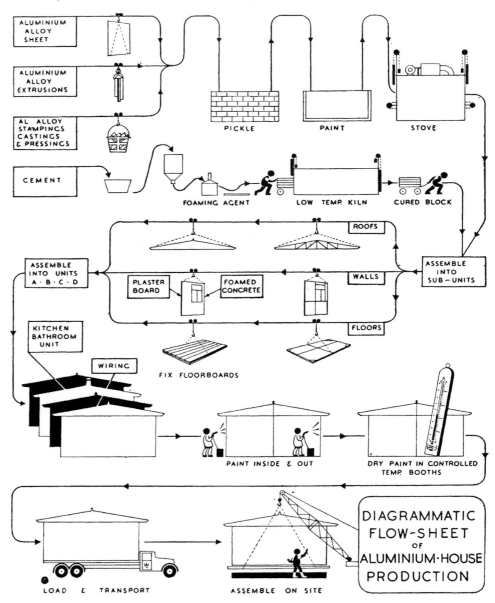

Figure 7.13 The production sequence for the aluminium bungalow (reproduced with permission from *Architects' Journal*, 1945, p.466)

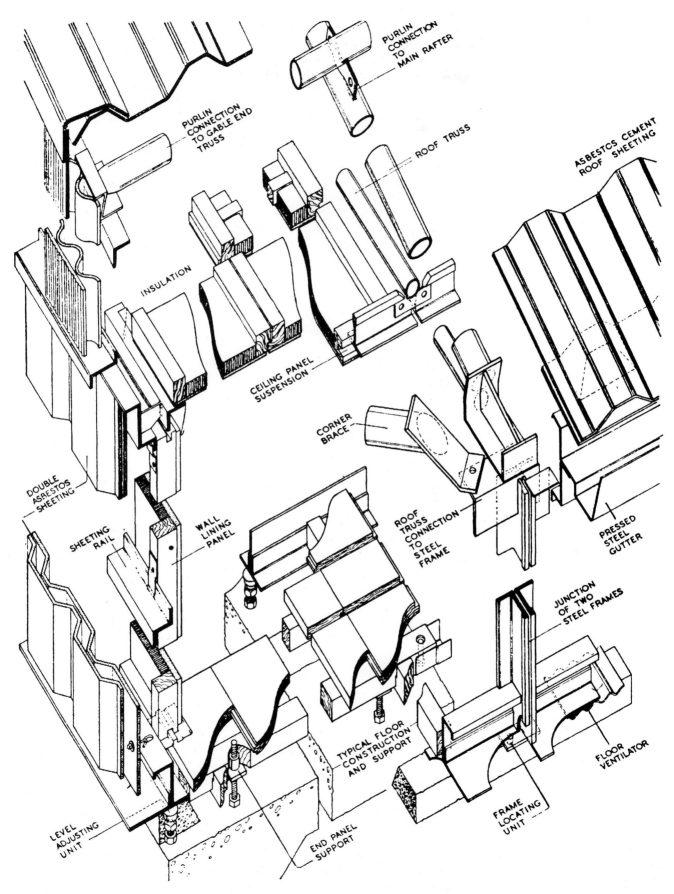

Figure 7.14 The construction of the *Arcon* (Mark V) house (reproduced from White, 1965, p.143, fig.3.2)

Figure 7.15 The *Arcon* (Mark V) house of the Temporary Housing Programme (reproduced from White, 1965, plate 3.1; courtesy Arcon, Chartered Architects)

the house could be erected, after some experience, by an unskilled gang in twenty hours of work.

The factory-based construction of the AIROH house stands in sharp contrast to that of the *Arcon* house (see Figures 7.14 and 7.15). For this, components were supplied by a variety of manufacturers to an 'agent', typically a large contractor, who would act as co-ordinator of supplies and delivery for sites in the locality. The basic structure of the final version of the *Arcon* house, Mark V, was a light steel frame of cold-rolled sections mounted on a steel foundation frame and roofed with a tubular steel truss (White, 1965). The external walls and roof were of moulded asbestos-cement sheeting used in a double thickness for the walls and a single thickness for the roof. Internally the walls were lined with storey-height plasterboard panels, backed with insulation and bonded onto light timber frames; together, this form of wall construction gave thermal insulation comparable to that of a 27.5 cm cavity wall. Internal partitions were also of plasterboard, the ceiling of fibreboard, and the floor of tongue-and-groove boarding nailed to battens set in a cement screed laid on a damp-proof membrane (DPM) over the site slab. The *Arcon* house, too, used the standard Ministry of Works service core located on the entrance side of the house to create an elegant plan which offered a kitchen overlooking the road and a marginally larger hall than was possible with other types of temporary bungalows. Assembly started with the levelling of the foundation frame and with this in position the main steel frame was welded together, along with the surrounds for windows and doors. This was followed by the cladding and the bolting in position of the windows and doors. Finally, with electrical circuits connected to the service core, the house was lined and partitioned internally. Assembly could be completed by a team of thirty-five labourers (with between six and nineteen on site at one time) in eight hours, longer than the time needed for the AIROH house, but balanced by savings in factory production time.

7.6 *The use of new forms of construction, 1945–50*

What had government programmes achieved as the first, most intense, phase of reconstruction was coming to an end in 1948? With the economy struggling to recover from the war and with labour and most building materials in short supply, it was soon clear that wartime hopes for a massive housing drive and the speedy reconstruction of war-damaged cities were impossible to achieve. In quantitative terms, far fewer houses had been built than the government had planned. The Temporary Housing Programme had provided 156,623 houses (against a target of 158,100) but had taken far longer to produce this number than originally planned: a year after VE day, fewer than 30,000 dwellings had been completed; two years after the end of the war – the target date for the completion of the programme – the number had increased but still stood at only 97,000 (Ministry of Works, 1945, 1948a). Less successful still was the Permanent (non-traditional) Housing Programme (Bowley, 1966). The White Paper on housing in March 1945 held out the hope that 200,000 dwellings would be completed in the first two years after the war. By the end of 1946 only 2,769 houses had been built; two years after VE Day only 7,850 had been completed and a year later the number was still only 36,000. By 1950 only 117,359 permanent non-traditional houses had been built, a much smaller contribution to the post-war housing drive than had been envisaged at the time of the White Paper on housing in March 1945 (Ministry of Health, 1945).

These difficulties were due in large measure to rising costs and, after 1948, to an increasingly unfavourable comparison in terms of value for money with conventional housing. The cost of both the temporary and the permanent non-traditional houses was much higher than expected, and as labour and materials became more freely available, became uncompetitive with the standard and cost of traditionally-built housing. There were exceptions: Wates built more 'no fines' houses after 1950 than before, and the Cornish Unit system of building (closely akin to the Airey system) continued to be used after 1950 in remote rural areas (Bowley, 1966). But these were exceptions to the general rule. The results of investigation into the cost of the permanent non-traditional programme carried out by the Building Research Station (Ministry of Works, 1948b; Bowley, 1966) showed that of the fourteen systems examined in detail, all except two *in situ* concrete systems (whose shuttering costs were excluded) were more expensive than traditional brick construction, and this by a considerable margin: six of the fourteen were ten per cent more expensive, and one was over twenty per cent more. After examining the way in which the non-traditional programme had balanced savings in conventional materials and labour against off-site labour and equipment costs, the study concluded that 'investigations provide a striking illustration of the relative ease of economizing in particular scarce resources compared to the difficulty of providing cost reductions, and the absence of any *a priori* reason for the former to coincide with the latter' (Bowley, 1966, p.247).

But if non-traditional housing was not cheaper, what were its advantages as traditional materials and skills were no longer in short supply? Conceived at a time of shortages, the general thrust of design and innovation in non-traditional forms of building had been to find substitutes for traditional ways of building, to match the qualities of the conventional house, not to offer fundamental innovation. A form of construction designed to use materials such as sheet timber or aluminium as a substitute for traditional materials and skills was found to have peacetime costs that simply could not have been anticipated in wartime. A non-traditional house that could offer the standards of space and comfort of a conventional house might be viable at a slightly higher cost during the period of greatest austerity, but as costs rose, for authorities such as the London County Council (LCC) who were grudgingly willing to try, many of the new systems of building became too expensive to offset the advantages of availability of materials and speed of construction.

Moreover, non-traditional housing had other disadvantages. To the local authorities who were expected to take the financial risk of building this kind of housing, its durability was suspect (Bowley, 1966). Members of the Glasgow Housing Committee, invited to inspect and comment on the systems being reviewed by the Burt Committee, were understandably sceptical about whether this kind of housing would last the full sixty years of its financial life if occupied by traditional tenants.[9] The LCC Housing Committee shared this view, despite the urging of the county's architect, Forshaw. Pressured to use new ways of building

---

[9]   The local authorities' doubts about the durability of non-traditional housing reported by Bowley are documented in the detail in the files of the Burt Committee (PRO, HLG94/11).

housing, the LCC reluctantly agreed to try a number of Braithwaite houses on their out-County estates but generally favoured those systems such as the Airey system which corresponded most closely with traditional construction. Nor did the appearance of non-traditional housing have much to offer. Most sponsors chose not to exploit the architectural potential of new forms of construction and disguised their designs to look as much like conventional houses as possible. A few were frank, some brutally so, in expressing the form of their construction; but very few had the straightforward simplicity of the BISF houses designed by Frederick Gibberd. Contemporary comments in *Architectural Design and Construction* and the *Architectural Review* record a muted enthusiasm for what were generally regarded as a series of worthy rather than exciting designs.[10]

If non-traditional construction offered a poor substitute for traditional methods of building, housing built in this way was of little value to those trying to replan war-damaged cities. The huge programme of housing and communities being built for the US defence industry had excited those preparing British plans for reconstruction, but these lessons were difficult to apply to the most urgent of British problems, rehousing in the London area. Here the issue was not just to build housing but to do so in ways that were compatible with the existing fabric of the city and with the new plans being made for post-war development. Abercrombie's plans for the county of London and the Greater London area envisaged an increase in the population density in the existing built-up area with 100 people per acre[11] as the lowest density in the county and 200 people per acre in the central residential areas. Temporary bungalows which could at most be laid out at a density of forty people per acre and were thus of little use in the East End, or indeed almost anywhere within the county, and threatened to pre-empt the use of sites for the Permanent Programme. In London (and other big cities) their use was limited to small groups on playing fields, in parks or on housing sites too small for multi-storey housing. Even permanent non-traditional housing was restricted to suburban sites outside the county, for example at Oxhey and Debden (LCC, 1949). Designed with a maximum of two storeys, this form of housing was inadequate where the demand for housing was most acute: none of the systems of non-traditional construction had yet been adapted to multi-storey use. Often built on sites on the fringe of cities or on the unpopular out-County estates built by the LCC as far as thirty-two kilometres from Charing Cross (the notional centre of London), large estates of temporary and non-traditionally built housing came to acquire a negative image (Young and Wilmott, 1957). These estates were remote from where people worked, not necessarily well served by public transport, lacked the shops and pubs, as well as the clinics, libraries and other communal facilities to be found in the existing city, and looked at best like an ingenious, if utilitarian, response to the problems of austerity and a far cry from the best of the US achievements.

Looking back at the period of reconstruction, it may seem to us that government plans started from the essentially optimistic and simplifying assumption that technical ingenuity would provide a way of meeting post-war demands for housing. To the crowd watching the erection of the AIROH bungalow on Oxford Street and to those who had just come through a war in which the government had been expected to find solutions to wartime challenges such as maintaining supplies of food and munitions across the Atlantic and launching an invasion of Europe across the Channel, these plans must have seemed more plausible than they may to us. That the housing

---

[10] See, for example, Mardall and Vulliamy (1948) and the 'Housing Forum' articles in *Architectural Design and Construction* during the mid-1940s.

[11] 1 acre = 0.4047 hectares

programmes fell short of producing the number of dwellings expected is readily understood, taking into account the shortage of labour and materials that the country then faced. But that so many of these designs should have been found to compare so unfavourably with conventional housing suggests that the plans to meet housing demands in this way were flawed. In the world of post-war shortages, any dwellings were welcome, even ones that looked so different from the norm. But most non-traditional permanent houses and the temporary bungalows were not regarded as anything more than mere substitutes for conventional housing. They never came to be valued in their own right.

A number of the designs, for example Gibberd's BISF houses, were not so different in their appearance from traditionally-built houses, and some non-traditional houses were disguised to look like their conventionally-built neighbours. But those that looked as inescapably unconventional as their construction or could not be disguised were generally viewed askance. Those who lived in them might welcome the benefits of their comfort and their equipment and even come to view them with affection, but they never came to be valued for their 'newness' or their 'otherness'; unlike the Airstream Caravan or the Citroën 2CV they never established a new canon, an acceptable image of a 'machine for living in'.

In the late 1940s and early 1950s, as the economy expanded and consumer goods again appeared in the shops and showrooms, the gap between what people wanted and what was available (if you could afford it) began to close. Washing machines, refrigerators and cars could be bought and people were again coming to expect a choice of goods and of housing too. With high costs, limited use at higher densities and unconventional appearance, non-traditional housing could find no market with private enterprise and had no future with the local authorities. In the exceptional circumstances of the immediate post-war years, developing new methods of building houses was an important way of meeting the government's housing targets but in the changed climate of the late forties and early fifties, technical innovation was no longer enough. Designing and building homes that people might want to live in proved to be more complex than merely solving the technical challenge of avoiding the use of traditional skills and materials in building houses.

References

ADDISON, P. (1975) *The Road to 1945*, London, Jonathan Cape.

ALBRECHT, D. (ed.) (1995) *World War II and the American Dream*, Cambridge, Mass., MIT Press.

Architects' Journal (1945a) 'Temporary aluminium housing', vol.100, 23.4.45.

Architects' Journal (1945b) vol.101, 21.6.45.

Architectural Design and Construction (1943–7) 'Housing Forum' series.

Architectural Design and Construction (1944) vol.14 no.9, September.

Architectural Design and Construction (1945) vol.15 no.12, December.

Architectural Review (1944) Special number on US Wartime Housing, vol.96, August.

BOWLEY, M. (1966) *The British Building Industry: four studies in response and resistance to change*, Cambridge, Cambridge University Press.

BULLOCK, N. (1987) 'Plans for post-war housing in the UK: the case for mixed development and the flat' in *Planning Perspectives*, vol.1, no.2.

BURT COMMITTEE PRO HLG94/7 and 9.

CASSON, H. (1945) *Houses by the Million*, Harmondsworth, Penguin.

COUNCIL FOR THE INDUSTRIAL AND SCIENTIFIC PROVISION OF HOUSING (1943) *Housing Production*, First and Second Reports, London.

CULLINGWORTH, B. (1975) *Environmental Planning*, London, HMSO, vol.1.

FINNIMORE, B. (1989) *Houses from the Factory: system building and the welfare state 1942–74*, London, Rivers Oram Press.

GOLD, J.R. (1997) *The Experience of Modernism: modern architects and the future city 1928–53*, London, E. and F.N. Spon.

GROPIUS, W. (1935) *The New Architecture and the Bauhaus*, London, Faber and Faber.

HISE, G. (1995) 'The airplane and the Garden City: regional transformations during World War II', in D. Albrecht (ed.) *World War II and the American Dream*, Cambridge, Mass., MIT Press.

HUXLEY, J. (1943) 'TVA, an achievement of democratic planning', *Architectural Review*, vol.93, June.

KOHAN, C.M. (1952) *Works and Buildings*, London, HMSO and Longmans.

LONDON COUNTY COUNCIL (1949) *Housing: a survey of the post-war housing work of the London County Council 1945–1949*, London, London County Council.

MARDALL, C. and VULLIAMY, J. (1948) 'Towards an architecture: post-war housing in Britain' in *Architectural Review*, vol.104, October.

MINISTRY OF HEALTH (1944) *The Design of Dwellings*, Report of the Dudley Committee, London, HMSO.

MINISTRY OF HEALTH (1945) *Housing*, Cmnd 6609, London, HMSO.

MINISTRY OF HEALTH AND MINISTRY OF WORKS (1944) *Temporary Accommodation: memorandum for the guidance of local authorities*, London, HMSO.

MINISTRY OF HEALTH, SECRETARY OF STATE FOR SCOTLAND AND MINISTRY OF WORKS (The Burt Committee) (1944) 'House Construction, Interdepartmental Committee on House Construction, First Report', *Post-war Building Studies*, no.1.

MINISTRY OF HEALTH, SECRETARY OF STATE FOR SCOTLAND AND MINISTRY OF WORKS (The Burt Committee) (1946) 'House Construction, Interdepartmental Committee on House Construction, Second Report', *Post-war Building Studies*, no.23.

MINISTRY OF HEALTH, SECRETARY OF STATE FOR SCOTLAND AND MINISTRY OF WORKS (The Burt Committee) (1948) 'House Construction, Interdepartmental Committee on House Construction, Third Report', *Post-war Building Studies*, no.25.

MINISTRY OF LABOUR (1944) *Training in the Building Industry*, London, HMSO.

MINISTRY OF WORKS (1944) *Demonstration Houses: a short account of the demonstration houses and flats erected at Northolt by the Ministry of Works*, London, HMSO.

MINISTRY OF WORKS (1945) *Temporary Housing Programme*, Cmnd 6686, London, HMSO.

MINISTRY OF WORKS (1946) *Post-war Building Studies*, no.23, London.

MINISTRY OF WORKS (1948a) *Temporary Housing Programme*, Cmnd 7304, London, HMSO.

MINISTRY OF WORKS (1948b) *New Methods of House Construction*, National Building Studies Special Report No.4, London, HMSO.

MINISTRY OF WORKS (1948c) *Post-war Building Studies*, no.25, London.

REICHFORSCHUNGSGESELLSCHAFT (1929) *Bericht über die Versuchssiedlung in Frankfurt-Praunheim*, Sonderheft, Berlin, Rfg. für Wirkschaft lichkeit im Bau-und Wohnungswesen E.V.

RIBA (1943) *Rebuilding Britain*, London, Lund Humphries.

RIBA Journal (1944a) 'American housing in war and peace', pp.227–30.

RIBA Journal (1944b) 'Prefabrication', pp.163–9.

VALE, B. (1995) *Prefabs: a history of the UK Temporary Housing Programme*, London, E. and F.N. Spon.

WHITE, R.B. (1965) *Prefabrication: a history of its development in Great Britain*, London, HMSO (National Building Studies, Special Report 36).

YOUNG, M. and P. WILMOTT (1957) *Family and Kinship in East London*, Harmondsworth, Penguin.

Chapter 8:
TECHNOLOGY, SOCIAL CHANGE AND THE PLANNING OF A POST-INDUSTRIAL CITY: A CASE-STUDY OF MILTON KEYNES

by Mark Clapson

8.1   Introduction

The new town of Milton Keynes was born in 1967. It is located in North Buckinghamshire, and as Figure 8.1 shows, its construction converted much of that rural area into an urban one. But in what ways did the planners of Milton Keynes view technological change in relation to social change? Their interpretations, in the second half of the 1960s, of the likely impact of motorized transport, and of increasingly sophisticated and accessible telecommunications systems upon social relations in towns and cities, were given practical expression in the way the new city was constructed.

This construction process will be discussed in this chapter in relation to two of its key elements. First to be discussed is the system of road transportation devised for the city, for Milton Keynes has a unique grid-road system which was consciously designed to integrate the motor car into urban life. Second, the chapter will examine how planners' understanding of social change was reflected in the communications infrastructure of Milton Keynes; that is, in the provisioning of the city with a system of underground cabling for a variety of anticipated immediate needs and future uses, including the provision of electricity, television and the telephone. In order to understand why Milton Keynes looks the way it does, we need to revisit the exciting planning debates in the 1960s about the relationship between transportation and communications technologies and societal evolution.

Figure 8.1 Early scenes of the construction of Milton Keynes in North Buckinghamshire (reproduced from *Architectural Design*, 1975, p.738; © John Wiley & Sons Ltd; reproduced with permission)

8.2 Context

The geographical area of the new town of Milton Keynes was designated in January 1967, following the passage through Parliament of the North Buckinghamshire New Town Designation Order. Nine thousand hectares of largely rural land were identified in North Buckinghamshire, and the process of compulsory purchase of this land began. The shape of Milton Keynes is now something of an icon within the town-planning profession. This area was to house a target population of 200,000 by the completion date of the new city (see Figure 8.2).

In common with previous new towns, and also with its contemporary new towns such as Peterborough and Telford, Milton Keynes' population growth was dependent upon the successful dispersal of population and employment from existing towns and cities. The relief of London's population densities, especially, had been central to the rationale of earlier new towns in the south-east

Figure 8.2 Icon of designated area (reproduced from *Architectural Design*, 1975, p.734; © John Wiley & Sons Ltd; reproduced with permission)

of England, notably the eight new 'Mark 1' towns designated in the London region from 1947 to 1949. Further new and expanded towns, moreover, the 'Mark 2' models, were debated during the 1950s, but only one, Cumbernauld, was built. Milton Keynes was a 'Mark 3' new town. Figure 8.3 shows Milton Keynes' regional location. It is about 80 kilometres north-west of London, and enjoys good communications.

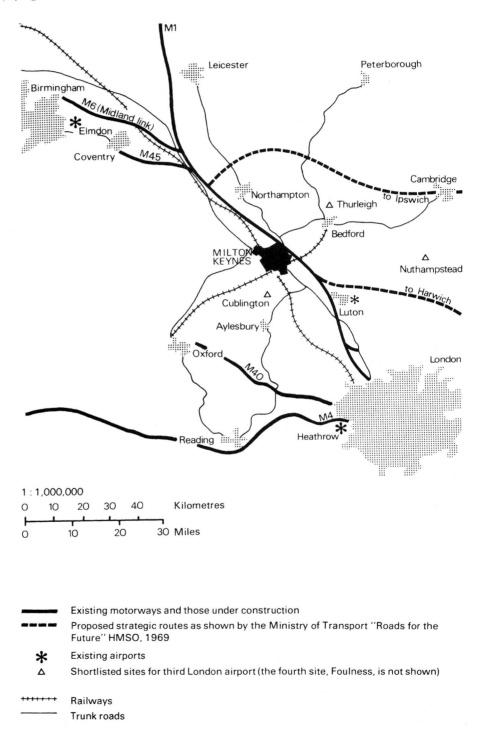

1 : 1,000,000

| 0 | 10 | 20 | 30 | 40 | | Kilometres |

| 0 | 10 | 20 | 30 | Miles |

━━━   Existing motorways and those under construction

■ ■ ■ ■   Proposed strategic routes as shown by the Ministry of Transport "Roads for the Future" HMSO, 1969

✳   Existing airports

△   Shortlisted sites for third London airport (the fourth site, Foulness, is not shown)

++++++   Railways

────   Trunk roads

Figure 8.3 The regional location of Milton Keynes (reproduced from Llewelyn-Davies *et al.*, 1970a, p.5)

There were some key differences between the 1960s new towns, and the earlier experiments. The latter had been planned and executed during the early post-war reconstruction, when there were far fewer cars and roads. Moreover, the earlier new towns were formulated at a time when manufacturing industry was the predominant basis of employment in the country. Hence, the early post-war planners operated within a context which viewed Britain, and England most extensively, as primarily an urban-industrial nation. Millions of workers, argued key members of the town-planning profession, were cooped up in overcrowded nineteenth-century towns and cities, and in part their salvation lay in dispersal to new, lower-density new towns modelled upon the Garden City experiments pioneered by Ebenezer Howard at Letchworth and Welwyn Garden City. Such thinking was strongly in evidence in the Reports of the New Towns Committee, a committee established by the incoming Labour government in 1945, and under the jurisdiction of Lord John Reith, a man better known for his contribution to British broadcasting. A healthy balance of town and country was envisaged, in planned urban environments. These would provide significant environmental improvements for the dispersed population (generally, see Hall, 1985, pp.3–12; Schaffer, 1972, pp.19–38).

However, the period from 1967 was a time when many people, planners among them, were aware that the urban-industrial structure and identity of Britain was undergoing, or more accurately, was beginning to undergo in heightened form, significant change. The early post-war planners of the new towns had failed to grasp adequately the changing nature of the British economy, and the growth of the service sector (Hall, 1985, p.8). To be fair, the importance of service-sector employment to new towns had certainly been recognized by the Reith Committee, but largely because the committee wanted to encourage the dispersal of white-collar work along with manual blue-collar employment, in order to provide a balanced and representative community of both middle- and working-class occupations in the new towns. This is evident in the *Reports of the New Towns Committee* (Ministry of Town and Country Planning, 1946, paras 119–22). However, the actual scale of growth of the service sector, notably in finance and banking, retail, and in the provision of commercialized leisure, had hardly been anticipated.

This structural movement away from manufacturing was evident in Milton Keynes. In 1968, the service sector accounted for about thirty-two per cent of full-time jobs in the area designated for Milton Keynes. By the late 1980s, this proportion had risen to over sixty-three per cent (Llewelyn-Davies *et al.*, 1970a, p.97; Milton Keynes Development Corporation (MKDC), 1987, p.9). This was partly due to the historical growth of the service sector in the first place, but also because Thatcherite economic policies since 1979 promoted service-sector employment, and allowed the ailing industrial sectors of the economy to decline (Halsey, 1989, pp.8–24).

If early post-war planners had not fully anticipated the growth of service-sector employment, then nor had they predicted what would become a quite remarkable growth in ownership of the motor car. This rose to sixty-two per cent of the population by the mid-1980s (Halsey, 1988, p.17). In common with the expansion of the service sector, this was a trend initiated during the inter-war years, but it was both accelerated and expanded during the era of full employment. These were the so-called 'golden years' of the British economy, a period that stretched from the mid-1950s to 1973. In that year, the oil crisis and industrial unrest brought this era to an end (Howlett, 1994, pp.320–39). Yet the momentous increase in the use of motor cars was well under way by then, and would continue. The motor car was a force that planners had to incorporate into their templates for new urban developments, and Milton Keynes, in common with other new towns of the later 1960s, was a city designed in no small part to facilitate the use, and to mould the impact, of motorized transportation.

Furthermore, a number of key thinkers in the 1960s were busy envisaging an urban society of the future wherein rapid interactive communication, increasingly facilitated by the telephone and the screen, would engender changes in the social organization of the city. They asked a seemingly simple, but extremely important and complex question: if we can converse or communicate instantly with people a few kilometres away, or even hundreds or thousands of kilometres away, then what happens to the nature of 'local' relationships? This led to a key issue facing planners: how far did they need to plan for the 'local' in relation to the forces of the motor car, and in relation to the possibilities of rapid communications technologies?

During the 1960s, influential planning thinkers such as F.J.C. Amos, J.B. Cullingworth and David Donnison were relating these considerations to the social fabric of Britain. They argued that increasing affluence enabled higher levels of car ownership. What had once been a largely aristocratic and middle-class luxury was becoming available to millions of working-class households. This, in turn, meant that people were more likely to have opportunities for a mobile lifestyle than had been possible in the older, and poorer, working-class areas of Victorian and Edwardian terraced housing. So one consequence of greater mobility – and of increased affluence – was the growing desire to move away from such housing areas to suburbs and to new towns. Since 1945, many millions of young working-class households moved away from those older proletarian heartlands in order to enjoy new and better housing in improved residential environments. This process was officially known as 'dispersal'. The role of new towns, argued many planners, was to concentrate dispersal spatially rather than allow it to flow outwards in *ad hoc* suburbanization.

In their suburban and new town destinations, moreover, people were able to maintain relationships with family and friends over wider spatial distances because of the car and the telephone. This understanding would find expression in the two-volume *Plan for Milton Keynes* (Llewelyn-Davies *et al.*, 1970), and in a variety of other planning materials. Both the road template of Milton Keynes and its system of underground cabling and wiring were intended to be flexible systems which could accommodate, and even embrace, current and future trends in transport and telecommunications.

8.3   Planning Milton Keynes (1): 'Pooleyville'

The *South East Study*, published in 1964 by Ministry of Housing and Local Government (MHLG), argued that the population shift to the south-east of England from the declining industrial areas of Britain required the consideration of further planned urban developments. And because these would be planned, they would be able to prevent at least some spontaneous urban sprawl brought about by the proliferation of roads, itself engendered by the growing use of the motor car. The *South East Study* argued that a number of locations, including the relatively under-developed north of Buckinghamshire, might be suitable for new towns (MHLG, 1964).

From that time, therefore, Buckinghamshire County Council lobbied continuously for a new town in the north of the county. A number of other local authorities, notably Bletchley Urban District Council, also wanted to develop a new town in the area for the economic growth it would engender in what was considered to be a relatively backward part of the county. Within Buckinghamshire County Council, the most important individual keen to establish a new town was Frederick Pooley, the chief architect and planner of that council. It was Pooley who is credited as the driving force in the negotiations between the council and the MHLG which led to the birth of the new town in 1967. And it was Pooley who would ultimately be disappointed by the nature of the *Plan for Milton Keynes* which Milton Keynes Development Corporation (MKDC) would produce (Bendixson and Platt, 1992, pp.20–21, 62–5).

In the years leading up to designation of the area for development, Pooley outlined his own ideas for the urban design and layout, the land uses, and the transportation facilities, of his hoped-for new town, or 'city', as he usually referred to it. His ideas would be rejected by the incoming MKDC. However, they are worthy of consideration, simply because their rejection was based upon a clear feeling that Pooley was out of date and too limited in his conception of how cities functioned in relation to transportation.

Pooley's ideas were proposed in a book he produced with Buckinghamshire County Council, and in articles in town-planning journals. In both words and illustrations, he gave expression to a version of the new city that would become known as 'Pooleyville' (Pooley, 1965, pp.281–3; Pooley, 1966). 'Pooleyville' was an explicitly Modernist city. In terms of layout, it was to be zoned into areas of high-density housing consisting of blocks of flats and town houses. Its residential areas, which Pooley termed 'townships', were to be zoned separately from employment areas. There would also be separate but easily accessible areas for shopping and leisure. In terms of transportation, the city was intended to deal with the motor car not by actively embracing it, but by supplying a public transport alternative to it in the form of a monorail. The monorail was to loop, or circuit, the city in two directions from a station located in the city's centre. Monorail stops would serve each township. The city, then, would resemble a butterfly, whose centre was the thorax, and whose adjacent wings, heavily dotted with townships and industrial zones, were given life by the monorail. This was the 'circuit linear' city, to use Pooley's terminology (see Figure 8.4 overleaf). Pooley was well aware of the growing use of the motor car. What he wanted to do, however, was to prevent '100 per cent motorization' of the new city, and provide a convenient alternative to the car which would bring people into the city centre, and hence provide a focal point for a large and growing urban area (Pooley, 1966, pp.47–54).

Pooley's vision was undoubtedly a bold one, but it was not to find favour with the planning team of Llewelyn-Davies *et al.* who were preparing the *Plan for Milton Keynes* on behalf of MKDC. As Walter Bor has since argued, the monorail was impractical and costly, because it required heavy initial investment:

> We tried to envisage this monorail (which can't be built in bits and pieces of course, it has to be built in one go), and how it would be built in relation to the development of the city as a whole. And we came to the conclusion that there was no specific time which would favour such a big investment … we reckoned that you needed at least 80,000 to 100,000 people to even justify it. By that time a complete pattern of movement would [already] have been established, and there would be no call for it, and we couldn't envisage anybody investing in the monorail at that point.
>
> (quoted in Percy, 1996, p.76)

David Donnison, who was a consultant to MKDC during the making of the *Plan*, has also recalled the 'necessary destruction' of Pooley's monorail system. It was spatially limiting, and as such it also restricted choices by circumscribing, as it would have done, much of the mobility of the local population by a rigid overhead rail system:

> Apart from the economic reasons, it would have completely cut across and contradicted the whole philosophy that the consultants and increasingly MKDC brought to the task, by imposing a single dominant mode of transport with dominant routes that people would have to follow. The philosophy taking shape [was] centred on choice, on creating a framework which would enable people in the course of time to build their own new town, and to renew it, as time went by, in their own ways.
>
> (quoted in Percy, 1996, p.76)

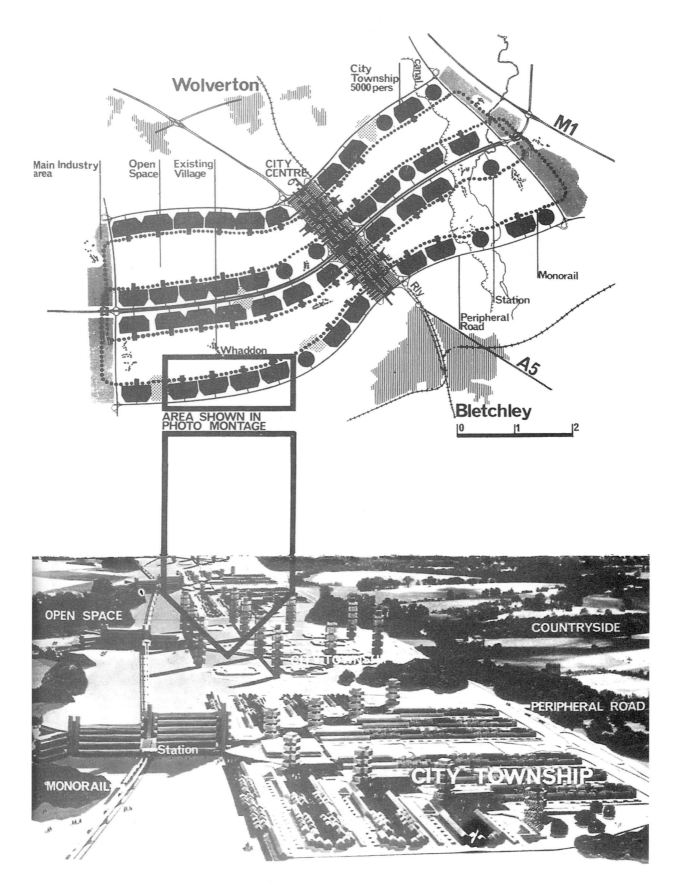

Figure 8.4 'Pooleyville' (reproduced from Pooley, 1966, p.7, fig.4)

Moreover, there was a spectre looming large over the erstwhile Pooleyville, a Californian spectre which was half-man, half-metropolitan region. The man? Melvin Webber, a professor of town planning at the University of California, Berkeley. The metropolitan region was Los Angeles, whose lifeblood was the automobile running along the freeways. In 1967, Llewelyn-Davies *et al.* invited Webber to become a consultant to the *Plan*, and he briefed them in a series of seminars about his interpretation of the urban society of the near future. Webber argued that because the nature of urbanity was changing so rapidly as a consequence of information technologies, and because much British planning was stuck in the 'middle-industrial' era, a new experiment such as Milton Keynes had an opportunity to break free from the past and become 'the spearhead in this changing face of urban civilization' (Webber, 1967).

8.4 Planning Milton Keynes (2): Melvin Webber

The planners of Milton Keynes began to demonstrate an enthusiasm for a marriage between the English Garden City tradition and a pseudo-Californian cityscape based upon the motor car. Hence, Milton Keynes would eventually come to be called the 'little Los Angeles in Bucks' by one writer on town-planning, and Webber's influence explains this (Mars, 1992). Derek Walker, who was the chief architect of Milton Keynes from 1970 to 1976, has argued that Webber more than anyone deserves the accolade of 'the father of Milton Keynes' (Walker, 1981, p.8). This is because Webber's writings during the 1960s were influential upon many town-planners who were trying to reformulate their ideas about what towns should look like, and about how towns were functioning, in an era of rapid technological change. Webber's ideas and arguments were complex, and are summarized, in a necessarily abbreviated form, in the following paragraphs.

Webber took San Francisco and Los Angeles as paradigms for his interpretation of the changing nature of towns and cities in relation to technological advances. Using one city or region from which to read off wider patterns is not a particularly unique or dangerous thing to do. Urban historians frequently do it in retrospect. Manchester, for example, has been viewed as paradigmatic of many changes occurring in the industrial towns of nineteenth-century England (Briggs, 1990, pp.88–138; see also Chapters 1 and 2 in this volume, on the industrial north).

For Webber, Los Angeles revealed much about the social and technological forces that were endemic to post-war towns and cities as they moved 'beyond the industrial age'. The aforementioned notion of dispersal was central to an understanding of Los Angeles' urban dynamic. People were moving away from older and often decaying urban centres to the burgeoning Los Angeles suburbs. They wanted a nice new home in the sun. Moreover, corporate America was tempted to move away from the inner cores. Service-sector places of employment – for example, comprising retail operations, offices and leisure facilities – were lighter and easier to move to cheaper peripheral sites than heavy industrial plant, or such fixed industries as ship-building and coal mining. As a consequence of such relative lightness and mobility, employment dispersal was gathering pace during the 1960s. Generally, increased opportunities of mobility through speedy motorized travel enabled a greater range of residential and locational choices.

However, the spatial dispersion of towns, argued Webber, did not unravel social and economic relationships: these were held together by instant communications via the telephone, and news and information was passed on through screens, the television screen, and increasingly the computer. In tandem, these developments also widened and stretched personal relationships and business networks spatially, far beyond the small scale implied by the term 'local'. People still bonded and interacted with each other, he observed, but they did not necessarily have to live cheek by jowl with each other any more. This was what he termed 'community without propinquity'. Webber argued strongly, therefore, that town-planners should liberate themselves from essentially Georgian, and certain continental European, ideas that a town centre with high-density developments close to it was an ideal form of town life.

His insights can be most clearly understood through the key words and terms which he used to portray his interpretation of the dynamic and changing relationship of urban form to social change in post-war California. Webber's terminology emphasized 'plurality', 'diversity', 'individualism', 'mobility', 'affluence', 'interest communities', the aforementioned 'community without propinquity', 'the non-place urban realm', 'disparate spatial dispersion', 'the knowledge explosion', and the culmination of all these social forces in the 'post-city age'. In synthesis, such forces provided a difficult social, economic and spatial context for highly localized notions of community to survive in. To put it another way, the 'classic slums' (as the title of Robert Roberts' influential book termed them; Roberts, 1973) and tight-knit urban communities of the industrial era were unravelling as advanced capitalist societies such as Britain and the USA entered the 'post-industrial era'.

There is no need to worry overmuch about the term 'post-industrial'. Anyone who argues that Britain still has a significant industrial base, clearly has not read the American sociologist Daniel Bell, the man who coined the term 'post-industrial' in the first place, during the 1950s. Unlike Webber, Bell never wrote directly about Britain, but his ideas may be viewed as relevant to British socio-economic change. For Bell, as countries moved away from goods production to an increasingly service-based economy, certain things would happen. For example, he predicted the fairly obvious growth in service-based employment. He also predicted the growth of 'intellectual technologies' and the rise of a more information-based society as communications became more sophisticated, but accessible (Bell, 1992, pp.250–51). Webber held broadly similar views about the onset of post-industrialism.

Webber's work was cogent, but conjectural. He was not a technological determinist. He made no focused predictions about the future of urban societies. Instead, he saw himself as 'trying to foresee latent qualitative consequences before they become manifest' (Webber, 1968–9, p.181). In this spirit, Webber argued that town-planners needed to embrace the spread of urban dispersal as a spatial expression of social forces which were working in harness with technical and economic forces.

Webber, then, stood against those pessimists who saw dispersal and suburbanization in negative terms. Many American and British writers had argued in the 1950s that the suburbs were essentially bland areas of mass-produced houses providing accommodation for millions of atomized and privatized people whose consciousness of status and cult of domesticity meant that they were forgetting how to be truly communal. What the US sociologist David Reisman called the 'suburban malaise' or the 'suburban sadness' was seen to result from the alleged isolation of suburbia (Reisman, 1950). In Britain, sociologists based at the Institute of Community Studies in Bethnal Green, London, compared the older, supposedly more vital, urban communities of close and extended kinship networks, with the fate of the bored and lonely suburbanite in their ostensibly nice but spiritually desolate new home. For example, Peter Willmott and Michael Young's *Family and Kinship in East London* (1979) was a social study of Londoners who had moved from the Cockney vitality of Bethnal Green to the dormitory suburbs of Essex. Here, it was argued, they huddled round the television set, became pathetically competitive with their neighbours, and communicated with relatives and old friends down the telephone wires, whereas once they had popped in for a chat. Willmott and Young castigated planners for bringing about this alleged social degradation through planned dispersal policies (Willmott and Young, 1979, pp.198–9).

Webber was altogether more optimistic than such maudlin appraisals of social evolution. People were 'realizing expanding opportunities for learning new ways, participating in more diverse types of activities, cultivating a wider variety of interests and tastes, developing greater capacities for understanding, and savouring richer experiences' (Webber, 1970, p.28). In this scenario, the role of place as a basis for stimulating social interaction required reconsideration. It was only one of many variables which brought people together, and an increasingly less important one (Webber, 1971, pp.496–501). Hence the significance, then, of phrases such as 'interest communities' and 'taste communities'. The planners of Milton Keynes broadly shared this view of the nature of social interaction, and they set about devising an urban framework which would facilitate it. To this end, a grid system was adopted, comprising fast roads which served a pattern of dispersed but easily accessible residential settlements. As Derek Walker has argued, the scale and nature of urban society had changed, and social communication was stimulated by many more variables than mere proximity. In consequence, the physical nature of the city had to be planned to reflect freedom of movement and of choice. Hence, he described the grid as 'an open matrix for selection' for the individual citizens of the new city (see Figure 8.5). This conception would find its physical expression in the gridiron of roads which would be spread, like a net, over the designated area of north Buckinghamshire.

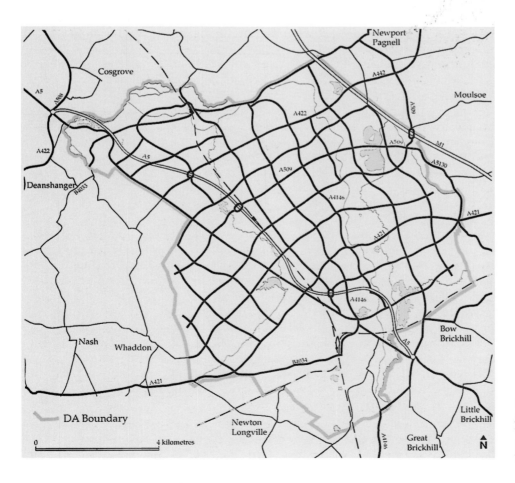

Figure 8.5 Milton Keynes grid-road system (reproduced from MKDC, 1992, p.42; used by permission of the Commission for the New Towns)

8.5   *Motorization and the road grid*

In the section on 'Transport' in the *Plan for Milton Keynes*, the goals of the Development Corporation reflected the input of Webber in the seminars held by MKDC. The following principles were emphasized:

1   a high degree of accessibility between all activities and places making up the city: homes, jobs, education, health, shopping, recreation, etc.;

2   freedom of choice between public and private methods of transport;

3   provision for the use of the car unrestrained by congestion.

MKDC were also aware of the problems of pollution and noise which cars produced, and they allowed for a generous system of banked earth and landscaping to shield the residential grid squares from the noise, fumes and views of main traffic routes.

While affirming the centrality of the motor car as a key mode of mobility, they also extended the principles of choice and freedom of movement to public transport and to pedestrians. A number of public transport experiments were to be tried which were intended to be flexible, and thus to bypass the limitations of fixed routes. There was also 'provision for free and safe movement as a pedestrian' (Llewelyn-Davies *et al.*, 1970b, p.279).

The grid was and remains a network of main roads spaced at roughly equal distances from each other. The roads provided for rapid, cross-city movement between the grid squares, and key 'A'-road arteries connected the city to the regional and national road system. A number of original consultants to the *Plan* favoured traffic-lights to control traffic flow on the grid roads, and to aid personal navigation at main road intersections. Roundabouts as opposed to traffic-lights, however, were introduced, and for three main reasons. First, as MKDC's budget was squeezed during the financial crisis of the 1970s, the capital cost of roundabouts was lower than for installing lights. Second, it was felt that the traffic-lights at large dual-carriageway intersections, involving left, right and straight-on options, would take up more valuable land than roundabouts would. Third, roundabouts with greenery on them looked more attractive than lights (Clapson *et al.*, 1998, pp.21, 28–30).

The *Annual Reports* of the MKDC provide statements on the continuing extent and cost of the road-building programme as the new city grew. By Easter 1973, for example, contracts worth £4.6 million had been awarded for the construction of principal and non-principal city roads, and over seventeen kilometres of main grid roads had been opened to traffic (MKDC, 1973, p.239).

The roads were intended not just for private vehicles but for buses, too. However, the public transport experiments in the new city were, basically, failures. For example, the Dial-a-Bus system, introduced early in the 1970s, was intended to provide a computerized bus-taxi which would pick people up whenever they wanted, and drop them off wherever they wanted, within the city boundaries. This was found to be too costly and unwieldy, and was soon scrapped. The regional bus companies were the only motorized public transport serving Milton Keynes, with a limited number of routes, until 1987, when de-regulation of the bus system worked to the new city's advantage, providing many small shuttle buses which weaved their way through and between residential areas and the shopping centres of the city (Llewelyn-Davies *et al.*, 1970a, p.35; Clapson *et al.*, 1998, p.13).

The emphasis upon pedestrian mobility was to find expression in the so-called 'green grid', a system of walkways and cycle routes which lace across the city adjacent to, passing over, or undercutting, the main roads. Although they were originally to have been called 'pedways', as in 'pedestrian ways', a typing error – allegedly – led to their being called 'redways', and ever since these motor-free paths have been made with pinkish-red tarmac (Mars, 1998, p.121). However, despite such pedestrian provision, a city primarily designed to facilitate the movement of the motor car has perhaps contributed to a level of car ownership and car usage that is higher than the national average. By 1988, seventy per cent of all households in Milton Keynes owned one or more cars, compared with the national statistic of sixty-five per cent (Banister *et al.*, 1997, pp.133–4; MKDC, 1990b, p.43). This raises an important question. Was the higher level of car ownership a result of the physical determinism of the grid and the dispersed nature of its grid squares which made car usage convenient?

Research related to such a question implies that the road grid, and the relatively dispersed low-density settlements, together contributed to a higher level of car ownership and usage when compared with other established towns. However, because traffic moved relatively freely around the grid roads, Milton Keynes was no more energy-intensive than established towns in terms of petrol consumption and exhaust emissions (Banister *et al.*, 1997, pp.133–4, 140; Clapson *et al.*, 1998, pp.28–9).

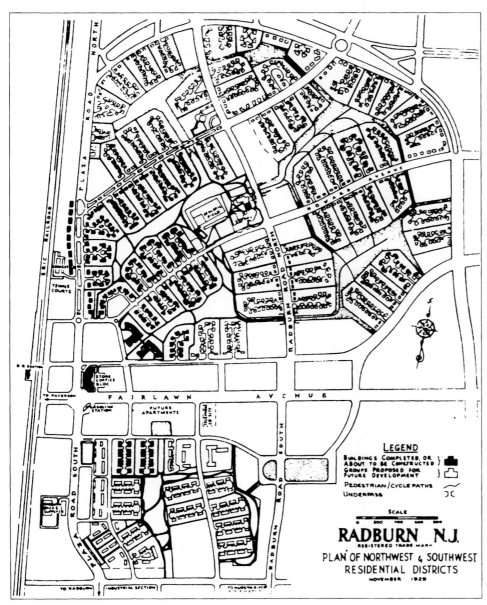

Figure 8.6 Radburn, New Jersey (reproduced from Bendixson and Platt, 1992, p.94)

Historically, the planners of Milton Keynes were not without precedents for their attempts to integrate motorization into town life. Planning for motor cars as well as for people had been rehearsed in some previously innovative town-planning experiments elsewhere. Historians of planning almost always point to Radburn, New Jersey, where the first attempt to directly facilitate car movement around towns was made, incorporating careful planning to minimize the visual intrusiveness and noisiness of the automobile. Radburn was influential upon both American and British town-planning (Birch, 1980, pp.424–36; Cherry, 1996, p.83; Stern, 1981, pp.84–5; Tiratsoo, 1990, p.81).

Hence, many of the attempts to accommodate the motor car in Milton Keynes had antecedents in Radburn. This is particularly in evidence in MKDC's *Milton Keynes Planning Manual*, published towards the end of MKDC's institutional life, and intended not simply as a guide to the past planning of the city, but also to influence its future development. At city-wide level, similarities to Radburn principles are evident in the above mentioned separation of fast roads from housing areas by various attractive landscaping measures. Moreover, pedestrians are kept largely separate from roads by the redways and by subways and other such motor-free walkways and cycleways. Figure 8.6 depicts an original drawing of a layout at Radburn, while Figure 8.7 depicts two American underpasses influenced by Radburn, and a similar underpass in Fullers Slade, Milton Keynes.

Figure 8.7 *Top* Underpasses and overpasses, at Radburn (reproduced from Birch, 1980, p.431; photographs: © Eugenie Ladner Birch); *Bottom* Underpass from Fullers Slade, Milton Keynes, 1971 (drawing reproduced courtesy of John Seed)

This separation of pedestrians from busy streets, however, has not been without its critics. For while the notion of separate walking routes has provided some attractive landscaping, it has also given rise to pathways and underpasses which are out of view of cars and houses. Feminist critics of patriarchal planning designs have emphasized that women have been more vulnerable to attack and harm in such circumstances, because the issue of female safety was one of many which were never on the 'male' planning agenda (Greed, 1994, p.133). The planners of Milton Keynes have been made aware of this and now acknowledge that redways are dark and lonely in many places, and accordingly design or adapt the redways in terms of heightened personal safety (Clapson *et al.*, 1998, p.127).

8.6 *Motorization and grid-square design*

Two examples will suffice to demonstrate very different grid squares, both built largely for rent by tenants of MKDC. One was the repudiation of some of Radburn's key tenets, in Netherfield. The other represents the relatively successful and more generous incorporation of Radburn principles in the grid square of Eaglestone.

We begin with Netherfield. This grid square was built between 1972 and 1977; therefore it ranks as one of the earliest in the new city. As Bendixson and Platt argue, its architects self-consciously moved away from 'stuffy old Radburn' (Bendixson and Platt, 1992, p.96) and its Garden City traditionalism. They worked within a modernist movement which followed architects such as Le Corbusier and Gropius, who had been influential in designing both industrial buildings and also workers' flats in Weimar Germany, flats which have since been viewed as the epitome of rational and modern design. Such housing was 'non-bourgeois', rejecting the suburban middle-class taste for romantic motifs such as 'Tudorbethan' window detailing and Dutch gable ends, both of which harked back to a pre-industrial past. And it was 'non-bourgeois' in its repudiation of the *rus-in-urbe* mentality which had stimulated so many wealthy middle-class households to evacuate the city for the rural fringes since the mid-nineteenth century. Modern designers wanted rid of all that. They favoured function and simplicity, and they wanted landscapes which looked not suburban, but regular and urban (Bendixson and Platt, 1992, pp.95–6; Pevsner, 1987, pp.211–17).

Hence, the grid square was to contain repetitive worker housing which was redolent of the proletarian terraces of the nineteenth-century heartlands of the Industrial Revolution. The houses were built in long straight terraces of one-, two- and three-storey dwellings. Every roof line in the grid square was flat, and exactly level with the others. But these workers' terraces were updated with the industrialized building methods of the later twentieth century, and instead of brick, a laminated metal sheeting was the main exterior finish and fibreglass fins acted as mock walls to separate each house from those next door (Bendixson and Platt, 1992, pp.95–6). The use of such materials reflected an experimental spirit among many architects and planners of the time (see Figure 8.8 overleaf).

As for the road layout of Netherfield, there were no culs-de-sac and banjos and curving avenues. The roads were straight, and thus inadvertently allowed for speeding cars, a consequence that appears counter to the spirit of Radburn. However, in keeping with the principles of the *Plan for Milton Keynes*, Netherfield enjoyed a generous provision of open green spaces. Private gardens, however, were tiny.

Figure 8.8 Netherfield housing (photograph: courtesy of Carl Vivian)

What does that august publication, Pevsner's *Buildings of England*, have to say about Netherfield in its Buckinghamshire volume? The authors describe it as, no less, 'the most immediately impressive housing development in the city', whose architecture has 'the straightforwardness of industrial building' (Pevsner and Williamson, 1994, pp.539–40). That was the view from above, as it were, the view of the architectural aesthete. But what of the view 'from below', from those people who lived in Netherfield? It is clear that the majority of the residents of Netherfield did not like the design of their accommodation. Two comments from the oral history Milton Keynes Living Archive publication, *What's Netherfield To You?*, demonstrate this point:

They're tin cans. We don't like living in tin cans.'

'There's not a lot we like about these houses. [The] exterior of the houses is not very pretty.'

(Living Archive, 1997, pp.2–3)

Further proof of such views is evident in the local campaigns by residents' associations to get the appearance of their houses altered to more traditional-looking designs, and to improve the performance of their interior workings, too. In addition to oral testimony, a study of the local newspapers throughout the new city's history gives ample evidence of the struggles of locals to improve their grid squares. For in its official publications, MKDC only rarely presented its experimental housing in a critical light.

The above-mentioned problems in exterior and interior finishing were not limited to Netherfield but also occurred in, for example, Beanhill, another grid square of straight roads and metal-clad houses. The heating and ventilation systems in Netherfield and Beanhill were plagued with problems, as protesting residents were keen to tell the local press. They also criticized their flat roofs, and called for pitched ones (*Milton Keynes Express*, 22 April 1977; 22 October 1981; 11 March 1982; *Milton Keynes Gazette*, 21 June 1985). So it was clear that such grid square housing was all very modern and mould-breaking in theory, but was not, in practice, particularly efficient or pleasurable for its users to live in. Because of this, one writer on architectural matters described Netherfield as 'a sort of TGV image of slickness technology with an Intercity 125 reality of noise and diesel and belching. Netherfield is a 125 in TGV drag' (Mars, 1998, p.124).

Another Milton Keynes grid square, Eaglestone, designed by the notable architect Ralph Erskine, and also built during the 1970s, saw no such campaigns to change the fundamental appearance of its housing. Eaglestone possesses 'clusters of houses which back onto a central green. [Its] density is comparable with contemporary low-rise urban housing' (Pevsner and Williamson, 1994, p.520). Here, the houses were designed to look like a clustered fishing village and hence were more traditional and cottagey in size and shape (see Figure 8.9). They were constructed from brick, which in parts was covered with weather-boarding. It was, therefore, more in keeping than was Netherfield with the Garden City tradition in domestic architecture. The housing at Eaglestone does appear to have been more popular with its residents than was the housing at Netherfield. One couple told MKDC they had not wanted to live in terraced housing, 'but they had been persuaded to change their minds by the individuality of Erskine's short rows' (Bendixson and Platt, 1992, p.102).

Figure 8.9   Eaglestone housing (reproduced from *Architectural Design*, 1975, p.770; © John Wiley & Sons Ltd; reproduced with permission)

Eaglestone was more explicitly 'Radburn' in layout, with a variety of short streets and culs-de-sac, parking bays and separate garages. Much of the grid square was, moreover, encircled by a service road from which the smaller roads led. When the personal history of the architect is considered, this is not surprising. Pevsner says that Erskine's first job had been for Louis de Soisson, a leading architect of Welwyn Garden City (Pevsner and Williamson, 1994, p.520).

So, when considering the design history of these two grid squares the urban historian gets some idea of the strength of popular preference for tradition in housing, and hence of the unpopularity of technical designs which look modern. Yet this relationship between tradition and experiment was nuanced. For Radburn itself, when it was built in the 1920s, was both a Garden City design and also a departure in urban planning which mediated the effects of motorization. It appears that the imprint of Radburn was more successful, therefore, in more traditional garden-suburb experiments.

There is a further implication for these findings. For as residents came together in Netherfield to combat unpopular housing designs, they affirmed the role of a measure of propinquity in social relationships within the so-called non-place urban realm. The planners of Milton Keynes had been aware that this might continue to be the case but, as noted above, they had not expected closely localized relationships to be paramount. They were, on the contrary, exercised by the demise of neighbourhood as it had been envisaged in the early post-war years. And the demise of neighbourhood, argued Webber, was due to the widening opportunities for social connection. Hence – and as noted on p.284 above – the *Plan for Milton Keynes* allowed for a more flexible and variegated social interaction, an interaction which could be sustained across greater distances than the neighbourhood. So Milton Keynes had been 'wired up' to the cable from 1970, thus historically presaging the fact that Britain in general would be increasingly wired up over the coming decades.

8.7 Cabling for the 'non-place urban realm'

This section discusses the provision of the electricity supply, and of the television and telephone services. It must be said that when these issues were discussed in the *Plan* and in other planning papers, they were usually treated in a rather matter-of-fact way. The social goals of the *Plan* are dealt with extensively in sections on social development, and in sections on community-building through the principles of voluntary association. However, these social goals are not discussed directly in relation to the communications infrastructure. So when electricity and telecommunications were treated, they were seen largely in terms of problems of operational delivery. This becomes understandable when it is recognized that not all planners were concerned with intellectual debates about the nature of the urban fabric. Many of them, as lists of personnel in the *Plan* show, were engineers, technicians and construction site managers. They were concerned with the nuts and bolts of the job in hand.

The electrical wiring of the city had to contend not simply with the heavy Oxfordshire clay upon which much of the new city was to be built, but also with the fact that MKDC had insisted upon a low-density, low-rise city in keeping with its relatively dispersed pattern of settlements. No tall structures were to be built, and that ruled out pylons. The alternative of cabling, however, was costlier than overhead power lines. Many in MKDC felt that those pylons inherited from earlier developments in North Buckinghamshire were unsightly, and negotiations were undertaken with the East Midlands Electricity Board to promote cabling wherever possible, and to replace pylons where they were particularly intrusive (Llewelyn-Davies *et al.*, 1970b, p.345; Bendixson and Platt, pp.178–9).

MKDC's aesthetic consideration extended to the television service. The televisions of the new city's households were also to be supplied through cable rather than through roof-top aerials. During the 1950s and 1960s there had been a good deal of criticism within architectural and planning circles of the appearance of unsightly TV aerials on the roof-tops of new housing developments. MKDC, with its concern for an uncluttered sky-space in the city, shared this distaste. However, they favoured the cable for other reasons, too, notably its adaptability to other uses. Some of these uses were known by 1970, others were to be anticipated and allowed for. Both concerns, aesthetic and practical, were in evidence in the *Plan for Milton Keynes*:

> In order to avoid proliferation of television aerials, the possibility of using a cable distribution network for [the city] has, of course been investigated. At the same time, the opportunity of incorporating other communications systems in the same network was explored.
>
> (Llewelyn-Davies *et al.*, 1970b, p.348)

These other uses included facsimile, remote meter readings, burglar alarms, fire alarms, closed-circuit television, and 'other such communications', notably the 'videophone', a telephone with a screen attached to provide visual as well as verbal communication between the callers (Llewelyn-Davies *et al.*, 1970b, p.348). Some of these anticipated uses have since been adopted, notably the facsimile. There was, however, no mention of computers and their potential uses in the *Plan*. By 1988, however, Milton Keynes demonstrated a higher household penetration rate of home computers than the national average, thirty-two per cent compared with twenty-seven per cent respectively. These figures may represent regional consumption levels as much as the uniqueness in the planning and provisioning of Milton Keynes. On a different level, however, they do at least demonstrate that MKDC was monitoring the ownership of computers and other items, in order to keep abreast of consumption trends and their requirements, in a rapidly growing new city.

It is possible to trace the development of the provision of the television cable and its associated facilities through the *Annual Reports* of the Development Corporation and other publications. For example, by March 1972, the franchise for the television broadcast relay system, 'an advanced UHF–VHF hybrid co-axial cable which will reach into all city dwellings', had been awarded to the Post Office Corporation, and work on its construction was 'well under way' (MKDC, 1972, p.242; Barrett, 1975, p.759). The television signals that fed into the cable were received by a specially built transmitter near the city centre, and then sent underground to the television sets of Milton Keynes via a cable and duct relay system (MKDC, 1992, pp.182–3).

Yet MKDC did not want the citizens of the new city to be merely passive viewers of their cable televisions. They also expressed their desire to explore the possibilities of electronic screen-based communications 'in community development'. To this end, as the Post Office began work, MKDC funded a small-scale production studio and a number of video cameras. MKDC's social development workers were to help any individuals or groups in the city who wanted to use the new equipment, in the hope that 'the cable system will allow people in the new city to communicate directly with others via their living-room TV sets' (see Figure 8.10). A number of experiments took place (Barrett, 1975, p.759). However, community television in Milton Keynes did not develop along these lines.

Figure 8.10 Residents at Greenleys using video equipment (photograph: reproduced from *Architectural Design*, 1975, p.759; © John Wiley & Sons Ltd; reproduced with permission)

Computers now provide, for those who can afford them, the remote but interactive screens in Milton Keynes homes.

The importance of telephones, too, in serving the social needs of an incoming population was evident in the *Plan* and in other MKDC papers. MKDC recognized that immediate access to the local and national telephone network would be vital for the local economy, and for the success of new companies establishing themselves in, or relocating to, Milton Keynes:

> The Post Office Corporation and the Development Corporation are determined that an effective and efficient telecommunications service will be provided for the city and to this end, exchange and underground cable capacity will be designed to provide for 100 per cent telephone usage by all residential, industrial and economic development ...
>
> (Llewelyn-Davies *et al.*, 1970b, p.345)

Five telephone exchanges served the designated area in 1970, but the expansion of the population required the building of two more state-of-the-art exchanges. Trunk cabling was run underground from these exchanges, alongside the main grid roads, beneath the grass reservations which were, in part, provided for this purpose. At any given time, not just motors but conversations would be speeding along the main corridors of the city (MKDC, 1992, p.182).

By 1988, the penetration rate of telephones into households was eighty-seven per cent in Milton Keynes, compared with eighty-one per cent nationally (MKDC, 1988, p.2). These statistics may reflect regional consumption differences, as the Milton Keynes area is on the margins of the south-east of England, a more affluent area than some areas of the north of England. Yet they also demonstrate that the flexibility and responsiveness which was intended for the infrastructure of Milton Keynes, and which was provided to homes in the form of the cable, has been able to accommodate the changes in the consumption of communications technologies since 1970. At that time, a subterranean communications infrastructure came into being, and grew to service the needs and desires of a mobile population above ground, a population whose numbers increased rapidly, year after year.

Moreover, the cable possessed an economic dimension. The extension of the cable network to businesses was also promoted by MKDC as an extra communications advantage largely unique to companies who chose to relocate to the new city (Firnberg and West, 1987). Furthermore, the cable was explicitly projected as complementary to Milton Keynes's image as a centre for telecommunications, robotics and automation, and satellite communications (MKDC, 1990a, p.5). In both these economic and social spheres, however, Milton Keynes is still evolving. At the time of writing, it has not even met its original population target. So this is a history still in the making, and it is a history which reveals a great deal to us about changes in British society in the final third of the twentieth century. The rejection of 'Pooleyville' symbolized a desire on the part of MKDC to provide a responsive urban template which would keep abreast of both the known and also the latent trends in the complex relationship between technology, social change and urban form. The recent history of Milton Keynes, therefore, reveals much to us about planners' understanding of this multi-faceted relationship.

References

ARCHITECTURAL DESIGN, December 1975, vol.XLV.

BANISTER D., WATSON, S. and WOOD, C. (1997) 'Sustainable cities: transport, energy and urban form', *Environment and Planning B: Planning and Design*, vol.24, pp.125–43.

BARRETT, M. (1975) 'Cable: the electric grapevine', *Architectural Design*, December, vol.XLV, p.759.

BELL, D. (1992) 'The coming of the post-industrial society' in C. Jencks (ed.) *The Post-Modern Reader*, London, Academy, pp.250–66.

BENDIXSON, T. and PLATT, J. (1992) *Milton Keynes: image and reality*, Cambridge, Granta.

BIRCH, E.L. (1980) 'Radburn and the American planning movement: the persistence of an idea', *Journal of the American Planning Association*, vol.46, no.4, pp.424–39.

BRIGGS, A. (1990) *Victorian Cities*, Harmondsworth, Penguin Books.

CHERRY, G. (1996) *Town Planning in Britain since 1900*, Oxford, Blackwell.

CLAPSON, M., DOBBIN, M. and WATERMAN, P. (eds) (1998) *The Best Laid Plans: Milton Keynes since 1967*, Luton, Luton University Press.

FIRNBERG, D. and WEST, D. (1987) 'Milton Keynes: creating an information technology environment' in W.H. Dutton, J.G. Blumler and K.L. Kraemer (eds) *Wired Cities: shaping the future of communications*, London, Cassell, pp.392–408.

GREED, C. (1994) *Women and Planning: creating gendered realities*, London, Routledge.

HALL, P. (1985) 'The people: where will they go?', *The Planner*, vol.71, no.4, April, pp.3–12.

HALSEY, A.H. (1988) 'Statistics and social trends in Britain' in A.H. Halsey (ed.) *British Social Trends since 1900*, London, Macmillan, pp.1–35.

HALSEY, A.H. (1989) 'Social trends since World War Two' in L. McDowell, P. Sarre and C. Hamnett (eds) *Divided Nation: social and cultural change in Britain*, London, Sage, pp.8–23.

HOWLETT, P. (1994) 'The "Golden Age", 1955–1973' in P. Johnson (ed.) *20th Century Britain: economic, social and cultural change*, London, Longman, pp.320–39.

LIVING ARCHIVE (1997) *What's Netherfield to You?*, Milton Keynes, Living Archive.

LLEWELYN-DAVIES, WEEKS, FORESTIER-WALKER AND BOR FOR MKDC (1970a) *The Plan for Milton Keynes*, vol.1, Milton Keynes, MKDC.

LLEWELYN-DAVIES, WEEKS, FORESTIER-WALKER AND BOR FOR MKDC (1970b) *The Plan for Milton Keynes*, vol.2, Milton Keynes, MKDC.

MARS, T. (1992) 'Little Los Angeles in Buckinghamshire', *Architects' Journal*, 15 April.

MARS, T. (1998) 'Milton Keynes: a view from exile' in M. Clapson, M. Dobbin and P. Waterman (eds) (1998) *The Best Laid Plans: Milton Keynes since 1967*, Luton, Luton University Press, pp.117–26.

MINISTRY OF HOUSING AND LOCAL GOVERNMENT (1964) *The South East Study, 1964–1981*, London.

MINISTRY OF TOWN AND COUNTRY PLANNING (1946) *Final Report of the New Towns Committee*, London, Cmd. 6876.

MKDC (1972) *Fifth Annual Report for the Year Ended 31 March, 1972*, London, HMSO.

MKDC (1973) *Sixth Annual Report for the Year Ended 31 March, 1973*, London, HMSO.

MKDC (1987) *Employers' Survey Report, 1987*, Milton Keynes, MKDC.

MKDC (1988) *Milton Keynes Household Survey, 1988: IT and media technical note*, Milton Keynes, MKDC.

MKDC (1990a) *The Dynamic Environment for Business*, Milton Keynes, MKDC.

MKDC (1990b) *The Milton Keynes Population Bulletin*, Milton Keynes, MKDC.

MKDC (1992) *The Milton Keynes Planning Manual*, Milton Keynes, MKDC.

PERCY, M. (1996) 'The best laid plans', *Town and Country Planning*, vol.65, no.3, pp.75–82.

PEVSNER, N. (1987) *Pioneers of Modern Design*, Harmondsworth, Penguin Books.

PEVSNER, N. and WILLIAMSON, E. (1994) *The Buildings of England: Buckinghamshire*, Harmondsworth, Penguin Books.

PLANNING EXCHANGE (1997) *The New Towns Record, 1946–1996: 50 years of New Town development*, Glasgow, The Planning Exchange.

POOLEY, F.B. (1965) 'Buckinghamshire new city', *Ekistics*, vol.19, no.1, pp.281–3.

POOLEY, F.B. (1966) *North Buckinghamshire New City*, Aylesbury, Buckinghamshire County Council.

REISMAN, D. (1970) *The Lonely Crowd: a study of the changing American character*, New Haven, Connecticut (first published 1950).

ROBERTS, R. (1973) *The classic slum*, Harmondsworth, Penguin Books (first published 1971).

SCHAFFER, F. (1972) *The New Town Story*, London, Paladin.

STERN, R.A. (1981) *The Anglo-American Suburb*, London, Architectural Design.

TIRATSOO, N. (1990) *Reconstruction, Affluence and Labour Politics: Coventry, 1945–60*, London, Routledge.

WALKER, D. (1981) *The Architecture and Planning of Milton Keynes*, London, Architectural Press.

WEBBER, M.M. (1967) 'The urban society of the future' in Llewelyn-Davies, Weeks, Forestier-Walker and Bor, 'Proceedings of the first general seminar held at the Park Lane Hotel, Piccadilly, London, N1, on 4 and 5 December, 1967', unpublished manuscript.

WEBBER, M.M. (1968–9) 'Planning in an environment of change, part 1: beyond the industrial age', *Town Planning Review*, vol.39, pp.179–95.

WEBBER, M.M. (1970) 'Order in diversity: community without propinquity' in L. Wingo (ed.) *Cities and Space: the future use of urban land*, Baltimore, Johns Hopkins Press, pp.23–54.

WEBBER, M.M. (1971) 'The post-city age' in L.S. Bourne (ed.) *Internal Structure of the City: readings on space and environment*, New York, Oxford University Press.

WILLMOTT, P. and YOUNG, M. (1979) *Family and Kinship in East London*, Harmondsworth, Penguin Books (first published 1957).

Part 3
URBAN TECHNOLOGY TRANSFER

Chapter 9: CITIES IN RUSSIA

by Colin Chant

9.1 Introduction

In Part 2, it became clear that city authorities in western and central Europe frequently adopted similar policies towards the implementation of some of the principal technologies of the Second Industrial Revolution. It could be plausibly concluded from those chapters that, apart from certain subtle differences, the form and fabric of the cities of that period reflected a planning and technological culture of increasingly international scope. That interpretation will now be scrutinized more closely. The influence of what might loosely be called the Western city-building tradition – which after 1870 included technological and planning innovations from the USA – is examined in Russia and colonial India, which provide markedly different political and economic contexts for the deployment of urban technologies in the post-1870 period from those thus far considered.

First, Russia. Russia's geographical relationship with Europe is inherently ambiguous. From the fifteenth century, the rulers of the land-locked principality of Muscovy began to accumulate what became a territorial empire of unprecedented size in world history, covering a sixth of the earth's land surface, from the Baltic Sea in the west to the Pacific Ocean in the east and from the Arctic Ocean in the north to the Caspian Sea in the south. The bulk of this territory lay beyond the Ural Mountains, in Asia. The ambiguity of Russia's geographical position is mirrored in the ambivalent attitudes of its political and cultural elite towards the West. From the time of Tsar Ivan the Terrible (1533–84), Western expertise and technologies began to be adopted, albeit usually with xenophobic suspicion; that is, until the reign of Peter the Great (1682–1725), the first great Westernizer of Russian history. Peter's reforms included the founding in 1703 of St Petersburg, a Baltic port built from nothing on marshlands at the mouth of the River Neva by the forced labour of tens of thousands of peasants, and which Peter made his capital in 1712. Peter's Westernizing policy was continued by the German-born Catherine the Great (1762–96), a devotee of the French Enlightenment, and of European ideas about urban administration and town-planning. Among her urban

The author of this chapter is grateful for detailed advice and corrections from Anthony French, and for the comments of colleagues at the Open University. The overall approach adopted, however, is the author's own.

reforms was the requirement that every city should have a standard development plan approved by the crown (Hittle, 1979, pp.196–7). For much of the subsequent imperial period, the Russian political and intellectual elite was riven by a debate between those anxious to emulate Western achievements, including its planned, comparatively hygienic, cities, and those who wished Russia to pursue its own Slavic, Orthodox path, untainted by the horrors of Western factory production and urban insurgency. Ironically, after the revolutions of 1917 had brought to power the Marxist heirs of the Westernizing intellectual tradition, the ambivalence towards the West persisted as the Soviet Union attempted to build a socialist alternative to the Western capitalist economy. This chapter will consider how the various resolutions of this perennial conflict over the West and its technologies were translated into the form and fabric of Russian cities.

9.2   *Technology and the tsars*

By the mid-nineteenth century, a pre-industrial urban system with distinctive features had developed under the Russian tsars. Settlements were generally smaller and more scattered than those of Western Europe. These characteristics reflected both the overwhelming predominance of agriculture and forestry in the Russian economy, and also its relative unprofitability. Apart from the 'black earth' region of southern Russia that was destined to become the granary of Europe in the second half of the nineteenth century, the soils in much of the Russian empire were less fertile than those in the regions of Europe that were the first to industrialize. Consequently, rather than investing their time and resources in their estates or in local government, absentee noble landowners looked for social advancement and for a share of the imperial tax revenues to state service, whether administrative, military or ecclesiastical. Because of the preponderance of subsistence farming, and consequent lack of internal demand for manufactured goods, Russia's middle classes lacked the vitality that spurred urbanization in Western Europe (Hittle, 1979). The Russian merchant class – a smaller proportion of the population than in Western Europe – traditionally relied on the granting of state monopolies and on protection from foreign competition through import tariffs; they were concerned, above all, with brokering the exchange of Russian raw materials for imported luxury goods for the nobility.

This socio-economic configuration underpinned a notably lopsided pre-industrial urban system. The domination of primate cities has already been remarked upon in earlier chapters on London and Paris. In the Russian case, this disparity among urban settlements was even more pronounced, except that primacy was shared between Moscow, the first capital of the medieval Muscovite state, and Peter's created capital, St Petersburg, the population of which overtook that of Moscow early in the nineteenth century. The only other cities with a population of more than 100,000 in the mid-nineteenth century Russian empire were Warsaw in captive Poland and the Black Sea port of Odessa, founded in 1794 on territory seized from the Turks (Hamm, 1986, pp.2–3; Herlihy, 1986). Moscow and St Petersburg were magnets for people of talent, aspiration and rank from the provinces. The two capitals (as historians of Russia usually refer to them) have often, with good reason, been seen as the two sides of Russia's Janus face: St Petersburg on the periphery as Russia's 'window on the West', cosmopolitan and alien, a showcase of Western urban planning and architecture; Moscow, Russia's traditional heart, mostly, beyond its formal centre, a 'semi-Asiatic big village' of one-storey wooden dwellings (Bradley, 1985, p.60).

Outside the two capitals, Russian cities displayed another kind of duality: urban settlements that had grown organically as market centres of an agricultural economy, settlements typified by wooden buildings and tortuous layouts of ring roads and radial connecting streets, were absorbed into the system of imperial administration.

To emphasize their acquired status as provincial and district outposts of tsarist dominion, the imperial bureaucracy imposed city plans on the old organic layouts; these plans enforced social-class segregation and strict land-use zoning, with noxious industries located on the periphery. Formally, they were characterized by geometrical street patterns focused on a central square; the square was overlooked by brick-built public and private buildings with a limited variety of prescribed plaster façades imitating Classical or Baroque masonry. But the ideal of imperial town-planning was hard for Russian provincial cities to realize; with municipal funds usually scarce, only the squares and adjacent planned streets could be paved and lit, if at all, and the prescribed building façades often fell into disrepair.

As it happens, 1870 marks a turning-point in Russian urban history, as the year of a statute granting, for the first time, a measure of local self-government to urban settlements. More profoundly, the urban system was beginning to change character around this time as nascent Russian entrepreneurial activity in cotton textiles multiplied, and private operators constructed a railway system in European Russia during the 1860s and 70s (see Figure 9.1 overleaf). The most vociferous proponents of the railways were entrepreneurs involved in the export of Russian agricultural produce (above all, grain). That rapidly growing trade was straining the capacity of the traditional deep-water barges and the new steamboats of the extensive river and canal system through which the tsars had linked the Volga to the Baltic Sea. The Russian autocracy, entrenched in hostility towards the innovations of Western Europe ever since the invasion of Napoleon in 1812, was nervous about the social effects of the railways; but after the exposure of Russia's military inferiority during the Crimean War (1853–6), the tsarist administration became less inhibited about the adoption of Western technology. The building of the railways ensured that some unconnected cities declined, while others in strategic locations experienced rapid growth. Many of the fastest-growing cities were Black Sea ports, directly involved in the grain trade; the population of Odessa, over 115,000 in 1861, had jumped to more than 313,000 by 1890, and reached 630,000 by 1914 (Herlihy, 1986, p.234). But the railways also spurred the growth of industrial regions. The outstanding example is the Yekaterinin railway, built in southern Russia during the 1880s. This line connected the rich iron ores of Krivoy Rog with coal deposits in the basin of the Donets, a tributary of the River Don, thereby fuelling the growth of the new industrial cities of the Ukraine metallurgical industry (Solovyova, 1984). Another key line, opened in 1883, linked the Black Sea to Baku on the Caspian coast, stimulating the expansion of the Caucasian oil industry, which the Nobel brothers had pioneered in the 1870s.

The railways affected the pattern of urbanization in a number of ways. They invigorated the urban economy, creating a demand for labour; and through third-class carriages filled with peasant migrants, they helped to satisfy the demand. In 1897, 95 million passenger-journeys were made by rail at a time when the first Russian census revealed a population of a little over 126 million (Brower, 1990, p.51; Mitchell, 1998, p.7), a ridership consistent with the railway's use for special journeys rather than routine commuting. The railways, therefore, were partly instrumental in the absolute growth of the urban population from 6.1 to 18.1 million between 1863 and 1913, and its relative rise from ten to fifteen per cent of the total (Falkus, 1972, p.34). The arrival of a railway also influenced the morphology of a given settlement. Railway stations became focal points of industrial and commercial growth, sometimes at some distance from the existing economic centre. In Moscow, the hub of the national system, the nine main railway termini were located on the outskirts, and the neighbourhoods of wooden dwellings that sprang up about them were among the most congested in the city (Bradley, 1985, p.58).

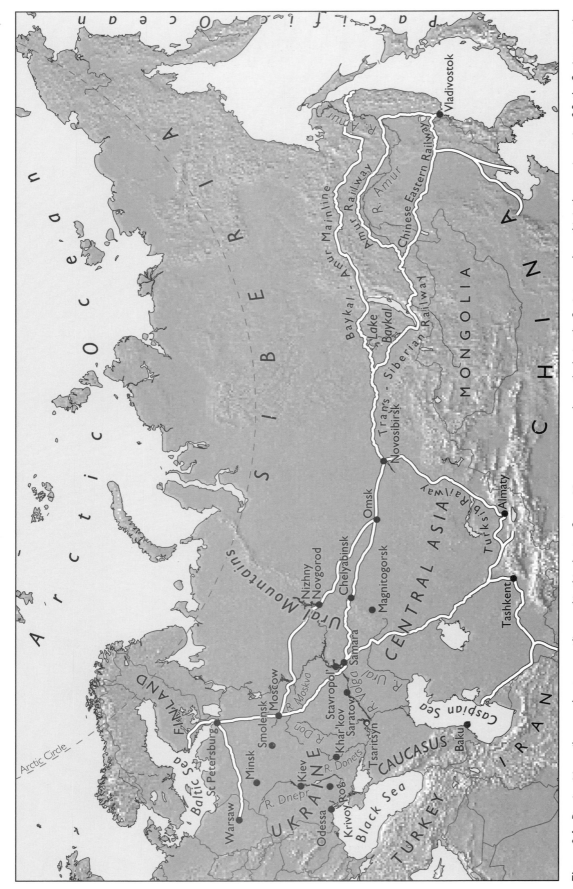

Figure 9.1 Russia: cities, railways and waterways discussed in this chapter. Some cities were renamed, mostly during the Soviet era; where this is the case in section 9.3, the Soviet name is followed by the original name in brackets. After the end of Soviet rule in 1991, these cities reverted to their original names. The Trans-Siberian Railway, which began at Chelyabinsk, was originally planned to follow the River Amur to Vladivostok, but for military reasons the Russian government decided instead to drive a section (the Chinese Eastern Railway) across the Chinese territory of Manchuria. Construction of the Amur railway was resumed in 1908 and completed in 1916 (Marks, 1991, pp.174, 205)

Following the demise of imperial planning, with its prescribed façades and layouts, 'railroad engineers and topographers largely replaced the state in giving shape to the city' (Brower, 1990, p.52). Too much emphasis, however, can be placed on the railways; more generally, change in the urban built environment was being driven by the liberalization of the economy, and in particular, by the growth of property speculation. The requirement that private buildings adhere to authorized model façades was lifted as early as 1858 (Brower, 1990, p.95).

Private enterprise, however, was nothing like as vigorous as it was in Western Europe. In the late 1880s and 90s, it was the state that stepped in and engineered a dramatic acceleration in the process of Russian industrialization. It became the main agent of railway construction and encouraged a flood of foreign investment, particularly from France and Belgium, into textiles, electrical engineering, metallurgy, mining and oil. As a result, industrial growth in the 1890s leapt to an average of eight per cent per annum. This industrial spurt was identified above all with Count Sergey Witte. Witte, whose name reflected his father's Dutch origins, was a former railway manager, and served as Minister of Finance from 1892 to 1903. He convinced the Russian autocracy to take a gamble on the idea that unleashing the forces of modern technology into the economy would strengthen, rather than jeopardize, a political system founded upon the privileges of a hereditary nobility. Indeed, Steven Marks has argued that the main aim of Witte's prize project – the 8,600-kilometre Trans-Siberian Railway, built between 1891 and 1901 – was not economic but political: to extend tsarist control over whole empire and counter external threats. Marks, though, had his own political agenda; he attributed the railway's technological shortcomings to state entrepreneurship, and saw the project as harbinger of subsequent failures of Soviet planning (Marks, 1991).

Witte himself was evidently something of a technological determinist:

> the railroad is like a leaven, which creates a cultural fermentation among the population. Even if it passed through an absolutely wild people along its way, it would raise them in a short time to the level prerequisite for its operation.
>
> (quoted in von Laue, 1974, p.191)

But the effects of the great industrial drive of the late tsarist era in Russian cities were beyond the calculation of Russia's new generation of technocratic officials and entrepreneurs. The first major experience of industrialization in Russian cities was even more traumatic than in the first decades of the Industrial Revolution in Britain. This was partly because in the Russian case, the technologies of the First and Second Industrial Revolutions were imported almost simultaneously. State-of-the-art metallurgical and electrical engineering facilities were transplanted, either directly into the countryside, where they acted as nodes for new urban settlement, or into existing cities devoid of the physical and social infrastructure of their late nineteenth-century Western industrial counterparts. One effect of this conjunction of technologically sophisticated plant and an abundance of unskilled Russian peasant migrants was an unprecedented degree of industrial concentration; by 1914, some forty-one per cent of Russian industrial workers were found in factories with more than 1,000 employees, more than twice the proportion in the USA (Rogger, 1983, p.113). One of the best known of these factories was the Putilov works in St Petersburg, where 12,000 metal-workers were employed at the turn of the century. This phenomenon was surely a catalyst for the unrest among urban industrial workers that led to the revolutionary overthrow of the autocracy in 1917, and, later in the same year, of the provisional government that replaced it.

The urban effects of the rapid industrialization of the 1890s were particularly acute in St Petersburg, where there was heavy foreign investment in munitions, shipbuilding, railway construction, mechanical engineering and electrical engineering, the last of which was led by the German companies, AEG and

Siemens. During the half-century of industrialization leading up to the 1917 revolutions, St Petersburg was reckoned to be the least healthy city in Europe. The population of the city had risen from 667,000 in 1870 to 1,962,000 in 1910, by which time a third of the city's inhabitants had no running water, and there was an average of seven people per apartment. This compared with 3.6 per apartment in Berlin, and 2.7 in Paris in the same year (Bater, 1976, p.329; 1980, p.18; Mitchell, 1998, p.75). In 1908, there was a cholera epidemic, and forty-seven per cent of all deaths were caused by infectious diseases; nevertheless, no steps were taken to improve the city's water supply and sewerage systems until 1914 (Bater, 1976, pp.350–53).

There was some investment in public transport in Russian cities; by the end of the century, apart from the horse-drawn cabs patronized by the best-off, there were horse trams in all major cities and most provincial capitals, but as in the West, these were beyond the pockets of the pedestrian poor. Moscow's trams delivered 45 million passenger journeys in 1895, a small fraction of the ridership in contemporary London; two-thirds of the Muscovite passengers took the option of travelling first class (Brower, 1990, p.53). As early as 1892, Belgian entrepreneurs built the first Russian electric tramway in Kiev, the capital of Ukraine province. In 1896, in order to dispel foreign perceptions about the backwardness of Russian life outside the capitals, the government sponsored a showcase tramway, and electric lighting, for the 'city on a hill' designed for the All-Russian Fair at Nizhny Novgorod. But the reality was that the traditionally weak municipalities, squeezed between the tax demands of central government, and the stinginess of local merchants, had much smaller revenues per head of population than their Western European counterparts, and most could not afford electrification. Only in 1907 did electric trams start to replace horse trams in the centre of St Petersburg. Even though public transport trips there averaged 150 per inhabitant in 1912, this was still well below the average in most West European cities; St Petersburg was still a walking city for most factory workers (Bater, 1973[1]; Bater, 1976, pp.268–77, 284).

Building construction from the 1860s followed the Western pattern, in that architects became professionalized, and increasingly undertook commercial and industrial briefs, including shopping arcades, large enclosed markets, banks, theatres, hotels and railways stations, using iron columns and beams where unobstructed spaces were needed, as well as the traditional stucco and brick, or wood. As in the West, architects had recourse to an eclectic mix of historical styles for building façades. An example is Alexander Pomerantsev's rebuilding of Moscow's Upper Trading Rows (1889–93). Its façade was in the style of the Russian Revival, favoured by those so-called Slavophile intellectuals who sought to combat the influence of the West; a style deemed necessary to fit in with the historic buildings of Red Square. But behind the façade, reinforced concrete was deployed possibly for the first time in Russia for the walkways spanning the second and third levels of the interior of the building (see Figure 9.2). By the turn of the century, architects were engaging more self-consciously with modern building technology and architectural styles. A landmark in Russia's so-called *style moderne* was William Walcot's Hotel Metropole in Moscow (1899–1905); the complex mix of materials and styles of the façade concealed the structural use of reinforced concrete (see Figure 9.3). But the use of load-bearing walls – usually of brick – persisted; despite its Gothic styling, probably the first example in Moscow of a fully load-bearing reinforced-concrete frame with a suspended exterior facing was

---

[1] An edited version of the article by Bater can be found in Goodman (1999), the Reader associated with this volume.

Figure 9.2 Central passage of the Upper Trading Rows, Moscow, on a site facing the Kremlin. Pomerantsev's design was based on the *galleria*, an up-market shopping feature of a number of nineteenth-century European cities, in particular Milan. But with over 1,000 shops, the Upper Trading Rows were far bigger, and modern construction technologies were necessary to provide access, light and ventilation. The design of the arched skylights by Vladimir Shukhov 'ranks among the remarkable achievements of civil engineering in Russia during the nineteenth century' (Brumfield, 1993, p.415). The structure is now the city's well-known GUM department store (reproduced from ibid., p.417; photograph by William Brumfield)

Figure 9.3 Hotel Metropole, Moscow. Despite the large arched panel in the centre, and other decorative devices, the façade dispenses with the 'illusionistic supportive elements' of neo-classical and other historicist styles. The large glass bays of the first and second floors have a structural as well as decorative function. The building revealed 'a tension between structure and decoration characteristic of the early phase of the *moderne*' (Brumfield, 1993, pp.426–7; reproduced from ibid., p.426; photograph by William Brumfield)

Figure 9.4 Muir and Mirrielees department store, Moscow. Beneath the decorative Gothic details (including pinnacles and a rose window) of the roof line and corner tower, the reinforced-concrete frame is clearly expressed around the generous plate-glass windows (Brumfield, 1993, pp.440–1; reproduced from ibid., p.442)

Roman Klein's department store for the British firm, Muir and Mirrielees (1906–8; see Figure 9.4). More outspoken examples of framed construction followed, though mainly for the administrative headquarters of Moscow's business district. Reinforced concrete was also a feature of some multi-storey apartment blocks for the wealthy, in districts of Moscow opened up by extensions of the horse-tram network, and of some suburban mansions that sprang up following the construction of the Moscow Ring Railway in 1908, partly along the line of the former customs barrier. But such buildings were exceptional in the otherwise dire and stressed landscape of Russian housing before the First World War (Brumfield and Ruble, 1993; Brumfield, 1993, pp.429, 436). Reinforced-concrete-framed corporate headquarters and up-market apartment blocks were also erected in St Petersburg, but to a lesser extent than in Moscow, because of strict regulations on building heights in the capital.

Before the Russian Revolution, the overall urban picture is one of the breakdown of imperial zoning restrictions and architectural controls in the face of rapid industrialization and massive in-migration. The municipalities had neither the power nor the resources to provide the necessary infrastructure. Nor was there any established tradition of town-planning; Ebenezer Howard's Russian admirers set up a Russian branch of the International Garden City Society in St Petersburg in 1913, but what few suburban and satellite developments it spawned – including Prozorovka, a railway workers' settlement near Moscow – were limited by the inadequacy of both interurban and intra-urban transport (Bater, 1980, pp.20–21; French, 1995, pp.24, 31).

9.3 *Engineering the socialist city*

Much of the rapid and chaotic urban growth of the late tsarist period was reversed by the economic dislocations and calamities associated with heavy Russian casualties on the Eastern Front during the First World War, the revolutions of 1917, and subsequent civil war and famine. A systematic depopulation of the cities followed, due to death, conscription and a mass return to the countryside: though reliable statistics are lacking, the populations of Moscow and St Petersburg seem to have dropped by a half and two-thirds respectively (Lewin, 1985, pp.211–12). The urban population had been only a sixth of the total in 1917; but after a rapid rate of growth of some four per cent per annum in both the 1930s and the 1950s, the number of urban inhabitants reached a half by the start of 1961, and two-thirds by the 1989 census (Bater, 1980, pp.60, 66; French, 1995, p.70). This demographic transition occurred later than in the leading industrial nations of north-west Europe, but at a similar point to that of several other industrializing European nations. The dramatic rate of increase took place despite the urban catastrophes of the Second World War, during which an estimated 25–30 million people perished, and a sixth of the national housing stock was damaged or destroyed (Bater, 1980, p.63). The increase testifies above all to the Soviet regime's burning ambition to match the military and industrial might of the leading Western nations, an ambition born of actual invasion during the aftermath of the October Revolution of 1917 (and subsequently by Nazi Germany), and of the leadership's sense of being encircled by hostile capitalist powers. At the Eighth All-Russia Congress of Soviets in December 1920, the Bolshevik leader Vladimir Il'yich Lenin famously declared:

> *Communism is Soviet power plus the electrification of the whole country* ... Only when the country has been electrified, and industry, agriculture and transport have been placed on the technical basis of modern large-scale industry, only then shall we be fully victorious.
>
> (Lenin, 1966, p.516; original emphasis)

Having wrested power, the Bolsheviks – the dominant faction of the revolutionary Social Democratic party, and soon to proclaim themselves the Communist Party of the Soviet Union (CPSU) – showed themselves at the outset to be far more committed to science and technology than were their tsarist predecessors during the century leading up to the First World War. But their ideological hostility to private enterprise precluded direct investment by Western companies, and great efforts were made to encourage indigenous inventiveness; by the mid-1930s, the Soviet Union was devoting a higher proportion of national income than the USA to scientific and industrial research (Lewis, 1979). By then, starting with the first of three Five-Year Plans in 1928, the Communist Party under the leadership of Joseph Stalin had embarked on an industrialization drive every bit as dramatic as the tsarist spurt of the 1890s. Between 1928 and 1940, annual industrial growth has been estimated at ten per cent (Blackwell, 1994, p.130). The fruits of the Russian research drive were slow to mature, and although direct foreign investment was out of the question, the Soviet regime had to buy in Western expertise and technology. Much of this was from the USA: the Ford Motor Company supplied a car manufacturing plant at Gor'ky, electrical engineering equipment was provided by General Electric, and RCA was involved in radio installations (Sutton, 1968–73).

The industrialization of the Plan era was partly based on the existing urban areas of Moscow, the Donets basin and St Petersburg, which by then had undergone two name changes: first to Petrograd in 1914, and then to Leningrad in 1924, the year of Lenin's death (Ruble, 1990). Great urban-industrial agglomerations were created in the east, notably in the vast Urals–Kuznets Basin region, which included one of the many new industrial settlements founded in the Soviet period: Magnitogorsk, built to exploit an outcrop of rich iron ore, known as the Magnetic Mountain (Kotkin, 1995). Other new cities based around deposits of coal and iron ore sprang up in Siberia and the Islamic republics of Central Asia. After the Second World War, further industrial urbanization reactivated the growth of these settlements; whereas in 1959 only three cities – the two capitals, plus Kiev, the capital of the Ukraine – had populations of more than a million, by the 1989 census, this figure had risen to twenty-three, despite the desire of Soviet town planners to avoid urban agglomerations (French, 1995, pp.72–3). Five of these cities (Tashkent, Novosibirsk, Omsk, Chelyabinsk and Almaty) were situated in Asiatic Russia; by this time, there were several cities in Siberia other than Novosibirsk and Omsk with at least half a million inhabitants, though from the 1960s, the state resorted to income differentials to attract workers to these remote and often inhospitable locations.

Despite the notable spatial extension of urbanization during the Soviet era, the post-war trend, as in Western Europe, was for the biggest urban agglomerations to take an even bigger share of the urban population. The primacy of Moscow in particular was enhanced by the Bolsheviks' decision in 1918 to restore it as Russia's centre of government, and the consequent expansion of the bureaucracy required for a socialist, centrally-planned economy (Colton, 1995). The Communist leadership also required a great deal of heavy industry to locate in or around the restored capital, which before the Revolution had been associated above all with cotton textiles. New engineering establishments were set up, and old ones expanded; the Likhachev Vehicle Works was transformed into the largest plant in Moscow, and along with housing for its employees, formed a new district on the south-eastern outskirts:

> It is difficult to think of any other major city outside the Soviet Union, let alone a metropolis, where the growth of heavy industry began to play a significant part at such a late stage in its development.
>
> (French, 1984, p.368)

Consequently, although at 1.76 million in 1914 the population of Moscow was slightly less than that of St Petersburg, it pulled well ahead during the Soviet era, and by some estimates became the most populous conurbation in Europe; in 1991, the final year of Soviet rule, the population reached its highest point of just over 9 million (Colton, 1995, p.758).

Urban planning

In theory, the Soviet political system, based as it was on a hierarchy of elected councils (Soviets) was a model of devolved democratic decision-making. In practice, Soviets at all levels were dominated by their executive committees, and executives at lower levels deferred to those above. The entire political system was controlled throughout by the CPSU, a political elite whose members occupied all key posts. The principal operative feature of this system – and the one that set it apart from Western nations – was the replacement of economic mechanisms for the distribution of resources by centralized planning. The near elimination of the market went hand in hand with the nationalization of land; henceforth land-values, among the most powerful determinants of the form and fabric of west and central European cities, counted for nothing in decisions by Soviet planners about the location of industry and housing, the placing of transport networks, the function of city centres, and so on. These radical changes had tumultuous implications for the urban system and the internal structure of settlements; in principle, town planners had free rein to create entirely new socialist cities, based on the deployment of technologies towards equitable and hygienic ends.

Figure 9.5 Central Union of Consumers' Co-operatives (Tsentrosoyuz) Building, 1929–36, Moscow. One of the last Modernist works to be completed in Moscow, the eight-storey building consisted of a central slab block set on pilotis (subsequently enclosed), flanked by two perpendicular blocks. Two of the blocks were clad in glass framed by red tufa, and the interior components were connected by spiral ramps (Brumfield, 1993, p.475; reproduced from ibid., plate 77; photograph by William Brumfield)

The opportunity was there to build an urban system and individual cities that were fully consonant with a socialist ideology, rather than one based on the divine right of kings and aristocratic privilege, or on the maximization of profit. But what kind of city was a socialist city? To start with, cities were looked upon with favour. Marx and Engels had famously mocked the 'idiocy of rural life', and Lenin regarded cities as the main motors of progress: on the face of it, the dominant urban ideology was the diametric opposite of the bourgeois semi-rural idyll associated with the suburbanizing flight from city centres in Britain and the United States. But the revolution that Marx predicted for mature capitalist societies actually took place in one that was still predominantly rural, and the Soviet version of Marxism reflected this anomaly. As a result, there was a great deal of emphasis in Marxist–Leninist thinking on parity between the peasantry and the industrial proletariat, and on fulfilling the call by Marx and Engels in the *Communist Manifesto* (1848) for the gradual abolition of the difference between the city and the countryside.

During the late 1920s, the tension in Soviet ideology over the city and the country found expression in a debate between two groupings commonly classified (though this may be to over-simplify the issues) as 'urbanists' and 'de-urbanists' or 'disurbanists'. The de-urbanists envisaged a townless socialist landscape consisting of ribbon developments set in green fields; communication between them would depend on buses and cars. Their urbanist antagonists were more sanguine about nucleated settlements but, echoing Ebenezer Howard, saw 50,000 as a population limit that struck a balance between a sense of community and the efficient allocation of resources. They also favoured high-rise housing set in green spaces, and so were advocating a mixture of Howard's Garden City and the Le Corbusier's Radiant City. Le Corbusier was among the early admirers of the new Soviet regime. He visited Russia three times between 1928 and 1930, came up with a plan for the reconstruction of Moscow every bit as radical as his Voisin Plan for Paris,[2] and designed a major Modernist building for Moscow, the headquarters of the Central Union of Consumers' Co-operatives (see Figure 9.5).

---

[2]   Discussed in Norma Evenson, 'Paris and the Automobile' in Goodman (1999).

Some of the earliest examples of Soviet urban plans actually implemented drew on both the urbanist and de-urbanist approaches. Most notable were the plans utilizing the concept of a linear city, first put forward for Madrid by Arturo Soria y Mata (1844–1920), and advocated in the Soviet Union by N.A. Milyutin, a Constructivist (see p.317 below). The linear city consisted of parallel transport corridors and industrial and residential zones, with the residential zone segregated from the other functions by a green buffer. These ideas were incorporated into the plans for the industrial cities of Volgograd (originally Tsaritsyn, subsequently renamed Stalingrad) and, in theory at least, Magnitogorsk (see Figure 9.6). Magnitogorsk, situated to the east of the southern tip of the Urals, was a brand-new industrial city, incorporating 'the latest word in world technology', as Stalin's henchman Ordzhonikidze boasted in 1935 (quoted in Kotkin, 1995, p.70): a gigantic steel-making plant, along with a dam on the Ural River for the site's water supply, a power station, and an iron-ore mine, commissioned in 1929 from Arthur McKee and Co. of Cleveland, Ohio. Despite great difficulties between the us contractors and the Soviet administration, the first steel issued from the open-hearth shop in 1933. The population of the site, entirely dependent upon long-distance rail for supplies, equipment and people, was put at 250,000 in the early 1930s, most of them deportees from European Russia (ibid., pp.72–3, 85). The site epitomized the priority accorded to industry. The first workers arrived at a railway halt, lived in tents and then in mud huts

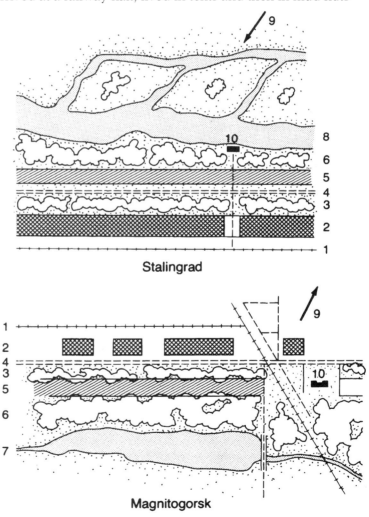

Figure 9.6 Milyutin's linear city: Stalingrad and Magnitogorsk. The main point of the parallel lines of development was to enable a short pedestrian journey to work. Milyutin's plan for Stalingrad was modified by Vladimir Semenov, and his scheme for Magnitogorsk by the German socialist architect Ernst May (reproduced from Bater, 1980, p.25, with the kind permission of Edward Arnold; adapted from Parkins, 1953, p.22 fig. 5 by kind permission of Chicago University Press)

1, railway; *2*, industrial zone; *3*, green zone; *4*, thoroughfare; *5*, residential zone; *6*, park; *7*, Ural River; *8*, Volga River; *9*, prevailing wind; *10*, House of Soviets.

Figure 9.7 Magnitogorsk: the gulf between state-of-the-art imported technology and its Russian context. *Top* US-designed blast furnaces from which pig-iron was poured for refining in the steelworks, photographed in 1932. *Bottom* Workers building their own mud-brick houses, as Sumerian craftsmen might have done at the beginning of the Urban Revolution; photographed in 1929 (Novosti Photo Library)

often shared with livestock, or in row upon row of filthy, overcrowded barracks; these were one-storey, wooden structures covered with white-washed stucco (see Figure 9.7). There was only one paved road, little street lighting, an inadequate sewage system, a contaminated water supply, a handful of buses and, from 1935, a single tramcar. The attempt to add to the steelmaking plant a permanent linear city with high-rise buildings was first made by the internationally acclaimed Modernist architect Ernst May, though his plans fell foul of the Soviet bureaucracy. A linear city of a sort emerged, though it was less a planned socialist city than an outcome of the industrial process, the local topography and the imperfections and imbalances of the Stalinist plan era (ibid., pp.141–5).

By the 1930s, Stalin had committed the Soviet Union to rapid industrialization, and notions of balanced town-planning slipped down the political agenda; indeed, in 1931 the Central Committee of the CPSU expressly terminated the debate on the socialist city, branding it as 'utopian' (French, 1995, p.42). But there was soon a repetition of the urban problems of the late tsarist era, as redoubled mass in-migration outstripped the provision of housing and municipal services; there was a desperate need to extend planning from the economic to the urban sphere. The first notable product of Soviet urban planning was the 'General Plan for the Reconstruction of Moscow' (Genplan) of 1935 (Richardson, 1991; Ruble, 1994). The Moscow plan, like those for London and Paris in the 1930s, saw the seemingly indefinite expansion of the capital as undesirable; a population limit of 5 million was specified, and further industrial development was to be directed out to satellite cities beyond a green belt. As it turned out, Moscow's post-war population far exceeded the numerical limit, and the establishment of a green belt (the Forest-park zone) had to await the Genplan of 1971. The policy of containment had some effect: even though the metropolitan population grew by twenty-five per cent between 1959 and 1975, that of Moscow's satellite cities increased 4.7 times (Bater, 1980, p.81). But by the 1980s, the growth of the capital had encroached upon the Forest-park zone itself (see Figure 9.8).

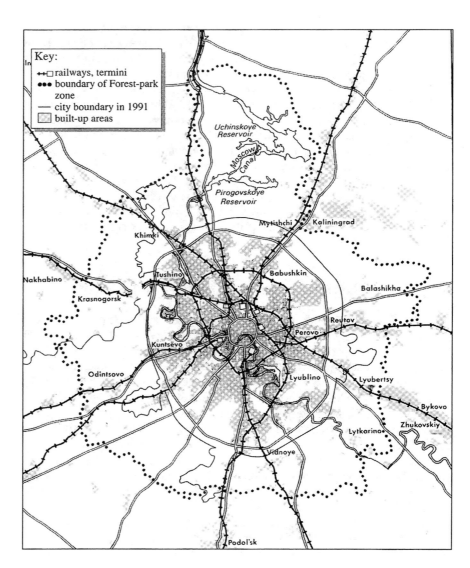

Figure 9.8 The expansion of Moscow into the Forest-park zone. There was considerable urban sprawl as a result of new housing on the city outskirts; during the decade 1974–84, 600 hectares of the designated green belt beyond the orbital motorway were built over. The city boundary, set in 1960 at the motorway, has since been correspondingly extended (French, 1995, p.103; reproduced with the kind permission of the publisher from ibid., p.104)

Moscow's plan of 1935 has been described as a 'dead letter', made unfeasible by Stalin's obsession with industrialization (French, 1984, p.369). Little was achieved, other than the building of two new ports and a granite embankment for the river Moskva, a new ship canal dug by Gulag prison labour to the Volga, and the widening of Red Square and some streets, including the Sadovoye (Garden) Ring, on the line of Moscow's sixteenth-century fortifications. One of the few successes of the period, albeit begun before the 1935 plan, was the famously ornate Metro, the interior decoration of which included marble stripped from the city's cemeteries; twenty-six kilometres of the underground network had been constructed by 1938, with advice on designs and equipment provided by representatives of the London Underground (French, 1984, p.370; Colton, 1995, pp.327–8; Robbins, 1997[3]). Despite the failure of the plan to curb Moscow's expansion in the post-war period, it incorporated some general features that would become enduring goals of Soviet planners: limitation of the journey-to-work by public transport to a maximum of forty minutes; land-use zoning, involving the use of green buffers to segregate housing from industry and heavy traffic, the allocation of the city centre to mass political demonstrations, rather than the high-rise administrative and distributive buildings fostered by Western urban economics; and the organization of residential areas into complexes with their own shopping centres, parks and schools, and consisting of a number of 'micro-regions' (*mikrorayony*) of between 8,000 and 12,000 residents. The constituent unit of each *mikrorayon* was the 'super block' (*kvartal*), officially an area of land of some six hectares with roads on all sides and street façades, intended to house some 1,000–1,500 people (Bater, 1980, p.102; French, 1995, pp.62–3).

Soviet cities were undoubtedly unlike those in Western Europe because of different planning desiderata, but the reality diverged from the ideal in many important respects (Bater, 1980; Shaw, 1991). With regard to the urban system, the increasing pressure on the leadership later in the Soviet era for increased industrial productivity was at odds with some of the main egalitarian planning tenets: the dispersal of industry and the elimination of the inequalities between the republics that made up the Soviet Union. As for individual cities, a number of the planning goals proved elusive. One was the limit on city size, which could certainly have been enforced, as an official permit (*propiska*) was required for residence in a city. But such controls were limited in their effects, as the drive for increased industrial production took precedence over planning goals. The ideal of the 1930s – a settlement of 50–60,000 – had continually to be revised upwards: to 150–200,000 in the 1950s and again to 2–300,000 in the 1960s (Bater, 1980, p.78; French, 1995, pp.72–3, 99–100). Another influential ideal prefigured in the 1935 Moscow plan was that of a socialist city organized into residential micro-regions, separated from industrial zones by green buffers. Again, the actual outcomes were much less ordered and efficient. Ironically, this was because the supposedly monolithic Soviet political and planning system was, in practice, full of the kind of rivalries and conflicts that characterized the planning process in liberal, capitalist societies.

At various times during the history of the USSR, attempts were made to invest the local Soviet (council) with authority sufficient to enforce a general plan ensuring the most efficient and hygienic integration of housing, industry, transport and services. In practice, the city Soviet, which had a much smaller staff of town planners than an equivalent Western local authority, was up against a number of regional and national organizations (especially ministries, and state-run industrial enterprises), all of which had a stake in the industrial development of particular locations, and which often built their own housing and provided their own services, including transport. Some housing developments had no services at all, apart from electricity; in the east, a Soviet

[3] See Goodman (1999) for an edited version of this article.

publication reported in 1975 that barely fifty per cent of houses were connected to the public water supply, the sewage system or the centralized heat and power networks which were among the Soviet regime's more positive environmental contributions (Bater, 1980, p.109). This 'departmentalist' approach, conjoined with the absence of a market in urban land, led to a wasteful and ill-integrated industrial land-use pattern in Soviet cities; according to one estimate, urban industry occupied fifty per cent more land than the equivalent in US cities. Housing was also deeply affected. In Saratov on the Volga, apart from the municipal authorities, eighty other enterprises and organizations of ministries and departments constructed housing in the 1970s, and more than seventy were involved in Kiev, the capital of the Ukraine. Only 35.3 per cent of state housing stock in the Russian republic was controlled by local authorities, according to a Soviet publication of 1978 (Bater, 1980, pp.49, 96, 131).

Transport

Railway construction continued into the Soviet period: the Turkestan–Siberian (Turksib) railway was completed in 1930, and during the late 1970s and 1980s, the Baykal–Amur Mainline was built, further reinforcing the industrialization and urbanization of Asiatic Russia (see Figure 9.1). As well as these inter-urban rail links, the Soviets also invested in civil aviation, as might be expected for such a vast territory. The road network was, as a consequence, much less developed than in the West; this even went for intra-urban roads – according to a Soviet source published in 1973, only forty per cent of the streets of Gor'ky (formerly Nizhny Novgorod), Kuybyshev (formerly Samara) and Novosibirsk were hard-surfaced (Bater, 1980, p.113). Freight-haulage by road was consequently mainly confined within cities, in which lorries dominated traffic until the recent rise in car ownership. Towards the end of the Soviet period, the increased use of 'juggernaut' lorries from East Germany, Hungary and Bulgaria in the place of lighter Soviet vehicles was proving too much for the physical road structure (French, 1995, p.160).

The Soviet planning ideal was that people should be within walking distance of their jobs and of amenities such as shops, schools and recreational spaces. This, and the planning goal that journeys by public transport should take no more than forty minutes, proved increasingly hard to realize, not least because of the inefficiencies of land allocation and use thrown up by the decision-making process. The industrialization drive also diverted investment away from public transport; even though trolleybuses were introduced to Moscow in 1933, and its tramway system increased from 408 kilometres in 1928 to 532 kilometres in 1934, the population grew by more than fifty per cent in that time (French, 1984, pp.359, 370). Another consideration unlikely to reduce commuting times was the egalitarian flat-rate fare policy on urban public transport. Public transport use was certainly heavy; by 1977, there were more than 48 billion passenger-journeys on urban trams, trolleybuses, buses and what were then seven operational metro systems, and in 1980, the Moscow Metro was the most heavily used in the world (Bater, 1980, p.112; Colton, 1995, p.519). Although twelve underground railway systems were in place by the end of the Soviet era in 1991, the workhorse of Soviet urban transport was the notably inefficient bus, on which the great majority of Soviet cities relied entirely, and which accounted for 58.2 per cent of public transport journeys in 1976. Trolleybus ridership increased fivefold between 1970 and 1990, overtaking that of trams; nevertheless a small absolute rise was recorded in tram passenger-journeys in that period (Grava, 1984, pp.188, 191; French, 1995, pp.161–2).

The favouring of public transport meant that Soviet cities were for many decades almost free from private motor vehicles. Even before the end of the Soviet era, however, this aspect of urban life was changing, as restrictions on car ownership were relaxed. In 1970, car ownership was two per 100 families, but by 1988 it had risen to seventeen; in the same year Soviet production of cars had risen to 1.3 million, and that of motor cycles and mopeds to 1.1 million, the bulk of these vehicles being manufactured at a huge plant built by Fiat on the Volga at Tol'yatti (formerly Stavropol', renamed after Togliatti, the secretary of the Italian Communist Party). In the 1980s and early 1990s, Russian citizens used their cars for pleasure rather than for commuting, and car parks, petrol stations and garages were thin on the ground (French, 1995, pp.101, 159, 164, 169–73). But the process of 'automobilization' (*avtomobilitatsiya*) was under way, with inevitable consequences for the urban fabric and the quality of air, which, in Russian cities outside the heavily-polluted regions of heavy industry, compared favourably with that of many a Western city afflicted with traffic-induced smog.

Building construction

The conclusion of Section 9.2 was that before the Revolution, the built environment of the two capitals was only slightly affected by Western construction innovations. But during the 1920s, there was a great ferment of conflicting ideas and designs among rival associations of Russian architects, many of whom had contacts with the Western avant garde, notably the Bauhaus group, Le Corbusier and the US architect, Frank Lloyd Wright. Much exciting 'paper architecture' resulted, exploring and debating the potential of new technology, but little of this was realized, as the designs were often way beyond the capacity of the infant Soviet economy and building industry. Indeed, among the earliest practical approaches to building in the immediate aftermath of the revolution was the revival of the medieval Russian practice of building houses from pre-cut, standardized wooden components (a practice reflecting the superabundant forest resources of Russia's central and northern regions). Among those avant-garde projects that got off the drawing board were a number of buildings designed by the Constructivists, the group most identified with the view that architecture should express the functional properties of new materials and construction techniques. This approach is clear from the angular framing and glass infills of Grigory Barkhin's 1927 building for the newspaper *Izvestiya* (see Figure 9.9), and in the more

Figure 9.9 *Izvestiya* building, Moscow. The off-centre glass shaft to the left is a stairwell. The prominent vertical and horizontal lines are stressed by asymmetrical balconies, and are juxtaposed with the full circles of the top-storey windows (Brumfield, 1993, pp.471–2; reproduced from ibid., p.471; photograph by William Brumfield)

Figure 9.10 Zuyev Club,
1927–9, Moscow. As in the
Izvestiya building, square and
circle are juxtaposed. The large
corner cylinder encloses a
stairwell. The curved corners of
Golosov's design were typical
of the ways Constructivist
architects exploited the
versatility of concrete and glass
(reproduced from Brumfield,
1993, p.472; photograph by
William Brumfield)

Figure 9.11 Lenin State
Library, 1928–40, Moscow,
designed by the 'traditionalist'
architects Vladimir Shchuko and
Vladimir Gelfreikh. The statuary
above the cornice, the frieze
above the main portico, and the
bronze reliefs of noted thinkers
between the pylons of the main
façade were all additions to the
original design, and show how
far was the progression from
Modernism to conservative
architectural forms over the
building's protracted period of
construction (Brumfield, 1993,
p.481; reproduced from ibid.,
p.483; photograph by William
Brumfield)

ambitious contours of Ilya Golosov's version of that distinctive Soviet building type, the workers' club (or 'house of culture' or 'palace of labour'), a more politically correct alternative to the public house (see Figure 9.10).

But in 1932, Stalin abolished the avant-garde architectural associations, and Modernist designs for public buildings began to give way to the heavy, monumental, often neo-classical façades that would also characterize Fascist architecture during the 1930s. This can be seen in the 'stripped' classical colonnades and porticos of the Lenin State Library (see Figure 9.11). The irony of this embrace of traditional architectural styles during a period of rapid industrialization is patent:

> Even as the functional, streamlined architecture of industrial production – in Marxist terms, the base – took shape with the help of western engineers and specialists in the Urals and on the Volga (particularly in Stalingrad), the blossoming of surface decoration on the buildings of the new administrative and cultural 'superstructure' reflected a diminution of functionalism in favor of a blatant display of power.
>
> (Brumfield, 1993, p.489)

Another kind of reconciliation between the exploitation of modern technologies and the age-old affirmation of centralized political power was manifest in the post-war period of Stalinist rule. To begin with, the authorities were preoccupied with the wholesale rebuilding of destroyed cities, notably Stalingrad, Smolensk, Minsk, Kharkov and Kiev. Then, from the late 1940s until Stalin's death in 1953, 'Stalinist Gothic' tall buildings with spires sprang up in many Soviet cities, at sites judged to have the greatest visual impact on the urban landscape. Of eight such tall buildings intended for Moscow, seven were completed; they were mostly office and apartment buildings and hotels, though the most dominant was the central building of Moscow State University, set on the Lenin Hills overlooking the city (see Figure 9.12).

Figure 9.12 Moscow State University, 1949–53, designed by Lev Rudnev, Pavel Abrosimov and Aleksandr Khryakov. This, and the other 'tall buildings' were partly inspired by New York's early neo-Gothic skyscrapers, such as the Woolworth Building. Their spires, reminiscent of medieval belfries and the spires of St Petersburg, reflect the reversion of the Stalinist regime to traditional Russian authoritarian rule (Brumfield, 1993, p.490–1; reproduced from ibid., p.491; photograph by William Brumfield)

After Stalin's death, there was a return to Modernist architecture. Some of the most controversial examples in Moscow dated from the 1960s: the gigantic Rossiya hotel, for which more than half the area of Moscow's medieval commercial quarter, the Kitay-gorod, was cleared; and the high-rise office and apartment blocks of a new street, Kalinin Prospect, again driven through a historic part of inner Moscow (French, 1995, p.186; see Figure 9.13). More functional building designs were also necessary to tackle the massive housing shortage. Urban housing had been an intractable problem from the regime's earliest years. Despite Stalin's revival of the tsarist passport system in order to control internal migration, Russia's urban population mushroomed by some 25 million during the first ten years of the Plan era. Because the lion's share of resources was poured into heavy industry, and in particular into defence, rather than urban amenities, existing housing capacity was overwhelmed, and many of these peasant migrants made do with tents, underground dugouts, mud and industrial scrap hovels and barracks; even in 1950, there was still an average of four individuals to a room (Blackwell, 1994, pp.104–5). Effectively, an apartment with three or four bedrooms would have to accommodate that very number of families.

Figure 9.13 Central Moscow, with the Moskva river in the foreground, and the Kremlin, the fortified medieval citadel, in the centre. On the right is the Rossiya hotel, built on a site originally intended for a Stalinist skyscraper, an example of which, on Ploshchad' Vosstaniya (Revolution Square), dominates the centre background. In the left background are the tower blocks of Kalinin Prospect (reproduced with the kind permission of the publisher from French, 1984, p.377)

Figure 9.14 West façade of communal apartment house for the People's Commissariat of Finance (Narkomfin), designed by Moisey Ginzburg and Ignaty Milinis, 1928–30, Moscow. Intended for 200 residents, the building represented a view, suppressed after Stalin's rise to power, that the family was a bourgeois institution that should be replaced under socialism. The main block, which rested on pilotis (now enclosed), included both apartments and communal dormitory rooms. At one end of the main structure was a block of communal services; on top was an open-framed solarium (reproduced from Brumfield, 1993, p.475; photograph by William Brumfield)

The preferred ideological solution to the housing shortage was the state-owned communal high-rise apartment block with nominal rents, rather than the privately-owned US and British detached family dwelling, with its associated bourgeois notions of family life. Housing policy also reflected the state's commitment to reduce income differentials and regional disparities, and to avoid the residential segregation that was so marked under the capitalist system. In practice, high-rise housing in the interwar period was limited, because of the priority accorded to heavy industry. One example is the apartment house for the People's Commissariat of Finance, designed by Moisey Ginzburg, the leading proponent of Constructivism, and Ignaty Milinis. This Modernist structure has features of the work of Le Corbusier, with whom Ginzburg was personally acquainted (see Figure 9.14). Another example, one which shows the rising tendency to monumentalism in public buildings, was the massive Government House, designed by Boris and Dmitry Iofan using a reinforced-concrete frame with brick infill (see Figure 9.15). This complex, which occupied about three hectares, contained some 500 apartments, as well as a library, gymnasium, club and shops; it was intended for the upper echelons of the government and the CPSU, and is a sign of the privileges that the élite had begun to accumulate.

The prominent structures of the interwar years were unrepresentative; only some 350 million square metres of state-owned or state-assisted housing were built between 1918 and 1950. This figure pales into insignificance beside the 1.55 billion square metres of floor space built between 1960 and 1975, allowing fully two-thirds of the population to be re-housed (Andrusz, 1984; Bater, 1980, pp.99, 102). Much of this expansion was achieved by industrialized, high-rise construction, standardized to a degree unknown in the West.

Figure 9.15 Government House, designed by Boris and Dmitry Iofan, 1927–31, Moscow. The lack of ornament and uniform fenestration was relieved somewhat by set-back roof structures. Behind the massive towers were three interior courtyards connected by enclosed passages raised on pilotis. The use of colonnades reflected the shift from Modernist to classicizing architectural forms that was getting under way at the time. On the left is the *Udarnik* ('Shock-worker') cinema (Brumfield, 1993, p.480; reproduced from ibid., p.482; photograph by William Brumfield)

Figure 9.16 Construction of high-rise blocks of prefabricated sections, Minsk, capital of the Belorussian Soviet Socialist Republic. Increased output of lifts was also necessary for heights to increase beyond the standard five-storey block of the Khrushchev period (1957–64). By the end of the Soviet era, probably nearly all new public-sector housing was of standardized prefabricated reinforced-concrete panels (French, 1995, pp.77–9; reproduced from ibid., p.79)

It involved the assembly on site of identical reinforced-concrete panels and other prefabricated modules or precast concrete components (see Figure 9.16). To avoid the expense of clearing inner-city sites, high-rise blocks were usually located on the fringes of cities, as near as possible to existing gas pipes, electricity and telephone cables, water mains and sewers. High-rise construction was adopted not, as often in the West, because of the scarcity or cost of land, but for speed and cheapness of construction, and to minimize the cost of services and transport links (French, 1995, pp.75–6, 80). Apartment heights rose from an average of 5.1 storeys in 1963 (reflecting the preponderance in the early years of the housing drive of brick-built five-storey buildings avoiding the expense of lifts) to an average of 7.8 storeys in 1971 (Bater, 1980, p.108). In 1974, a minimum of twelve storeys was stipulated for large cities, and in the 1970s, some blocks reached twenty-five storeys (French, 1995, p.79). Budgets remained tight, and the seams and cracks attendant upon the construction process added shoddiness to the pervading drabness of such developments.

Figure 9.17 Tropar'yevo housing area, 1970s, Moscow, designed by A. Samsonov, A. Bergelson *et al.* (reproduced from Brumfield, 1993, p.493; photograph by William Brumfield)

It was mass-production housing: 'architecture had been supplanted by engineering in the routinized production of buildings issuing from design bureaus' (Brumfield, 1993, p.492). In the 1970s and early 1980s, these methods gave rise to new 'housing massifs' in the outer zones of Moscow encompassed by the orbital motorway (completed in 1960), and connected to the central areas by new transport arteries; one of these developments, reminiscent of the *grands ensembles* of the Parisian suburbs, was the Tropar'yevo housing area (see Figure 9.17). In the Soviet era, partly because of the continuing housing crisis caused by the unceasing influx of new urban residents, there was nothing comparable to the economic, political and aesthetic disenchantment with the residential tower block that set in after the 1960s in Britain.

Electrification and communications

The quotation from Lenin at the beginning of Section 9.3 (p.309 above) evinced the special faith of the Soviet regime in the progressive implications of electric power. Lenin regarded the plan of the State Commission for the Electrification of Russia (GOELRO), which called for the building of power stations and of several new towns in association with them, as the 'second programme of our Party'; second, that is, to the political programme (Lenin, 1966, p.514). This is a clue that the policy of electrification was highly politicized: in an avowedly social-constructivist study, Jonathan Coopersmith emphasizes the political and ideological contexts of Russian electrification, in an attempt to combat any technological-determinist interpretation of social effects of technology transfer (Coopersmith, 1992). He documents the dependence of pre-revolutionary electrification on foreign investment, and the constraint placed on its development by inflexible tsarist institutions. He also shows how the technocratic ideology of non-Bolshevik Russian engineers paved the way for Lenin's GOELRO electrification plan. The plan resulted in the first major Soviet engineering project: the construction of a huge hydroelectric dam across the River Dnepr in 1927–32 (Rassweiler, 1988). But in many ways Lenin's expectations were not realized; his target of electrifying 3,500 kilometres of railway within fifteen years was not achieved until after the Second World War (Hutchings, 1976, p.96). Coopersmith attributed the failures of Soviet electrification to severe lack of resources, including lack of foreign investment capital.

The implementation of communications technologies made possible by the prior investment in electrification was also clearly shaped by politics. Rather than investing in telephony, which involved two-way communications, the authorities preferred the one-way information and propaganda route of radio. Regular radio broadcasting began in Moscow in 1924; by 1940, there were 1.1 million radio receivers and 5.8 million loudspeakers, connected to a sending station by telephone wires, and often mounted in clubs and meeting halls, and along streets. This method of transmission ensured that most listeners were urban dwellers (Hopkins, 1970, pp.90, 246, 248). The Soviet government also appreciated the propaganda value of cinema, especially as a way of influencing the illiterate; during the interwar period, after Stalin had banned imports from the Hollywood dream factory, the vast Russian cinema audience were fed on a heavily-censored diet of equally escapist historical drama, and 'socialist realist' tales of heroic workers succeeding against the odds (Kenez, 1985, pp.258–9; Kenez, 1992, pp.161–3).

9.4   Conclusion

A dimension hidden in this survey of the transfer of urban technologies from Europe and the United States to Russia is a secondary process whereby the colonizing Russians imposed their ways of city-building on the established urban cultures of the Caucasus region and central Asia. A prime example is Tashkent, the capital of Kazakhstan, where, partly assisted by the earthquake of 1966, the Russians laid their geometric street patterns and high-rise blocks over the winding streets and mud-walled courtyard houses of a typical Islamic city (French, 1995, p.76). The main emphasis of this chapter is on the noteworthy contrasts that Russian cities provide with the Western European cities covered in Part 2 of this volume. In the tsarist period the principal differences stem from the peculiar conflation of First and Second Industrial Revolution technologies, and the chasm between imported, state-of-the-art production technologies and the urban infrastructure needed to cope with their consequences. In the Soviet period, a completely different political and planning context profoundly affected the ways in which new industrial processes, transport innovations and new building technologies left their mark on cities. The absence of competitive retailing and of any market in urban land, the economic priority given to the development of defence and heavy industry over personal consumption, and the ideological preferences for communal forms of state-owned housing and public transport – all these features of the Soviet political system produced an urban landscape very different from that of the West. Among these characteristics were the more even spread of industrial premises throughout the urban fabric; reduced social segregation – where there was any, it was by apartment block only, rather than by urban sector, no part of the city being debarred by high rents or transport costs (French and Hamilton, 1979, pp.15–17; Andrusz, 1984, p.220); the absence of the trappings of consumerism, such as advertising displays, shopping precincts and self-service supermarkets; and the lack of many of the urban physical manifestations of widespread car ownership. As a result of the post-war housing drive, there was a higher level and flatter gradient of population density than in most Western cities; this was because flat-rate public transport fares and the absence of a land-market encouraged high-rise building in the suburbs. Overall densities nevertheless were somewhat lower in the outer areas, as there was probably as much land given over to privately-owned, wooden, one- or two-storey detached housing as there was to high-rise *mikrorayony* (Bater, 1980, p.109; see Figure 9.18).

The differences between Western and Soviet cities may not, however, be as great as they appear at first sight; it can plausibly be argued that well before 1991, the Soviet city was set on a path converging with that of the capitalist city (French, 1995, p. 203). The opportunity had clearly existed for Soviet planners to create a city quite different in kind from the pre-industrial and industrial cities considered so far in this series. The judgement of James Bater, an authority on Soviet urbanization, is that there was a degree of success:

> In the history of the city there is as yet no equivalent of the Soviet achievements in consciously manipulating the tempo and distribution of urban growth and in planning the city building process.
> (Bater, 1980, p.163)

And yet the main focus of Bater's survey is the widening gap between planning ideal and urban reality throughout the Soviet period; and the reality was perhaps closer to the Western city than it was to the ideal. Concerns about 'efficiency' increasingly over-rode ideals of regional equality, and of abolishing disparities between town and country, and urban agglomerations as big as those in the West emerged. More and more concessions to private

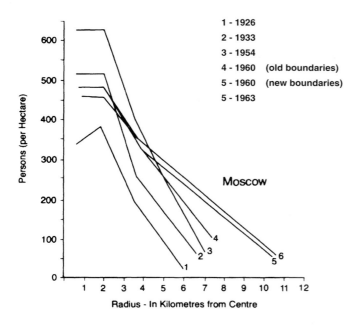

1 - 1926
2 - 1933
3 - 1954
4 - 1960   (old boundaries)
5 - 1960   (new boundaries)
5 - 1963

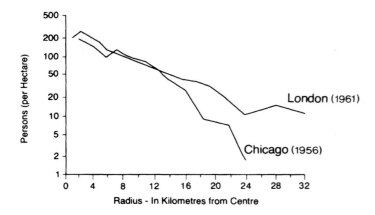

Figure 9.18 Population density gradients: Moscow, Chicago and London. The lines on the Moscow graph show a flattening of the density gradient over the period 1926–63, as the centre became less densely populated and the suburbs more so. Comparison of lines 3–6 with those given for London and Chicago (allowing for the differing scales of the two diagrams) shows much greater density in almost all areas of Moscow. The presence of two lines for 1960 in the Moscow diagram reflects the fact that in that year the city's boundaries were extended to the new orbital motorway (adapted from Gol'ts, in Bater, 1980, p.110, with the kind permission of Edward Arnold)

consumption were made in the post-Stalinist period, not least that of rising car ownership. In the post-1870 period as a whole, we find those Western innovations in public transport and building construction examined in Part 2 of this volume being applied to Russian cities experiencing common problems of rapid industrial and population growth. Even where architectural styles appear to have been ideologically specific, as in the monumental, neo-classical architecture favoured by Stalin, Hitler and Mussolini in the 1930s, it turns out that similar monumental styles were often adopted for contemporary Western public and commercial buildings (see Chapter 5). The nub of this study of urban technology transfer appears once again to be a tension between the generalized effects of diffusing technologies, and the shaping of those technologies by specific contextual conditions.

Readers at the start of the twenty-first century, of course, have the luxury of hindsight in knowing that the Soviet urban experiment would end in failure, and that Russian cities were to become exposed to the full glare of Western capitalist development. Ultimately it proved impossible for Soviet planning ideals to withstand international forces of private consumption and mobility. It is tempting to conclude that the Western technologies embraced by the Soviet regime proved its ultimate undoing, just as they helped undermine the old tsarist order. The reality is surely more complex than this. To the extent that technology is involved, the crucial consideration is perhaps the distortion of the Soviet economy by the political elite in an attempt to rival the United States as a nuclear superpower. As far as ordinary Russian citizens are concerned, it remains to be seen whether a Western-style urban landscape proves more congenial than a Soviet one.

References

ANDRUSZ, G.D. (1984) *Housing and Urban Development in the USSR*, London and Basingstoke, Macmillan.

BATER, J.H. (1973) 'The development of public transportation in St Petersburg, 1860–1914', *Journal of Transport History*, vol.2, no.2, pp.86–93.

BATER, J.H. (1976) *St Petersburg: industrialization and change*, London, Edward Arnold.

BATER, J.H. (1980) *The Soviet City: ideal and reality*, London, Edward Arnold.

BLACKWELL, W.L. (1994) *The Industrialization of Russia: an historical perspective*, 3rd edition, Arlington Heights, Harlan Davidson (first published 1970).

BRADLEY, J. (1985) *Muzhik and Muscovite: urbanization in late imperial Russia*, Berkeley, University of California Press.

BROWER, D.R. (1990) *The Russian City between Tradition and Modernity, 1850–1900*, Berkeley, University of California Press.

BRUMFIELD, W.C. (ed.) (1990) *Reshaping Russian Architecture: western technology, utopian dreams*, Washington DC and Cambridge, Woodrow Wilson Center Press and Cambridge University Press.

BRUMFIELD, W.C. (1993) *A History of Russian Architecture*, Cambridge, Cambridge University Press.

BRUMFIELD, W.C. and RUBLE, B.A. (eds) (1993) *Russian Housing in the Modern Age: design and social history*, Washington DC and Cambridge, Woodrow Wilson Center Press and Cambridge University Press.

COLTON, T.J. (1995) *Moscow: governing the socialist metropolis*, Cambridge, Mass., Belknap Press of Harvard University Press.

COOPERSMITH, J. (1992) *The Electrification of Russia, 1880–1926*, Ithaca, Cornell University Press.

FALKUS, M.E. (1972) *The Industrialization of Russia*, Basingstoke, Macmillan.

FRENCH, R.A. and HAMILTON, F.E.I. (1979) 'Is there a socialist city?' in R.A. French and F.E.I. Hamilton (eds) *The Socialist City: spatial structure and urban policy*, Chichester, John Wiley and Sons, pp.1–21.

FRENCH, R.A. (1984) 'Moscow: the socialist metropolis' in A. Sutcliffe (ed.) *Metropolis, 1890–1940*, London, Mansell, pp.355–79.

FRENCH, R.A. (1995) *Plans, Pragmatism and People: the legacy of Soviet planning for today's cities*, London, UCL Press.

GOL'TS, G.A. (1972) 'Vliyaniye transporta na prostranstvennoye razvitiye gorodov i aglomeratsii' in Yu. Pivovarov (ed.) *Problemy sovremennoy urbanizatsii*, Moscow, Statistika, pp.159–90.

GOODMAN, D. (ed.) (1999) *The European Cities and Technology Reader: industrial to post-industrial city*, London, Routledge.

GOODMAN, D. AND CHANT, C (eds) (1999) *European Cities and Technology: industrial to post-industrial city*, London, Routledge.

GRAVA, S. (1984) 'Urban transport in the Soviet Union' in H.W. Morton and R.C. Stuart (eds) *The Contemporary Soviet City*, Armonk, M.E. Sharpe, pp.180–201.

HAMM, M.F. (ed.) (1986) *The City in Late Imperial Russia*, Bloomington, Indiana University Press.

HERLIHY, P. (1986) *Odessa: a history, 1794–1914*, Cambridge, Mass., Harvard University Press.

HITTLE, J.M. (1979) *The Service City: state and townsmen in Russia, 1600–1800*, Cambridge, Mass., Harvard University Press.

HOPKINS, M.W. (1970) *Mass Media in the Soviet Union*, New York, Pegasus.

HUTCHINGS, R. (1976) *Soviet Science, Technology, Design: interaction and convergence*, Oxford, Oxford University Press.

KENEZ, P. (1985) *The Birth of the Propaganda State: Soviet methods of mass mobilization*, Cambridge, Cambridge University Press.

KENEZ, P. (1992) *Cinema and Soviet Society, 1917–1953*, Cambridge, Cambridge University Press.

KOTKIN, S. (1995) *Magnetic Mountain: Stalinism as a civilization*, Berkeley, University of California Press.

LENIN, V.I. (1966) *Collected Works: April–December 1920*, vol.31, Moscow, Progress Publishers.

LEWIN, M. (1985) *The Making of the Soviet System: essays in the social history of interwar Russia*, London, Methuen.

LEWIS, R. (1979) *Science and Industrialization in the USSR: industrial research and development 1917–1940*, Basingstoke, Macmillan.

MARKS, S.G. (1991) *Road to Power: the Trans-Siberian Railroad and the colonization of Asian Russia, 1850–1917*, London, I. B. Tauris/Cornell University Press.

MITCHELL, B.R. (1993) *International Historical Statistics: Europe 1750–1993*, London and Basingstoke/New York, Macmillan/Stockton Press.

PARKINS, F. (1953) *City Planning in Soviet Russia*, Chicago, University of Chicago Press.

RASSWEILER, A.D. (1988) *The Generation of Power: the history of Dneprostroi*, Oxford, Oxford University Press.

RICHARDSON, W. (1991) 'Hannes Meyer and the general plan for the reconstruction of Moscow, 1931–5', *Planning Perspectives*, vol.6, no.2, pp.109–24.

ROBBINS, M. (1997) 'London Underground and Moscow Metro', *Journal of Transport History*, vol.18, pp.44–53.

ROGGER, H. (1983) *Russia in the Age of Modernization and Revolution 1881–1917*, London, Longman.

RUBLE, B.A. (1990) *Leningrad: shaping a Soviet city*, Berkeley, University of California Press.

RUBLE, B.A. (1994) 'Failures of centralized Metropolitanism: inter-war Moscow and New York', *Planning Perspectives*, vol.9, no.4, pp.353–76.

SHAW, D.J.B. (1991) 'The past, present and future of the Soviet City Plan', *Planning Perspectives*, vol.6, no.2, pp.125–38.

SOLOVYOVA, A.M. (1984) 'The railway system in the mining area of southern Russia in the late nineteenth and early twentieth centuries', *Journal of Transport History*, vol.5, pp.66–81.

SUTTON, A.C. (1968–73) *Western Technology and Soviet Economic Development*, 3 vols., Stanford, California, Hoover Institution on War, Revolution and Peace.

VON LAUE, T.H. (1974) *Sergei Witte and the Industrialization of Russia*, New York, Atheneum (first published 1963).

Chapter 10: COLONIAL INDIA

by Michael Bartholomew

10.1 Introduction

Why does a chapter on Indian cities appear in a book apparently devoted to Europe? The simple answer is that, in decisive ways, the chief cities of India – Delhi, Bombay, Calcutta, Madras – *are* European cities. This is not to deny the vivid, unmistakably Indian feel of these places: plainly, the lives of their residents express powerful, enduring Indian cultural traditions. But the fact is that the lives of Indian city-dwellers have been shaped by urban fabrics that were devised by British soldiers, merchants and administrators, all of whom were

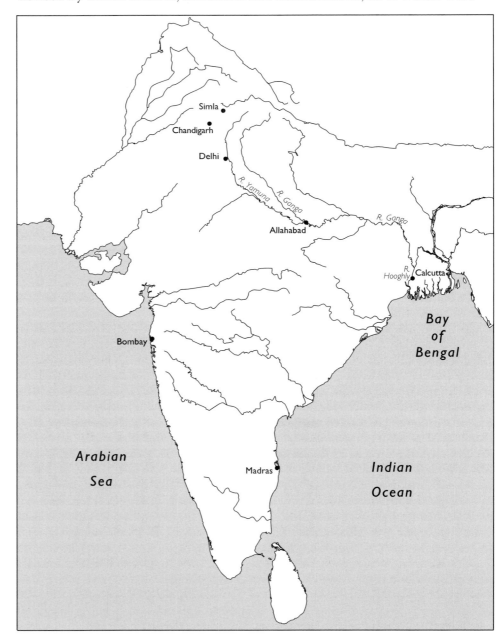

Figure 10.1 The locations discussed in this chapter. ('Delhi' indicates the location of both New Delhi and the old Shahjahanabad. Bombay is also known as Mumbai, and Madras as Chennai)

driven not by considerations of the steady development of indigenous Indian urban traditions, but by the twin overriding concerns of the imperial rule of the subcontinent, and their own personal safety, comfort and enrichment. In 1947, when India won its independence, an important part of the colonial legacy was a network of cities that had been devised according to British conceptions, for British purposes. One way of formulating the grim challenge of explosive urban growth presently facing Indian society – in some parts of the cities, people live at densities of c.150,000 inhabitants per square kilometre (adapted from Murphey, 1977, p.25) – is to say that Indians are having to cope with cities that they have inescapably inherited, but that they have had a very limited part in designing, and whose roots may not even have penetrated very deep into their culture.

Very general, unanswerable questions arise here. What would have been the 'natural' pattern of the subcontinent's urban development if there had been no British Raj? Were there inexorable demographic forces that – one way or another, Raj or no Raj – would have led to the emergence of giant cities, densely packed with people, most of whom would have been extremely poor? Traditionally, many cities in pre-colonial India had been sited at holy places, and had grown up to serve pilgrims. Rivers are especially important in Hinduism, and cities such as Varanasi (Benares) on the holy River Ganga (Ganges), and Allahabad (Prayag) at the confluence of the Ganga and the Yamuna (Jumna),[1] had long histories before the subcontinent was colonized by the British. Had there been no colonial intrusion, might the subcontinent have remained overwhelmingly rural, agricultural and inward-looking, with just a few small cities serving ceremonial and local administrative functions?

This last suggestion may seem bizarre. But it is significant that in 1931, when Mahatma Gandhi – fully aware of the potency of symbols – walked up the steps of the viceroy's imperial palace in New Delhi to begin his negotiations for India's independence, he had dressed himself not in the suit of an urbane lawyer but in the homespun cotton costume of the villager. He deplored what he saw as 'the waste of money on architectural piles'. The true values of India, he insisted, were embodied in her villages (Irving, 1981, p.351). We cannot put the clock back to a time before the British colonization of India and watch the unfolding of the alternative plot suggested by Gandhi's vision, but we might keep at the back of our minds the possibility that there may be some sort of misfit between the cultures of India and the cities left by the Raj – a possibility we would never entertain if we were considering the fit between Londoners, Parisians or New Yorkers and their cities.

The colonial imperative that governed the construction of Indian cities did not lead the British to adopt only one urban pattern. Indeed, there were great contrasts: Bombay, Calcutta and Madras were built for trade, whereas Delhi was built for government. Simla, a much smaller town, in the foothills of the Himalayas, was unique in having been built partly for the purposes of government (the administration of India shifted from Delhi to Simla during the hot season) and partly for pleasure – as a recreational resort for the British. Lastly, Chandigarh, a city commissioned as the regional capital of the Punjab by the newly independent Indian government in 1948, and intended to be a symbol of the new nation, none the less bears the heavy and, arguably, alien imprint of the North American and European Modernist movement. These diverse origins and original functions have left their stamp on the cities. But there are some common features of Indian colonial cities, and it may be useful to open them up here before we move to the study of particular places. Two preliminary points about the colonial history of India need to be made.

First, the colonization of India by Britain was not systematic. There was no grand government plan. British traders had been operating in India, along with the French, the Dutch and the Portuguese, since the early seventeenth century.

---

[1] The names in parentheses may be more familiar to some readers.

British interests were co-ordinated and administered by the East India Company, a trading company founded in 1599. Although the company existed solely for the purposes of trade, it steadily accumulated quasi-constitutional governmental, legal and military authority over the territories with which it traded. The British government eventually took a hand in the rule of the subcontinent, sending its naval ships and its armies. But not until 1857 – after an uprising known variously as The Indian Mutiny or The First War of Indian Independence – did the British government formally and finally end the East India Company's role. Twenty years later, in 1877, Queen Victoria was crowned Empress of India. In 1911 Delhi was proclaimed the capital of the Indian Empire, and an entirely new city – to be called 'New Delhi', and to be built next to the old city – was commissioned. Independence was won in 1947, when the Indian Empire was split into the two new states, India and Pakistan.

Second, as the colonial traders had moved deeper into the subcontinent during the seventeenth and eighteenth centuries, they were not moving into a single, pre-existing, unified state. No regime, until the British Raj, had ever controlled the whole subcontinent. There had been no single, pre-existing country called 'India'. Quite apart from European influences, the Indian subcontinent had been subject to invasions and regional and religious upheavals. The most significant, for our purposes, were the Muslim invasions from the north-west which had, from the thirteenth to the eighteenth centuries, spread Islamic influence deep into the subcontinent's indigenous Hindu regions, but which, even after 400 years, had led to little homogenization of the two religious cultures. An urban consequence of this was that Muslims and Hindus (and Sikhs and Buddhists) tended to live apart, even when their houses were separated only by an alley. The distinctive pattern of Indian urban building kept up significant barriers. And within Hinduism, the caste system further emphasized strict divisions in housing areas.

Most indigenous Indian housing is based on the *compound* – a plot of land on which the extended family lives.[2] The boundary of the compound is marked by a high wall, against which, on the inside, the accommodation is built, facing on to a central, private courtyard. To the passer-by in the street or alley, the only markers of the house will be a doorway, or perhaps a row of high, screened windows. This layout of alleys and high-walled compounds – especially when it departs from a grid pattern of streets, and branches crookedly into webs of narrower and narrower lanes and alleys, often terminating in dead-ends – can give to Western observers an impression of labyrinth[3] or even chaos. But the urban and architectural historian Spiro Kostof argues that such layouts of streets and alleys make sense:

> The labyrinthine medina proves to be quite rational after all. To cite Old Delhi as an example: the primary streets carry the bazaars with the large retail and wholesale outlets. Production, storage and service centers are immediately behind, set in the clearly defined residential neighborhoods or *mohallas* [sometimes transliterated as *mahallas*]. Secondary streets run as spines of commercial and residential activity throughout the *mohallas*, and can be closed off by doorways at their connections with the primary streets. Dead-ending tertiary streets, closed to general circulation, penetrate the cores of the *mohallas*. At the junction of two or more streets, modest expansions, called *chowks*, provide some breathing room in the crushing density of the town.
> (Kostof, 1991, p.63)

---

[2]   In this respect it is similar to the indigenous Chinese and African housing described in Chant and Goodman (1999), a companion volume in this series.

[3]   The labyrinthine nature of Muslim cities is discussed in Chapter 4 of Chant and Goodman (1999).

During the later nineteenth and the twentieth century, when British planners and sanitary reformers addressed some of the problems in the indigenous parts of Indian colonial cities, they rarely bothered to investigate the underlying logic of the settlements and, instead, drove straight roads into the middle of them.

At its best, this unit of housing (the compound), set in an intricate web of alleys within a densely packed city, offered (at least to the well-off and their dependants) peace and quiet, a measure of defensibility, and privacy – a particular requirement in a culture that included purdah for women. In an excellent book on the influence in India of Western conceptions of the city, on which I have drawn extensively in this chapter, Norma Evenson quotes a British physician's description, written in 1837, of wealthy Hindus' compound-houses in Calcutta. The houses, wrote the physician,

> are uniformly built in the form of a hollow square, with an area of from 50 to 100 feet [15–30 metres] each way, which on the occasion of Hindoo festivals, is covered over, and when well lighted up, looks very handsome. The house itself is seldom of more than two stories, the lower portion, on three sides of it, being used only for storerooms, or for domestics; on the remaining side, and that always the northern one, is to be found the Thakoor Ghur, or abode of the Hindu Gods … jutting out from this main building are situated the accommodations allotted to the females and family; they consist of smaller hollow squares, with petty verandas opening inwards, and some houses have two or three sets of these *zunnanahs*, with one or two tanks attached.
>
> (quoted in Evenson, 1989, p.67)

This convention in housing contrasts so sharply with the conventions of the colonizers that it is worth making a general point here. The traders of the East India Company were interested above all in making fortunes which they would ultimately spend back in Britain. They were not much interested in building in India permanent equivalents of the fine urban developments that were transforming cites such as Bath, Edinburgh, London and York during the eighteenth century. In Britain, and later in Haussmann's Paris, there developed an urban idiom of grand avenues, less grand side streets, squares and circuses of terraced housing, ranging from the monumental to the humble and basic. But in India, traders wanted individual *estates* of the sort that punctuated the British countryside. In the early years of the East India Company's settlements, employees and traders lived inside forts, built to defend the company's personnel against attacks both from Indians and from rival colonial powers. But as Britain's supremacy came to be acknowledged, residents of the forts ventured beyond the walls and set themselves up in greater style. Cheapness of land and labour in India meant that a trader could build an estate of a sort he could not have dreamed of, had he stayed at home.

These mini-estates were built within commuting distance of the warehouse, dockyard or mill managed by the trader. A consequence of this is that British sections of Indian towns tended to be characterized by miles of plots of pseudo-parkland, each with a house in the middle. Privacy, which was desired by the British as much as by the Indians, was achieved not by building around the perimeter of a small urban plot and facing inward, but by securing as large a plot as possible and hiding the house in its middle, shielded by a shrubbery and approached by a drive. This convention persisted into the twentieth century. When, as we shall see presently, New Delhi was laid out as the capital of India, the unit of housing for government staff, from the lowly clerk upwards, was not the apartment on a city centre boulevard, but the individual bungalow set in its own grounds. The sharp contrast between the British and indigenous parts of Indian cities is explicable largely by reference simply to wealth and power, but the cultural conventions governing house design have something to do with it too.

Technology transfer

The exploration of the explicitly technological aspects of Indian cities is difficult: little scholarly work has been done. There are, for example, plenty of studies of the planning and building, between 1911 and 1947, of New Delhi, the imperial capital; but information on the design and construction of the new city's water supply, or its drains, or its electric power, is hard to find. To take another, more general example: many writers have addressed the processes of urbanization in India, but – understandably enough, perhaps – they have tended to emphasize the formidable demographic and administrative aspects of cities, rather than the technological aspects. Furthermore, historians tend to be interested in change and innovation, and although both were in plentiful supply in Indian cities, there were no great innovations in urban technology: there is nothing in India equivalent to the invention of the railway, or the skyscraper. For all these reasons, the relationship between technology and the Indian city is often somewhat oblique. We find not radical invention, but an extraordinary mixture of traditional techniques (for instance, the woven basket on a woman labourer's head as the fundamental unit for the shifting of building materials), and the uneasy, problematic transfer of Western technology (for instance, the building of a highly expensive, extremely short, underground railway system in Calcutta, a city that tends to flood during the monsoon season, and whose chief problem is not commuting but homelessness; Evenson, 1989, pp.203–4).

We find also that the social contexts within which technology works are thrown into especial prominence. In Western cities, a set of technologies and distinctive technological ways of thinking – the unlimited supply of electrical power or safe drinking-water, and the growth of motor traffic – are so ingrained that they appear (or, at least, appeared until fairly recently) to be natural, inevitable, universal dimensions of urban life. But when they are transferred to other countries, and especially to developing countries, their taken-for-grantedness starts to look dubious. Consider, for instance, New Delhi – a city designed by an architect who despised Indian buildings and who turned for inspiration resolutely away from India to sources as diverse as Palladio, the English Garden City movement, and the civic centre of Washington DC. It is impossible to look at this city and conclude that the imperial transfer of conceptions of the city, and of the technological assumptions that underpinned them, was entirely appropriate. And even if we move to more recent, post-imperialist, decades, questions about the transfer of technologies from one culture to another are no less insistent. An editorial in an Indian architectural journal on the mania for building prestige high-rise buildings in Indian cities is illuminatingly quoted by Evenson. The editorial, written in 1984, asks:

> Do our social priorities and economic resources permit us to allow a handful of developers, in a nation of 700 million, to make fortunes and to use up colossal energy resources for these skyscrapers (air-conditioning, elevators, water pumps and generators) when millions of urban dwellers live in dark and dingy hovels, and ten times their number in our rural areas do not have electricity with which to run their tubewells? Are we prepared for the sake of these few developers and the new corporate-sector elites to allow India's oil import bills to keep mounting, or its other energy sources channelized towards these inappropriate, expensive, and impractical status symbols? If at present we are unable to supply even our single- and double-storied buildings with water and electricity, how will we supply 30-storied buildings and more with these facilities? … [Moreover, building services are often out of order. How would the occupants of such structures] survive in summer temperatures of 110° [43°C] and more during periods of power failures? How will they reach the top floors during power blackouts when the elevators do not work? [Could such buildings provide] all that is required for eventualities like firefighting and emergency evacuation?

(quoted in Evenson, 1989, pp.241–4)

Technology meets politics, meets culture, meets commerce. In these collisions, Indian people have usually been on the receiving end of Western political, military and technological solutions to urban problems. Even so, we should beware of lurching to the patronizing assumption that an authentically Indian solution would be bound to be limited to the application of only deeply traditional Indian technologies.

10.2   The early port cities: Bombay, Calcutta, Madras

Indigenous traditions of shipbuilding and seafaring, and of manufacturing and exporting textiles, predated the arrival of European traders. But these indigenous traditions had produced no large-scale port cities: India's cities tended to be inland, serving local needs. But from the eighteenth century onward, the East India Company's expansion of inter-continental, seaborne trade led to the rapid development of what had started out, in the mid-seventeenth century, as their trading posts. Bombay, on the west coast (see Figure 10.1), developed into a port specializing in the export of raw cotton. On the east coast Madras and Calcutta also grew significantly; they served both textiles and general imports and exports, with Calcutta also specializing in the export of jute, the vegetable fibre used for the manufacture of rough fabrics such as sacking, and for the backing of floor coverings such as linoleum. (Calcutta grew symbiotically with Dundee, the Scottish port city that specialized in the import and processing of jute.) These port cities reoriented India to seaward, and opened it up decisively to foreign influence.

Of the three ports, only Bombay had a natural harbour. The east coast, by contrast, offered no natural harbours and, in their absence, the siting of Madras and Calcutta seems to have been almost casual. In Madras, for example, the loading and unloading of merchantmen required a fleet of rowing boats to put out through the surf of the Indian Ocean for the hazardous business of the transhipment of cargoes and passengers to and from ships anchored in huge swells offshore (see Figure 10.2). The savage nature of the ocean defeated several attempts to build harbour walls out into the deep water, and it was not until 1895 that a successful harbour was constructed.

Figure 10.2 Madras beach in 1867, by William Simpson. A handsome set of port buildings had been constructed, but bales of goods were still having to be carried down to the water's edge and ferried in rowing boats to merchantmen out in the bay (reproduced from Evenson, 1989, p.3; Victoria Memorial, Calcutta)

Calcutta was sited not on the ocean but up the River Hooghly, a 'delta-stream' of the Ganga. The Ganga drains most of the north of India, spreading in a huge delta at the head of the Bay of Bengal. The siting of Calcutta was governed partly by considerations of how far ships of the draught used by the East India Company could sail upriver. Deltaic rivers such as the Hooghly are subject to silting, flooding and changes in course. Furthermore, steady increases in draught, as larger and larger ships were built, made deep-water wharves a necessity. Madras and Calcutta were founded by traders who barely looked beyond their immediate trading concerns, but the growth and momentum of trade determined that two permanent, large port cities would emerge – even though, in the case of Madras, it took the best part of 250 years to build a proper harbour.

On the west coast, the natural harbour at Bombay gave the city a clearer *raison d'être*. Furthermore, wider developments, both political and technological, stimulated the growth of the port in the nineteenth century. The Lancashire 'Cotton Famine' of 1861–5, when the American Civil War brought American cotton exports to a halt, gave a boost to Bombay's cotton trade. The Suez Canal, which opened in 1869, and the development of the steamship, together with the expansion of the British presence in the interior of India, especially in Delhi, further encouraged shipping and thus consolidated Bombay's position as the obvious port of entry to the subcontinent.

Although the existence and shape of these port cities were determined largely by the needs of the British traders, these traders did not invariably trample over Indian traditions. Shipbuilding and its associated urban features, such as docks, rope-walks, timberyards and slipways, were areas of activity in which traditional Indian techniques and craft specialists were readily employed. Indeed, as far back as the sixteenth century there had been interchanges in matters of ship design between Indian, Arab, Spanish and Portuguese mariners around the Indian coast. These interchanges became more formalized in the eighteenth century, when the East India Company, and eventually the Royal Navy, commissioned large ships from Indian shipyards. Indian timber and labour were cheaper than British, but this was by no means the only consideration. Indian teak was superior to British oak for some purposes, and an Indian method of jointing the edges of the planks that formed the ship's sides offered greater watertightness than European methods, and was thus readily accepted by British owners when they commissioned ships from Indian yards. Ships were built at yards all around the Indian coast and on the River Hooghly near Calcutta, by Indian shipwrights working initially under close British supervision but eventually with considerable autonomy. For our purposes, it is particularly interesting to consider the yards at Bombay, for they were in the hands of the Indian Wadia family which, for nearly a century and a half from 1740, oversaw shipbuilding and extensions to the port facilities, including the building of a large dry dock (Pacey, 1990, pp.66–8, 123–9; Sangwan, 1995; Wadia, 1957).

What did the cities themselves look like? The sharp distinction between British and indigenous Indian areas of the cities, and the presence of army garrisons, made these colonial cities seem very un-British. But in their patterns of development they were in many ways similar to the cities in Britain that were being transformed by the Industrial Revolution. That is to say, they developed rapidly, with scarcely any municipal regulation, attracting large numbers of migrants from the countryside to work in the docks and the mills that sprang up as India's own textile industry mechanized and urbanized.

Calcutta, with its substantial jute trade, is a good example of a city that industrialized. The upsurge in world trade in the eighteenth and nineteenth centuries led to a demand for jute sacks; and by 1850, nine million, woven on hand-looms, were exported from Calcutta each year. By that year, however, manufacture was passing to Dundee in Scotland, where steam-powered mills were being built for the processing of raw jute shipped from Calcutta. But

Figure 10.3 St Paul's Cathedral, Calcutta, designed by Major W.M. Forbes, built 1847 (photograph: Dinodia/Images of India)

Figure 10.4 Bombay University library, designed by G.G. Scott, built 1869–78 (photograph: Dinodia/V.H. Mishra/Images of India)

eventually the trade passed back to Calcutta as European merchants set up mills in the city. Raw material was close at hand, labour was cheap, and hours and working conditions were unregulated. By the early 1880s Calcutta was exporting 53 million sacks per year. By 1914 – by which time indigenous Indians, against stiff opposition from Europeans, were beginning to buy into the jute industry – there were fifty mills along the River Hooghly, extending about fifty kilometres north and forty kilometres south of the city (Sethia, 1996).

As in Britain, Indian cities became separated into industrial areas, areas of dense slums and areas of spacious, leafy suburbs. And, again as in any Victorian British industrial city, up went a variety of imposing civic and commercial buildings – churches, hotels, banks, libraries, town halls, railway stations. Indeed some of the building types were transferred directly from Britain to India, with regard neither to climate nor culture. Calcutta Cathedral and the library of Bombay University shown in Figures 10.3 and 10.4 are Victorian Gothic Revival buildings, straight from Britain.

One difference, though, is that in India buildings were often designed by military engineers, who thereby invaded what, back in Britain, would have been the jealously guarded professional province of the architect. Bombay University was designed by the highly acclaimed British architect George Gilbert Scott, designer of London's Albert Memorial and St Pancras Station. But Calcutta Cathedral in Figure 10.3 was designed by Major W.M. Forbes of the Bengal Engineers. Army engineers, trained to build fortifications and bridges, turned their hand to civilian buildings, using pattern-books of architectural style and ornament – a practice that sometimes led to clashes over the status of the

Figure 10.5 The Municipal Corporation Building, Bombay, built 1893 (photograph: Dinodia/Images of India)

professional architect. Not all large civic buildings, however, were uncompromising transpositions of British styles. Attempts were made to fuse European and Indian traditions, but the results often came out as unmistakably European buildings with a few gestural Indian motifs: the Bombay Municipal Corporation Building of 1893 (see Figure 10.5) is a good example. A more balanced fusion of styles was not easily achieved, given the one-sided, European function of the buildings: what, for instance, would an authentically Indian railway terminus look like?

Turning from grand civic buildings to the generally wretched houses in what the British airily designated the 'native quarters', the parallel with conditions in British cities holds, although the scale of poverty in India was much more formidable. But it is not fanciful to compare, say, the slums of Bombay with the slums of Manchester or Liverpool or Glasgow. Indeed, if we make the comparison with the areas of British cities in which Irish immigrants lived, a similar sense of uncomprehending foreignness between the well-to-do and the slum-dweller is apparent. Many of the people living in the overcrowded areas were single men whose aim was to send wages, earned in the mills and docks, back to their families in the rural heart of India. To exploit the needs of these men, landlords of properties (like landlords the world over) subdivided existing houses into smaller and smaller units, added lean-to extensions that covered what little space there was for circulation around the buildings, and erected further, rickety, timber-framed storeys on top of the existing buildings. Collapses, fires and epidemics were common. In Bombay, this dense pattern of housing led to the emergence of a distinctive tenement-type apartment building known as the 'chawl'. Chawls, like any building type, range from the decrepit to the sound and well maintained (see Figure 10.6).

Again, as in Britain, municipal improvement in Indian cities was slow in coming; and when improvement commissions and trusts were formed, they faced problems of the type familiar to municipal reformers in Britain, though much more fearsome in scale. But the comparison between Britain and India is still worth making, and has been illuminatingly used by the historian J.B. Harrison in a study of the sanitary history of Allahabad, a city roughly half-way between Delhi and Calcutta (see Figure 10.1). Harrison (1980) describes the Indian technological and cultural solutions to the universal problems of water supply and sewage disposal, and he makes the point that sanitary reform became a priority for colonial administrators – not, typically, out of altruism but because of the potential threat to the health of British troops stationed in the town. But he sets his study firmly in the context of the theory and practice of sanitary reform as it was developing in Britain, in the hands of reformers such as Edwin Chadwick. As a case-study of the sorts of challenge facing civic improvers in all Indian colonial cities, Harrison's work is of great interest.[4]

The parallels between Indian and British cities break down when we consider climate. First, the fierceness of the Indian climate ravaged the health of the British. In the middle of the nineteenth century, during peacetime, troops in Britain died

4   Harrison's article is reprinted in Goodman (1999), the Reader accompanying this volume.

Figure 10.6 Façade of a Bombay chawl, showing the highly decorated verandahs of the apartments (photograph: Dinodia/V.H. Mishra/Images of India)

at the rate of seventeen per 1,000. In India, the rate for British troops was fifty-eight per 1,000. The life expectancy of a peacetime soldier in Britain was fifty-nine years; in India it was thirty-eight (King, 1976, p.103). Second, and despite a good deal of stubborn transfer of unmodified British building types, the fierceness of the climate did lead to a number of minor adaptive responses in the design of buildings – adaptations drawn largely from Indian technology and building traditions.

Let us consider health. The three port cities under consideration here were all built in low-lying, swampy places, where fevers were endemic. It will be recalled from Chapter 3 of this volume that, in accounting for fevers, the prevailing medical theory during the nineteenth century often referred to miasmas – emanations from swamps and rotting vegetation. The term 'malaria' is derived from the Italian for 'bad air'. Not until 1897 did a doctor – significantly, a doctor serving with the army in India – demonstrate the role of the mosquito in transmitting the disease. Until then, the preventive measures taken against the fever were directed at improving the atmosphere, for some sort of connection between fever and damp, tropical environments had long been recognized. James Lind, an eighteenth-century pioneer of what we would now call tropical medicine, recommended in his *Essay on Diseases incidental to Europeans in Hot Climates* that indigenous labourers

> might be usefully employed in clearing the ground, draining the swamps, and either in burning down the woods and shrubs, or at least in opening avenues through them for purifying the air.
>
> (Lind, 1768, p.151)

Further, he recommended that Europeans should, wherever possible, live on headlands exposed to sea breezes, up and away from any stagnant water: 'the sea air', he wrote, 'affords a certain asylum in all hot and unhealthy countries' (p.153).

By the nineteenth century, Lind's common-sense recommendations were widely practised; but they were difficult to implement in Indian port cities. Bombay, built on a group of low-lying islands that had been bridged together, did have a breezy headland, Malabar Hill, and it became a fashionable suburb during the nineteenth century. But residents of Madras and Calcutta (both cities low-lying) had to make the best of the fever-ridden locations. Furthermore, the implications of a theory that associated fever with abundant, damp vegetation clashed with the desire to shade buildings from the fierce sun with trees and shrubs. An unhappy compromise had to be struck between the need for health and the need for shade. In his book on colonial cities in India, Anthony King describes elaborate regulations drawn up by the army to guarantee a prescribed volume of air space per soldier so as to maximize the free flow of whatever breeze might be blowing. Military cantonments were, wherever possible, built upwind of Indian towns, at a presumed safe distance from them. The free circulation of air through buildings, and therefore the cutting down of adjacent trees and shrubs, was a consideration that the military engineers who built them had to take into account (King, 1976, pp.108–14).

Turning now to the design of buildings, we certainly do not find much absorption by the British of Indian ways of doing things. By and large, the British sought to reproduce the buildings and fittings they had left behind at home, right down to the overstuffed Victorian furniture, now packed wall to wall in drawing rooms heated by the tropical sun to temperatures never contemplated in Surrey. But here and there designers acknowledged the efficacy of Indian technology, albeit in very restricted areas of building design. For example, an Indian device for improving ventilation was adopted by the British. Indian rooms had been cooled and mildly perfumed by the filling of window openings with frames filled with a local grass, cuscus. The cuscus was soaked with water, and the breeze – in passing through the screen (called a 'tatty') – evaporated the water and blew slightly cooler air, scented by the sweet-smelling grass, into the room.

Another device designed to stir the air in hot rooms (though incapable of reducing the room's temperature) was the punkah, an apparatus of Indian origin in whose technical development East met West (see Figure 10.7). Evenson gives the following account of this low-tech contribution to the comfort of occupants of oppressively hot rooms:

> Internal comfort in British building was heavily dependent on a swinging fan called a *punkah*. Although its origins are unclear, it appears to have been in use in Calcutta by the 1780s, and from there it spread to Madras and Bombay. In form it comprised a light wooden frame about fifteen feet long and four feet wide [4.6 x 1.2 metres], over which was stretched cloth or canvas, with a loose fringe or border on the lower edge. Suspended from the ceiling, it was swung by means of ropes and pulleys. Often the pulling rope passed through a wall aperture to a veranda where the *punkah-wallah* sat moving the rope up and down with his hands or, occasionally, lay on his back, moving it with his feet.
>
> After giving the matter serious study in the 1870s, British military engineers projected optimum dimensions for the punkah. The frame was to be twelve to eighteen inches wide [30–45 centimetres], with a heavy fringe eighteen to twenty-four inches deep. If several punkahs were used together, they needed to be rigidly connected so as to swing evenly, and for maximum effect should move through an arc of five feet with a velocity of two-and-one-half feet per second [76 centimetres/ second]. In spite of advancing technology, it was concluded 'that no machine had been brought to their notice equally effective and economical with a man's arm'. Although people unaccustomed to punkahs sometimes complained that the device gave them headaches, most of the British found them indispensable. It was observed that 'there is a *punkah* over the sleeper in bed; over the preacher … in the pulpit; over the party at dinner, whether on land or sea; over every man, woman, or child who wishes to breathe with any degree of ease'.

(Evenson, 1989, p.53)

Figure 10.7 Indian Victorian interior, with punkah (reproduced from Evenson, 1989, p.53)

Looking ahead a little, to the period when buildings could be designed on the assumption of a regular electricity supply, ceilings could be lower than the one shown in Figure 10.7, for the electric ceiling fan required less air space in which to operate.

India had powerful traditions of building with stone – as the Taj Mahal triumphantly testifies – and these traditions were pressed into service by the British, notably in New Delhi, but also in some of the grand civic buildings in the port cities. But often buildings were of brick, faced with stucco of various sorts. In Calcutta and Madras, a stucco called *chunam*, made from ground-up sea shells, was used. When applied to brick columns, for example, it could be burnished to resemble marble. But utter stability and permanence in buildings was difficult to achieve. The extremes of temperature, the instability of foundations in swampy ground, the attacks of termites on structural timber – all played havoc with grand schemes. Stucco flaked, walls cracked, joists rotted.

As the nineteenth century ended, Bombay, Madras and Calcutta were fully established industrial cities, with a range of industrial enterprises and civic amenities equivalent to those of cities in Britain. The social structures were different, though, and left their marks on the overall shape of the cities. Madras, for example, where land was cheap and where there were no physical barriers to expansion, developed a tripartite urban fabric. First, there were European traders who had moved outwards from the old fort near the harbour, in order to build low-density, individual bungalows set in large gardens. (The word 'bungalow' is Hindi in origin, meaning 'belonging to Bengal'.) Second, there were small agricultural villages which had been engulfed by the suburban expansion but which continued much as before. And third, there were densely packed urban areas of the destitute, of small traders and craft specialists, and of labourers whose jobs were in the city but whose roots were back in the rural regions from which they had migrated and to which they returned for rites of passage such as weddings, funerals and religious festivals.

The Indian context of these cities had plainly contributed to their character. A city will always have a distinctive character if, on the one hand, most of its most powerful citizens owe their chief allegiance to a home country thousands of miles away while, on the other hand, huge populations of subject, sometimes destitute people are sucked into the colonial cities to live lives that are in many ways alien to them. The cow is sacred to the Hindus and wanders freely in the countryside, causing no great problems. It is another matter, however, when the cow wanders freely in the tight alleys of a city, or ambles across an urban arterial road.

The pioneering British town-planner Patrick Geddes worked intermittently in India between 1914 and 1924, and taught for a while at Bombay University. He was highly unusual in valuing aspects of the congested Indian areas of cities, and recommended a more sensitive approach to their renovation than the practice of driving brutally straight roads into the middle of them – or, worse, simply knocking them down altogether. (While working at Edinburgh University he had chosen to move from a flat in the elegant late eighteenth-century New Town into a crowded, run-down tenement in the old town – good training for understanding the pains and pleasures of life in Bombay.) As a founder of the British Garden City movement, he sought to bring its ideas to bear on Indian cities. He wanted to unite Indian and British conceptions of the city:

> On the one hand the planner must strive to maintain the populous and gregarious nature of Indian life in village and town and yet abate its congestion and, at the same time, to lead more dwellings into garden villages without [outside, beyond] the town and provide more civil developments within. On the other hand, he must mitigate the Crusoe-like individualism[5] of the scattered and formless bungalow compounds and endeavour to build them up into coherent communities.
>
> (quoted in Gupta, 1986, p.250)

We should now turn to Delhi where – in the juxtaposition of the old, Mogul city and the new, British imperial city – we find scant recognition of Geddes' vision.

10.3   Delhi and New Delhi

The port cities Bombay, Calcutta and Madras were built virtually from scratch. By contrast, there had for centuries been cities in the Delhi area, which was 1,300 kilometres inland from both Bombay and Calcutta, on the western bank of the River Yamuna (Jumna). The city that preceded the arrival of the British was the beautiful Shahjahanabad, 'the mansion of Shahjahan', built by the Moguls. Shahjahan was the Muslim ruler who had founded his city in the mid-seventeenth century. Nineteenth- and twentieth-century images of Indian cities are so dominated by the British colonial presence, and by the poverty of the Indians and their settlements, that it is instructive to look at Shahjahanabad as it was at its peak, early in the eighteenth century, before the British presence was felt (see Figure 10.8).

Shahjahan's architects had laid out a planned, walled city with two focuses. At the eastern end, bordering the river, was the imperial palace complex, with its own moat and an open space separating it from the main city. In the centre of the city, on a low natural mound, was built one of the largest mosques in India, the Jama Masjid. A grid of streets had been established, but this grid had been modified, as *mahallahs* – neighbourhoods – had developed. The River Yamuna was subject to violent fluctuations in level, and tended to change course within its

---

[5]   Robinson Crusoe, the shipwrecked hero of Daniel Defoe's book of 1719, engaged in an energetic programme of building for himself a fortified, comfortable home on the island. This has been taken as a symbol of a primary impulse within Western individualistic capitalism.

bed. Accordingly, a steady and reliable source of water was needed for the
new city. Engineers responded by constructing a canal which branched off
from the river about 120 kilometres upstream and which, as it was led across
the dusty plain towards the new city, irrigated farms and the gardens of nobles
who had built estates beyond the city walls.

Muslim conceptions of both the house and the city place great stress on
formal gardens and courtyards with pools and fountains at their centres.
Shahjahanabad was well supplied with running water. When the canal reached
the city walls, it branched into a number of channels: some carried water to the
gardens of nobles who lived within the walls, some replenished the city's
wells, and one was led formally down the centre of the city's main
thoroughfare, the Chandni Chowk, in a marble-lined bed. The historian
Stephen Blake, who has described Shahjahanabad in its heyday, quotes an
eighteenth-century visitor who was clearly enraptured by the way in which the
city's engineers had brought water to supply both the decorative and sanitary
needs of the city. The visitor wrote:

> [The canal] brought greenness to Delhi. It ran in all of the city from lane to lane,
> and the wells became full from it … it flowed into the imperial fort and around
> the moat … having flowed to the mansions of the princes and amirs it flowed
> into the city – to Chandni Chawk [Chowk], to the Chawk of Saadullah Khan, to
> Paharganj, to Ajmiri Gate, to the grazing places, to the other *mahallas* and to all
> the lanes and bazaars of the city.
>
> (quoted in Blake, 1986, p.164)

The imperial palace was of extraordinary splendour, built from marble and the
local red sandstone (the ruins are now known as the 'Red Fort'). The city itself
included not only the mansions of the rich, but caravanserai (compounds for
accommodating visiting traders and their pack-animals) and many small
mosques and craft workshops. It also contained shops, some of which sold
up-market luxury goods – eyeglasses from China, for example. In a piece of

Figure 10.8 The walled Mogul city of Shahjahanabad (Delhi), viewed from an imagined aerial point to the east of the river; by A. Maclure, c.1857. In the centre is the great mosque, Jama Masjid. To the right of it is the central thoroughfare, the Chandni Chowk, down the middle of which runs a tree-fringed canal, which can be seen entering the city from the distance. The emperor's palace is by the river, and the British military lines are up on the ridge, well to the right of the walled city (reproduced by permission of the British Library)

self-advertisement, lines of poetry were inscribed around the arches of one of the buildings within the palace complex:

> If there is a paradise on the face of the earth,
> It is here, it is here, it is here.
>
> (quoted in Blake, 1986, p.174)

Lest we get too carried away, it is necessary to point out that the city contained not only stone and marble, but also areas of cramped mud-and-thatch huts that tended to be washed away in the monsoon. Not all was sweetness and light. But Shahjahanabad does give us an example of a coherent, designed city functioning within a powerful urban tradition, long before the British presence was felt in the middle of the subcontinent.

The Mogul empire, of which Shahjahanabad was the capital, declined as the eighteenth century wore on. By the beginning of the nineteenth century (in 1803), the ruler – the aged, blind Shah Alam – called in the British army to protect him and the remnants of his empire. The British moved in and were there for nearly 150 years, ousted during that period only once, briefly, by the Mutiny/War of Independence of 1857. The pretext that the British were protecting the fading Mogul emperor wore thinner and thinner, and in 1877 a great durba, or assembly, was arranged to publicize the proclamation of Queen Victoria as the empress of the whole subcontinent of India. In 1911 the British capital of India was moved from Calcutta to Delhi, and a completely new imperial capital city was commissioned.

Building New Delhi

The focus of this chapter shifts now to the design and building of New Delhi, but the fate of Shahjahanabad needs comment. The gracious city designed by Shahjahan's architects was very badly knocked about by the British. As Figure 10.8 shows, there was initially a decent distance between the old city and the British military lines, a distance established not out of regard for the beauty of the city, but by army health and safety regulations. The fate of Shahjahanabad has been described by Samuel Noe (1986). In brief, first there were substantial changes to the city inside the walls. After the 1857 uprising, the British occupied the imperial palace area inside the Red Fort and demolished most of the buildings, including royal pavilions. Then they cleared a defensive free-fire area between the fort and the city, demolishing further palaces in the process.

Making just as great an impact on the city as the military, came the railway, demolishing one of the city's old gates and a segment of the palace complex itself as it was laid out from west to east across the northern edge of the city, crossing the river at a new bridge close to the palace. The western edge of the city was hemmed in by further rail developments. Then a cordon sanitaire was cleared next to the walls along the southern boundary of the city to separate it from New Delhi, which was laid out to the south with hardly any organic links to the old city. The result of all these developments was that the city, which was bounded on its eastern side by the river, became hemmed in and compressed on its other three sides (see Figures 10.9 and 10.10).

As a consequence of these cramping physical developments, the elegant old city of Shahjahanabad suffered chronic

Figure 10.9 A small section of the area shown in Figure 10.10: this aerial photograph reveals a striking contrast between the dense *mahallas* of the lower half of the old city and the spacious avenues running south towards New Delhi's Connaught Place. Keeping the two areas strictly separated is a cleared area, or cordon sanitaire, alongside the old city wall (reproduced from Tyrwhitt, 1947; photograph: Indian Air Surveys)

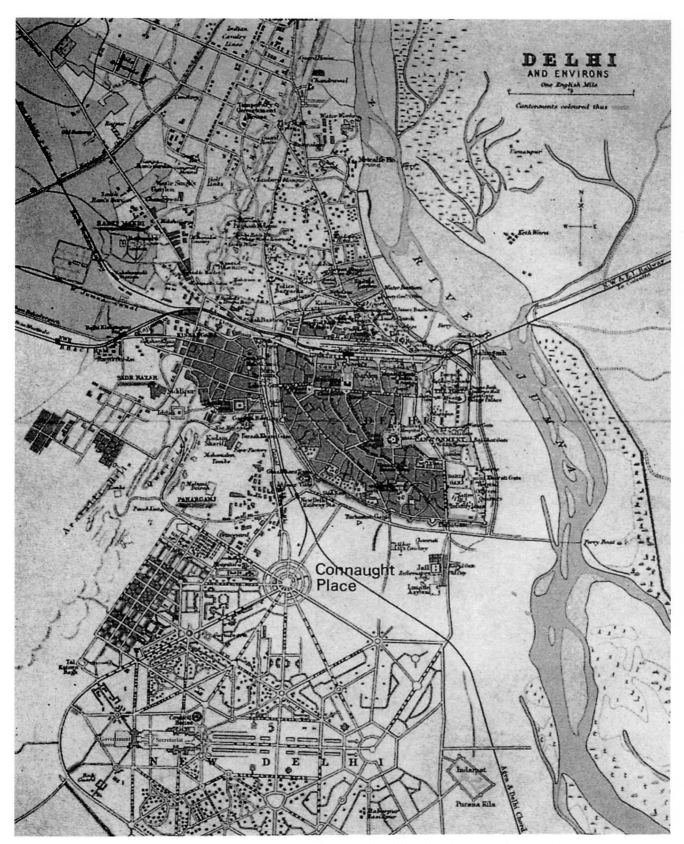

Figure 10.10 The old city of Shahjahanabad, in the centre of the map, and New Delhi to the south. Shahjahanabad has been hemmed in against the river by the railway lines running east–west and north–south. The radical disjunction between the old city and the spacious rectilinear streets of New Delhi is clear (reproduced by permission of the British Library)

problems of overcrowding. These problems were made worse when colonial India was partitioned in 1947: thousands of Hindu refugees migrated to the Delhi area.

The purpose of the new city of New Delhi, in the words of the planning committee which had been set up by the viceroy, and which reported in 1913, would be to 'convey the idea of peaceful domination and dignified rule over the traditions and life of India by the British Raj' (quoted in Irving, 1981, p.52). The words were carefully chosen. The new city was to express British domination, but it was to be peaceful and dignified: it was not to look like a garrison. Nor should it look like an indigenous Indian city: Indian traditions were to be ruled over, not harmonized with.

There was some dispute over this last point. The viceroy, Charles Hardinge, admired India's own monumental buildings (the Taj Mahal is only about 200 kilometres from Delhi), and favoured some sort of blending of styles, and an acknowledgement of Indian urban traditions. But Edwin Lutyens, the architect appointed to oversee the layout of the city and to design the viceroy's palace, would have none of it: he despised Indian architecture, dismissing it as veneers of highly ornamented surfaces stuck on to crude substructures. Not even the Taj Mahal was exempt from his criticism. Lutyens was perfectly in tune with the king's private secretary who, expressing the view of the king himself, wrote that 'we must now let him [the Indian] see for the first time, the power of Western science, art, and civilization' (quoted in Irving, 1981, p.73).

Figure 10.11 India Gate, New Delhi, nearing completion, 1920. This arch, designed by Edwin Lutyens, is also a memorial to the dead of the First World War. It stands at the eastern end of the long ceremonial axis of the King's Way (Rajpath) (reproduced from Irving, 1981, p.260; by permission of Mrs Marjorie Cartwright Shoosmith)

Note that in this particular formulation, the chauvinistic British tradition is transcended in favour of a notion of a 'Western' tradition. Lutyens was not at all interested in imposing specifically British architecture on India; there were to be no Tudor or English Gothic motifs in his buildings. He loathed the British Gothic buildings of Bombay as much as he loathed the indigenous Indian ones. He and his fellow architect on the New Delhi project, Herbert Baker, saw themselves as the inheritors of a *European* tradition running from the Greeks, through the Romans, on through the Italian Renaissance (especially in the work of Palladio) and through to the buildings of Wren. This, they argued, had emerged – refined, and with thoroughly understood principles – as a truly universal architecture that could be applied to any site in the world where monumental civic buildings were required.

Building in the new city started in 1914 and continued right up to Indian independence in 1947. The overall conception was of a city with imposing civic buildings, grand vistas, open spaces, wide streets, monuments. Figure 10.10 gives an indication of the ground-plan: near the bottom of the figure, running east–west and perfectly straight, is the grand ceremonial route, King's Way (Rajpath), nearly three kilometres long. It leads from a ceremonial arch (India Gate: see Figure 10.11), through formal parkland, slightly uphill, to Raisina Hill. Upon this small eminence at the western end stand the buildings that are the focus of the ceremonial King's Way –

From India Gate and Raisina Hill radiates a formal pattern of streets, defining the residential and shopping areas of the city. Connaught Circus (or Connaught Place), an arcaded circle of prestigious shops, lies to the north of Raisina Hill, towards the old city of Shahjahanabad, but keeping a respectable distance from it. New Delhi is urban planning in the Grand Manner.

the government secretariat buildings, and Viceroy's House. (It is called a 'house' even though it was the grandest of palaces, covering more ground than the palace of Versailles; on Figure 10.10 it is marked as Government House.) Figure 10.12 gives the view along King's Way towards Viceroy's House.

Figure 10.12 View from the top of the India Gate, westwards along King's Way toward Viceroy's House and the government secretariat buildings, 1945. The parade shown here marked the end of the Second World War (photograph: by permission of the British Library)

Although the layout was designed chiefly to impress, it is significant that the efficient functioning of the new city depended on developments in communications technology that had taken place in the early twentieth century. Distances within the new city, unlike the old, were too great to be covered routinely on foot. The bicycle, which had reached a high level of safety and reliability in the 1890s, and which was mass-produced in large quantities in Europe by the early twentieth century, became a standard form of urban transport for British and, eventually, Indian government clerks, who pedalled between bungalow and office. The telephone, linking government offices, meant that the offices themselves could be further apart. And lastly the motor car, slowly replacing the carriages that had been an essential part of the paraphernalia of British colonial social life, made rituals such as the trip to the fashionable shops in Connaught Circus more convenient.

The climax of the urban drama in New Delhi is the final approach up King's Way to Viceroy's House. This climax was botched. The small eminence of

Raisina Hill had to accommodate not just the palace, but two long secretariat buildings, flanking the approach, and designed by Lutyens' associate, Baker. The two architects fell out over the relative positions of palace and secretariat, and the result is that, as the grand ceremonial approach is made up King's Way, instead of the palace's appearing progressively larger, the bulk of the domed building dips temporarily below the horizon as the gradient of the avenue steepens, to reappear only when the final approach is made. As a demonstration of the superiority of Western urban design, this bit of the management of the principal architectural vista was a failure.

But the chief civic buildings themselves are undoubtedly impressive, even though tinged with a faint outlandishness. Their chief and visible concession to the climate is their use of deep, outwardly projecting stone mouldings designed to shade the windows below from the sun (see Figure 10.13).

The logistics of building the city were formidable. To supply the stone, the largest quarry in the world, employing 2,500 masons, was created. Temporary

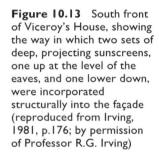

Figure 10.13 South front of Viceroy's House, showing the way in which two sets of deep, projecting sunscreens, one up at the level of the eaves, and one lower down, were incorporated structurally into the façade (reproduced from Irving, 1981, p.176; by permission of Professor R.G. Irving)

Figure 10.14 Brickfield; women labourers were also ubiquitous on the building sites (reproduced from Irving, 1981, p.334; by permission of Mrs Marjorie Cartwright Shoosmith)

railway lines were laid to transport the materials. Vast brickfields were opened (see Figure 10.14).

On the site of Viceroy's House itself, Lutyens ruled with a rod of iron, exercising an industrial discipline over the masons. Evenson (1989, p.108) records that 'He often recounted an incident of an Indian mason who tried to alter a template to fit a badly cut stone, and declared that no one could achieve good work in India "without the power of life and death" over the workers'. Details of the actual techniques used by the builders are hard to come by, but Robert Grant Irving, in his book on New Delhi, gives glimpses. Here, for instance, is his brief account of how the shallow saucer domes (designed by the British architects) were built by the Indian bricklayers:

> The plastered brick saucer domes … were constructed without temporary support or centering by a method ancient in Asia and known to the Byzantines. The Indian contractors' foremen would in each case organize a tight ring of laborers at the lip of the circular void they planned to close. Fresh mortar was already laid along the rim. At a given signal – often the beat of a drum which punctuated accompanying music! – every man placed a brick from each hand simultaneously, creating an instant circle. When the mortar dried, the workers repeated the procedure. So they raised a dome, course by course, until it stood entire, its last bricks held in place by callused fingers till the mortar set.
>
> (Irving, 1981, p.270)

The old Mogul city of Shahjahanabad, with its walls, courtyards and fountains, had been a traditional Asian response to the climate of an extremely hot and dusty plain. But the huge, shadeless open spaces and grand avenues of New Delhi defiantly disregarded the climate. Not until the viceroy's private gardens or the suburban streets of shaded bungalows are reached, do we find a more sensitive response to the fierce heat. In the suburban bungalows, scrupulously graduated in size and pretension precisely to fit the status of the thousands of government officers and clerks who ran the empire, and for whom they were built, the influence of the British Garden City movement of the early years of the century is felt. Surprisingly, though, it was never intended that residents of New Delhi – or at least their upper echelons – needed a city capable of making life tolerable during the summer, for the convention was for government to move wholesale on 1 May to Simla, a town in the cool foothills of the Himalayas, and to stay there until 1 October – by which time the heat of the plains had abated somewhat.

10.4 Simla

The contrast between imperial New Delhi, and cosy suburban Simla, is as sharp as can be. Simla is situated at over 2,000 metres in the mountains that terminate the plain upon which Delhi stands. It started, in the early nineteenth century, as a resort town, and it kept something of its feeling of ponderous Victorian gaiety even when it grew to accommodate the viceroy, who arrived each spring with his large staff to carry on the business of the government of the Indian Empire. Simla appealed to the British because it presented a site where they could build for themselves the sorts of house they associated with the Scottish Highlands, the Surrey hills or Switzerland. The result might be a mock-Tudor Swiss chalet with a corrugated iron roof (imported from Britain), adorned on the inside with the stuffed heads of beasts of the chase, in the manner of a Highland hunting lodge. Seasonal torrential rains and unstable bedrock meant that in the early days of the town, numbers of villas built into the hillsides ended up washed into ravines, but steadily a British provincial town was imposed on the mountainsides. The public buildings simulated those of a small British town – a Gothic church, a small racecourse, a theatre for amateur performances of Gilbert and Sullivan. As in other towns and cities in British India, the social fabric depended on armies of local servants. A special version of the rickshaw – adapted to the steep gradients in the town, and requiring four men rather than the single man customary in level cities – was developed for the personal transport of the mem-sahibs. Squads of garden servants kept the monkeys out of the precious rose gardens. Successions of bullock carts hauled all the appurtenances of British social and administrative life up the mountain track from the plain.

Figure 10.15 Viaduct on the narrow-gauge railway line connecting Simla, at c.2,000 metres, to Kalka on the plains (photograph: by permission of the British Library)

Initially the accommodation, for the viceroy and the government officers who ran India during the summer from this hill station, was improvised. Furthermore, since the initial purpose of the town was pleasure and recuperation, there was always a tension between the demands of government and the demands of pleasure-seeking. In order to get some government business done in peace and quiet, one of the viceroys, Lord Curzon, set up a tented camp even further into the mountains and sent his dispatches back and forth to the empire and to London, either by bearers or, at night, by flashing lights and, during the day, by a heliograph – a flashing signalling device using the sun's rays. By the 1880s the pressing needs of government were acknowledged. Official secretariat buildings were constructed, and in 1888 a new Viceregal Lodge, a vast mock-Renaissance pile set at the very summit of a hill, was opened. And finally, in 1903, Simla – this improbable site for the rule of empire – was connected to the rail network of plains India; a narrow-gauge line, built across terrain susceptible to landslips, and threading through more than one hundred tunnels, was opened. Figure 10.15 shows one of the viaducts across which the line passed.

10.5   *Chandigarh*

The Modernist movement in architecture, coming out of Germany, France and the USA in the early twentieth century, did not have much impact in British India. But immediately following India's independence in 1947, Nehru, the first Indian prime minister, put his weight behind a grand civic commission to build a new city, in the international Modernist style, at Chandigarh, in the Punjab region in the north-west of the country. At independence, British India had been partitioned into two brand-new states, India and Pakistan. India was predominantly Hindu, and Pakistan predominantly Muslim. The huge old province of the Punjab straddled the new boundary between India and West Pakistan. The old capital of the Punjab, Lahore, fell on Pakistan's side of the border, and India therefore needed a new provincial capital for the still-substantial province of the Indian Punjab. In the absence of a trained corps of home-grown town-planners and architects, Nehru's advisers turned to a US firm of architects. They in turn brought in a team of British architects, who in their turn brought in the Swiss arch-Modernist Le Corbusier.

From the early 1920s, Le Corbusier had been designing and proselytizing on behalf of a new style of architecture, appropriate for the machine age, the age of mass production. Houses, he said, in one of the most frequently quoted phrases in the history of architecture, are 'machines for living in'. An implication of this, in Modernist thinking, is that the house-machine needed by an Indian family is substantially the same as the house-machine needed by a French or a British family, for it was believed that nations everywhere were converging on a single industrial, democratic form of society which in turn entailed a universal way of living. Sensitivity to the ancient cultural patterns embodied in Indian cities was no more necessary to the Modernist Le Corbusier than it was to the Classicist Lutyens. As Le Corbusier put it, 'What is the significance of Indian style in the world today if you accept machines, trousers and democracy?' (quoted in Chandigarh Research Group, n.d., p.19).

Le Corbusier devised a plan for Chandigarh that would be recognized by the citizens of the new town of Milton Keynes in the UK. A rectilinear grid of high-speed traffic routes defines the town. Set among greenery within the grid are neighbourhood units – 'superblocks' – of housing, schools and shops. A linear park fringes the river which flows from a huge artificial lake. Industry is kept at a distance. A monumental civic capitol, containing the three buildings that symbolize democracy and good administration (a provincial parliament building, lawcourts, and a government secretariat), stands at one end of the town, as the head stands in relation to the body. Here is the City Beautiful, the humane application of technologically sophisticated theories of Western town-planning.

Evaluating the Chandigarh project

The extent to which this vision of Indian civic life of the mid-twentieth century has been realized in Chandigarh is a matter of dispute, but three things are fairly clear. First, the new city could never shake itself free from the influence of political and demographic pressures that it could never hope to control. Population pressure has been the hardest to absorb. Cities that have unusually large quantities of public money pumped into them for the building of good-quality houses are likely to attract far more families than there are houses. Units in Chandigarh designed for one family are sometimes occupied by three or four (although in India it is clearly inappropriate to conceive of 'the family' as two parents plus two or three children). A fringe city of huts quickly sprang up, accommodating the labourers who came to work on the huge building sites in the city itself. One result of the population pressure has been that building heights have gone up. This in turn makes serious demands on the technical infrastructure of the city. In the words of S.S. Virdi, the chief engineer of the Chandigarh administration:

> Le Corbusier's original concept for the dwelling units envisaged housing of only two storeys, and the system of water supply and the sewerage system was planned accordingly. Unfortunately, it was found necessary to increase housing density, and many four-storey blocks were subsequently constructed. The system is incapable of supplying water to the upper storeys however; boosters therefore have to be installed, and this can be extremely costly. Apart from the cost though, these boosters also disrupt the supply of water to one- and two-storey dwellings. Furthermore, the increase in the number of four-storey dwellings has led to a two- or even threefold increase in the population density in some areas, stretching water and electricity supplies to the limit. There is now an enormous gap between the norms accepted at the planning stage and those currently being adopted. The growth of unauthorized illegal labour colonies has worsened the situation. Apart from being blemishes on the face of the City Beautiful, these colonies also represent a significant drain on its resources.
>
> (Virdi, n.d., p.53)

Second, Le Corbusier's vision of a universal, machine-age city may have been flawed by an inadequate understanding of Indian culture, and perhaps by the questionable validity of the machine aesthetic itself. In some senses his response was bold and powerful. He recognized, for instance, that India was short on resources but long on human labour, so despite his allegiance to the machine, he rejoiced at the way the city was rising by the application of simple human, donkey and buffalo power, using cheap materials – chiefly concrete and local brick – that required no fancy and expensive finishes or supplementation by materials, such as seasoned timber, that were scarce and expensive. But in other ways, his Modernism was a dubious guide. For example, his design for the parliament building was in part inspired by the shape of concrete cooling towers at power-stations (see Figure 10.16). However, the acoustics in the resulting debating chamber are so poor that the chamber is rarely used (Chandigarh Research Group, n.d., p.36).

But from the point of view of Chandigarh as a whole, the most serious problem is the grand conception of the city's layout. The separation of the grid of high-speed traffic from the neighbourhood areas, however humanely intended, may have been misconceived. Chandigarh is not, nor is ever likely to be, like Milton Keynes. An assumption of near-universal car ownership is, to say the least, premature. Thus huge open areas of roundabouts, dual carriageways and one-way systems, intended for a volume of through traffic that has never materialized, have reverted to local use by pedestrians, bicycles, rickshaws and animals (see Figure 10.17). Avenues in the commercial and shopping areas, which presupposed the development of large shops and businesses, fail because Indian shopping tends to be very local, and businesses

Figure 10.16 Chandigarh Legislative Assembly, under construction; designed by Le Corbusier. The inspiration of cooling towers is plain. Note the donkeys, basic motive power on the construction site (reproduced from Evenson, 1966, plate 110; photograph: Professor Norma Evenson)

Figure 10.17 Street scene in Chandigarh: successful urban planning? (photograph: Evert Bloemsma)

tend to be family affairs, conducted from the family's home or from premises close by. An assumption of the wide spatial separation of working and living areas seems to be inappropriate in India.

Further, the provision of extensive green areas may seem to be an unproblematically good thing, but in a poor city, short of water, the green quickly becomes brown, and the civic open spaces of the planner's dream become empty, sun-baked, dusty urban deserts.

The indigenous Indian urban tradition produced, at its best, cities such as Shahjahanabad. They were small, and they warded off the relentless sun and the dust-storms by packing the buildings close together, along shaded alleys. Privacy, in this dense pattern, was achieved by building around courtyards. In the name of modernization, this tradition was abandoned in Chandigarh, but it is not at all clear that its abandonment has produced a better city.

Third, and all-encompassingly, there has never been sufficient money to carry the project through to completion. In the conclusion to her book on Chandigarh, published when the city's outlines had not been fully filled in, Evenson was pessimistic about the likely success of the project. For example, the basic layout of widely spaced clusters of buildings entailed a far more elaborate, costly and lengthy infrastructure of services such as water, sewerage and electricity than would have been needed in a more compact city. Furthermore, Chandigarh's water-needs may have been underestimated. The city's ground water is being depleted faster than the rain can top it up, and water is having to be piped in from great distances. Evenson's reflections are worth quoting at some length, for – in relation to the transposition of the Western city into India – they illuminate a number of the issues that I have tried to open up in this chapter:

> As one views the miles of wide, paved streets bordering unbuilt areas and contemplates the extended water, sewer, and electrical lines serving isolated settlements strung out over the vast expanse of the city, it is difficult to believe that Chandigarh represents a truly economical solution to the problem of urban planning in a poor country.
>
> Does Chandigarh in its overall conception represent the sort of city which India can really afford? There is no doubt that within the basic physical standards set up for the city there has been an earnest concern for economy in details of construction and finishing, but one wonders if the physical standard from the beginning has been too high. Chandigarh represents an attempt virtually to duplicate the public utilities and civic amenities of a modern Western city, and in so doing has made housing for really poor people impossible to provide. The minimum house established by the planners, providing two rooms, a kitchen, bathroom, and courtyard, and equipped with electricity, running water, and a flush toilet may represent a standard exceeding that which can be generally realized in much of India. Moreover, it seems increasingly evident that the extravagant provisions for parks and landscaping are not only in many ways inappropriate but are beyond the means of the city to maintain.
>
> … In presenting fundamental solutions to India's pressing housing needs, and in demonstrating town-planning solutions consistent with the Indian economy, climate and way of life, Chandigarh, unfortunately, has little to offer.
>
> (Evenson, 1966, p.98)

Evenson's gloomy conclusion was written several decades ago. Perhaps the Indian economy has performed much better than she imagined; maybe India will eventually have sufficient money to finance and carry through urban projects of the sort pioneered at Chandigarh. Furthermore, Evenson's unfavourable judgement might be tempered by the reflection that European and North American cities have only rarely been triumphantly successful in housing their own poor.

10.6 Conclusion

In this chapter, I have described a number of cities built in India by the British, and one city built on behalf of the independent Indian government by a European–American consortium of architects and planners. The suggestion has been that these cities may, in some ways, be alien to the subcontinent and to the cultures upon which they were imposed.

Save for Chandigarh, the original purpose of the cities was to serve the colonial needs of the British, and these needs have left their marks on the enduring urban fabric. But as the cities continue to develop, perhaps their colonial aspects will turn out, in the long run, to be rather superficial. Maybe, early in the twentieth century, the implications of the dynamic technologies originating in Europe and North America – skyscrapers, electric power, air-conditioning, telephones, cars – were beginning to be felt all around the world.

The implications, perhaps, are that fundamentally there will henceforth be only one sort of city. The skyscrapers of Chicago and New York, and the automobile-filled streets of Los Angeles, can and will be reproduced endlessly around the globe. In response, tenacious local cultures will certainly insinuate themselves into these new, universal cities, forcing them to yield the sorts of living and working spaces that the residents want, but they will be fighting rearguard actions against universal urban homogenization.

From now on, perhaps the only really big difference between cities will simply and brutally be the gap between those that have the money to keep spending on the renewal of the technological infrastructures necessary to keep the city running, and those that do not. The only essential difference between, say, Calcutta and Los Angeles will be the obvious, glaring difference – wealth.

References

BLAKE, S.P. (1986) 'Cityscape of an imperial capital: Shahjahanabad in 1739' in R.E. Frykenberg (ed.) *Delhi through the Ages*, Delhi, Oxford University Press, pp.152–91.

CHANDIGARH RESEARCH GROUP (n.d.) 'Birth and development of Chandigarh' in *Chandigarh: forty years after Le Corbusier*, Amsterdam, Architectura and Natura Press, pp.11–52.

CHANT, C. and GOODMAN, D. (eds) (1999) *Pre-industrial Cities and Technology*, London, Routledge, in association with The Open University.

EVENSON, N. (1966) *Chandigarh*, Berkeley and Los Angeles, University of California Press.

EVENSON, N. (1989) *The Indian Metropolis: a view toward the West*, New Haven and London, Yale University Press.

GOODMAN, D. (ed.) (1999) *The European Cities and Technology Reader: industrial to post-industrial city*, London, Routledge, in association with The Open University.

GUPTA, N. (1986) 'Delhi and its hinterland' in R.E. Frykenberg (ed.) *Delhi through the Ages*, Delhi, Oxford University Press, pp.250–65.

HARRISON, J.B. (1980) 'Allahabad: a sanitary history' in K. Ballhatchet and J.B. Harrison (eds) *The City in South Asia*, London, Curzon Press, pp.167–95.

IRVING, R.G. (1981) *Indian Summer: Lutyens, Baker and Imperial Delhi*, New Haven and London, Yale University Press.

KING, A.D. (1976) *Colonial Urban Development*, London, Routledge.

KOSTOF, S. (1991) *The City Shaped*, London, Thames and Hudson.

LIND, J. (1768) *An Essay on Diseases incidental to Europeans in Hot Climates*, London, Becket and de Hondt.

MURPHEY, R. (1977, 2nd edn) 'Urbanisation in Asia' in J. Walton and D.E. Carns (eds) *Cities in Change*, Boston, Allyn and Bacon, pp.19–31.

NOE, S.V. (1986) 'What happened to Mughal Delhi: a morphological survey' in R.E. Frykenberg (ed.) *Delhi through the Ages*, Delhi, Oxford University Press, pp.237–49.

PACEY, A. (1990) *Technology and World Civilization*, Oxford, Blackwell.

SANGWAN, S. (1995) 'The sinking ships: colonial policy and the decline of Indian shipping, 1735–1835' in R. MacLeod and D. Kumar (eds) *Technology and the Raj*, New Delhi, Sage, pp.137–52.

SETHIA, T. (1996) 'The rise of the jute manufacturing industry in colonial India: a global perspective', *Journal of World History*, vol.7, no.1, pp.71–99.

TYRWHITT, J. (1947) *Patrick Geddes in India*, London, Lund Humphries.

VIRDI, S.S. (n.d.) 'Supply of water and electricity' in *Chandigarh: forty years after Le Corbusier*, Amsterdam, Architectura and Natura Press, p.53.

WADIA, R.A. (1957) *The Bombay Dockyard and the Wadia Master Builders*, Wadia, Bombay.

Index

Page numbers in *italics* refer to figures

Acknowledgements

Grateful acknowledgement is made to the following for permission to reproduce material in this book:

pp. 70–71 Kellett, J.R. (1969) *The Impact of Railways on Victorian Cities*, Routledge.

pp. 86–8 Hobhouse, H. (1971) *Thomas Cubitt Master Builder*, Macmillan Press Limited; by permission of Hermione Hobhouse.

pp. 168–70 Howard, E. (1985) *Garden Cities of To-morrow*, Attic Books.

pp. 171–4 Le Corbusier, *The Radiant City*; © FLC.

Figures 3.18, 3.20, 3.21 Hobhouse, H. (1971) *Thomas Cubitt Master Builder*, Macmillan Press Limited; by permission of Hermione Hobhouse.

Figure 4.19 (left) Short, J.R. (1984) *An Introduction to Urban Geography*, Routledge.

Figure 4.19 (right) Johnson, J.H. (1972) *Urban Geography: an introductory analysis*, Pergamon; © the author.

Every effort has been made to trace all copyright owners, but if any has been inadvertently overlooked, the publishers will be pleased to make the necessary arrangements at the first opportunity.